Principles *and* Standards *for* School Mathematics

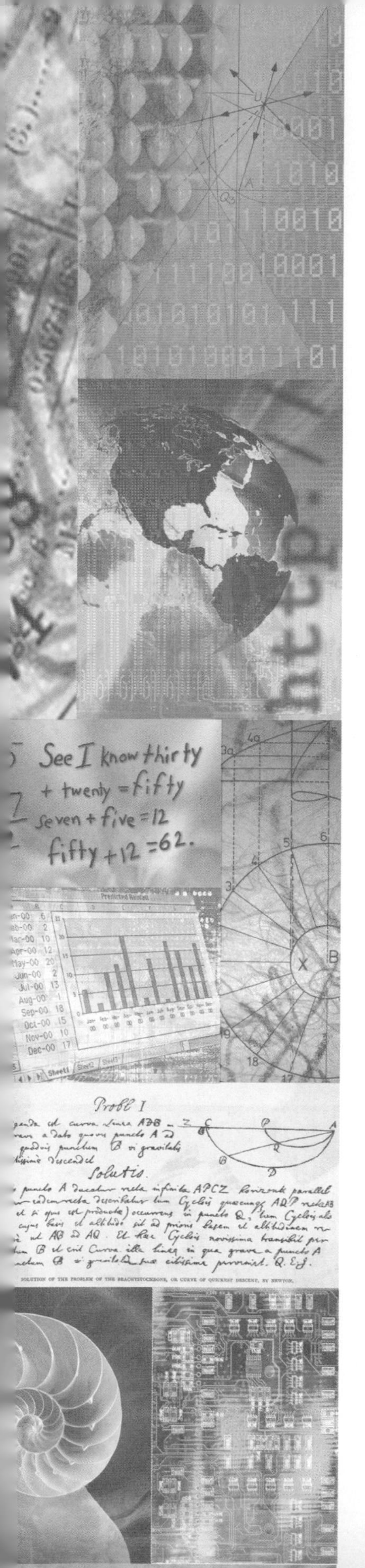

Principles *and* Standards *for* School Mathematics

National Council of Teachers of Mathematics

1906 Association Drive, Reston, VA 20191-9988
www.nctm.org

Library of Congress Cataloguing-in-Publication Data:

ISBN 0-87353-480-8

Photographs by Kathleen Beall

Printed in the United States of America

Table of Contents

Chapter 5

Chapter 6

Chapter 7

Chapter 8

Appendix

Standards 2000 Project

Writing Group

Joan Ferrini-Mundy, *Chair*
University of New Hampshire, Durham, New Hampshire (through 1999); Michigan State University, East Lansing, Michigan

W. Gary Martin, *Project Director*
National Council of Teachers of Mathematics

Grades Pre-K–2

Jeane Joyner, *Chair*
North Carolina Department of Public Instruction, Raleigh, North Carolina

Angela Andrews
Scott School, Naperville, Ilinois

Douglas H. Clements
State University of New York at Buffalo, Buffalo, New York

Alfinio Flores
Arizona State University, Tempe, Arizona

Carol Midgett
Southport Elementary School, Southport, North Carolina

Judith Roitman
University of Kansas, Lawrence, Kansas

Grades 3–5

Barbara Reys, *Chair*
University of Missouri—Columbia, Columbia, Missouri

Francis (Skip) Fennell
Western Maryland College, Westminster, Maryland

Catherine M. Fueglein
Webster Groves School District, Webster Groves, Missouri

Melinda Hamilton
Rosemont Elementary School, Orlando, Florida, 1998–99

Melissa Manzano-Alemán
Fort Worth Independent School District, Fort Worth, Texas, 1996–97

Susan Jo Russell
Education Research Collaborative, TERC, Cambridge, Massachusetts

Philip Wagreich
University of Illinois at Chicago, Chicago, Illinois

Grades 6–8

Edward A. Silver, *Chair*
University of Pittsburgh, Pittsburgh, Pennsylvania

Mary Bouck
Battle Creek Public Schools, Battle Creek, Michigan

Jean Howard
C. R. Anderson Middle School, Helena, Montana

Diana Lambdin
Indiana University Bloomington, Bloomington, Indiana

Carol Malloy
University of North Carolina at Chapel Hill, Chapel Hill, North Carolina

James Sandefur
Georgetown University, Washington, D.C.

Grades 9–12

Alan Schoenfeld, *Chair*
University of California at Berkeley, Berkeley, California

Sue Eddins
Illinois Mathematics and Science Academy, Aurora, Illinois

M. Kathleen Heid
Pennsylvania State University—University Park Campus, University Park, Pennsylvania

Millie Johnson
Western Washington University, Bellingham, Washington

Ron Lancaster
The Bishop Strachan School, Toronto, Ontario

Alfred Manaster
University of California—San Diego, La Jolla, California

Milton Norman
Granby High School, Norfolk, Virginia (through 1999)

Editors

Jean Carpenter
National Council of Teachers of Mathematics

Sheila Gorg
National Council of Teachers of Mathematics

Cover Design, Book Design, and Illustration

Debra G. Kushner
National Council of Teachers of Mathematics

Commission on the Future of the Standards

Mary M. Lindquist, *Chair*
Columbus State University, Columbus, Georgia

Shelley Ferguson
Standards 2000 Outreach Coordinator, National Council of Teachers of Mathematics

Fred Crouse
Annapolis Valley Regional School Board, Centerville, Nova Scotia

Portia Elliott
University of Massachusetts, Amherst, Massachusetts

Mazie Jenkins
Madison Metropolitan School District, Madison, Wisconsin

Jeremy Kilpatrick
University of Georgia, Athens, Georgia

Michael Koehler
Blue Valley North High School, Overland Park, Kansas

James R. C. Leitzel
University of New Hampshire, Durham, New Hampshire, 1996–98

Marilyn Mays
North Lake College, Irving, Texas

Richard Schoen
Stanford University, Stanford, California

Bonnie Hanson Walker
Lamar Consolidated Independent School District, Rosenberg, Texas

Electronic Format Group

Enrique Galindo, *Chair*
Indiana University, Bloomington, Bloomington, Indiana

S. Thomas Gorski
The Gilman School, Baltimore, Maryland

Beverly Hunter
Boston College, Chestnut Hill, Massachusetts

Eugene Klotz
Swarthmore College/Math Forum, Swarthmore, Pennsylvania

Nanette Seago
Video Cases for Mathematics Professional Development, San Diego, California

Len Simutis
Eisenhower National Clearinghouse for Mathematics and Science Education, Columbus, Ohio

Acknowledgments

Additional Writing

Cathy Kessell
Dawn Berk
Eric Hart

Editorial Assistance

Steve Olson
Sister Barbara Reynolds
Marie Gaudard
Bradford Findell
Marcia A. Friedman
Charles A. (Andy) Reeves
Kathleen Lay
Daniel Breidenbach
Joan Armistead
Harry Tunis
Charles C. E. Clements

Design and Production Assistance

Jo H. Handerson
Kevin Chadwick (logo design)

Staff Support

Edward T. Esty,
Staff Liaison, 1996–97
Edith Huffman, 1997–98
Sharon Soucy-McCrone, 1996–97
Alan Cook, Staff Liaison,1995–96

Research Assistance

Hope Gerson
Todd Grundmeier
John Beyers
Maria Alejandra Sorto
Stephen Hwang
Margot Mailloux
Gary Lewis
Cengiz Alacaci
Marjorie Henningsen

Clerical Support

Rebecca Totten
Ana Graef-Jones
Alexandra Torres
Denise Baxter
Aimee Spevak
Claire Williams
Erin Kelly
Alphonse McCullough
Ayn Lowry
Teresia Kiragu
Helen Waldo
Elizabeth Wallace
Susan Suchar

Appreciation must also be expressed to members of the NCTM Publications Department not listed above and to members of the many other NCTM departments that supported the efforts of the Writing Group and the Standards 2000 Project. Finally, a special thank-you to all the persons (too numerous to mention by name) who so generously provided their feedback throughout the preparation of the discussion draft and the final document, with particular thanks to the chairs and members of the Association Review Groups and to the commissioned reviewers.

Preface

Principles and Standards for School Mathematics is intended to be a resource and guide for all who make decisions that affect the mathematics education of students in prekindergarten through grade 12. The recommendations in it are grounded in the belief that all students should learn important mathematical concepts and processes with understanding. *Principles and Standards* makes an argument for the importance of such understanding and describes ways students can attain it. Its audience includes mathematics teachers; teacher-leaders in schools and districts; developers of instructional materials and frameworks; district-level curriculum directors and professional development leaders; those responsible for educating mathematics teachers; preservice teachers; school, state, and provincial administrators; and policymakers. In addition, the document can serve as a resource for researchers, mathematicians, and others with an interest in school mathematics. *Principles and Standards* has been produced by the National Council of Teachers of Mathematics (NCTM), an international professional organization committed to excellence in mathematics teaching and learning for all students.

The NCTM had previously produced a landmark trio of *Standards* documents—*Curriculum and Evaluation Standards for School Mathematics* (1989), *Professional Standards for Teaching Mathematics* (1991), and *Assessment Standards for School Mathematics* (1995). These three documents represented a historically important first attempt by a professional organization to develop and articulate explicit and extensive goals for teachers and policymakers. Since their release, they have given focus, coherence, and new ideas to efforts to improve mathematics education.

From the beginning of its involvement in proposing education standards, NCTM has viewed its efforts as part of an ongoing process of improving mathematics education. For standards to remain viable, the goals and visions they embody must periodically be examined, evaluated, tested by practitioners, and revised. In the early 1990s, the Council began discussing the need for monitoring and updating the existing NCTM *Standards*. These discussions culminated in the appointment of the Commission on the Future of the Standards in 1995. In April 1996, the NCTM Board of Directors approved a process for revising and updating the original *Standards* documents. This project, which was dubbed "Standards 2000," illustrates how the setting of standards can serve as a reflective and consensus-building mechanism for all those interested in mathematics education.

A number of structures were established within NCTM to initiate Standards 2000. First, the Commission on the Future of the Standards was appointed in 1995 and charged to—

- oversee the Standards 2000 project and related projects;
- collect and synthesize information and advice from within and outside NCTM throughout the development of the project;
- develop a plan for the dissemination, interpretation, implementation, evaluation, and subsequent revision of future *Standards* documents.

The Standards 2000 Writing Group and the Standards 2000 Electronic Format Group were appointed by spring 1997. Each included individuals—teachers, teacher educators, administrators, researchers, and mathematicians—with a wide range of expertise. The Writing Group was charged to establish standards that—

- build on the foundation of the original *Standards* documents;
- integrate the classroom-related portions of *Curriculum and Evaluation Standards for School Mathematics*, *Professional Standards for Teaching Mathematics*, and *Assessment Standards for School Mathematics*;
- are organized into four grade bands: prekindergarten through grade 2, grades 3–5, grades 6–8, and grades 9–12.

The Electronic Format Group was charged to—

- think of alternative ways to present and distribute the document that would result;
- envision ways in which technology-based materials could be incorporated in the *Standards*;
- keep the Standards 2000 Writing Group up-to-date on uses of technology;
- assist in the work of the Standards 2000 Writing Group by finding examples of appropriate uses of technology.

The primary work of the Writing Group was carried out in sessions during the summers of 1997, 1998, and 1999. Extensive efforts were undertaken to ensure that the Writing Group was informed by the best of research and current practice. The writers had access to collections of instructional materials, state and province curriculum documents, research publications, policy documents, and international frameworks and curriculum materials.

Additional input was sought for the Writing Group through a series of activities orchestrated by the Commission on the Future of the Standards. In February 1997, invitations were extended by the NCTM president to all the member societies of the Conference Board of the Mathematical Sciences to form Association Review Groups (ARGs) that would "provide sustained advice and information as it reflects on K–12 mathematics from the perspective of your organization." Over the course of the project, fourteen Association Review Groups were formed, and five sets of questions were formulated and submitted to these groups for their responses. (A list of the Association Review Groups, as well as a complete set of questions and publicly released responses from the ARGs, is available at www.nctm.org/standards/.)

NCTM's Research Advisory Committee commissioned a set of "white papers" summarizing the current state of education research in eight areas of mathematics teaching and learning to serve as background for the Writing Group. In addition, the Conference on Foundations for School Mathematics, held in Atlanta in March 1999 with support from the National Science Foundation, provided background to the writers concerning theoretical perspectives about teaching and learning. The papers written for this conference, along with the "white papers," are being published by NCTM as *A Research Companion to the NCTM Standards.* Two conferences, supported in part by the Eisenhower National Clearinghouse, also were held to inform the Writing Group about technology and advise it on the development of the electronic version of *Principles and Standards.*

A draft version of the *Standards*, entitled *Principles and Standards for School Mathematics: Discussion Draft*, was produced in October 1998 and circulated widely for reaction and discussion. Nearly 30 000 copies of the draft were furnished to persons interested in reading it, and many tens of thousands more accessed it from NCTM's Web site. Presentations and discussion sessions were held at all NCTM regional conferences in 1998–99, presentations were held at the conferences of many other organizations, and articles inviting feedback appeared in NCTM publications. In addition, 25 people were commissioned to review the draft from the perspective of their particular areas of interest. In total, reactions were submitted by more than 650 individuals and more than 70 groups (ranging from school study groups to graduate seminars to sessions held by NCTM Affiliates). The reactions were coded and entered into a qualitative database, resulting in the identification of a series of major issues for consideration. A synthesis of the issues and sample responses, as well as printouts of detailed feedback, was made available to the Writing Group for its work during summer 1999. Arguments on all sides of the issues were examined in the feedback. In light of the feedback, using the writers' best judgment, the Writing Group made careful decisions about the stance that *Principles and Standards* would take on each of the issues.

In response to a request from the NCTM Board of Directors and with funding from the National Science Foundation, the National Research Council formed a committee of experts from diverse backgrounds to review the process of gathering and analyzing reactions to the discussion draft, the plan to respond to the issues raised in the reactions, and the work of the Writing Group in carrying out that plan in the final document. The Writing Group was able to benefit greatly from the committee's guidance on responding to the comments and suggestions from reviewers and the field, and the document was improved as a result.

Principles and Standards reflects input and influence from many different sources. Educational research serves as a basis for many of the proposals and claims made throughout this document about what it is possible for students to learn about certain content areas at certain levels and under certain pedagogical conditions. The content and processes emphasized in *Principles and Standards* also reflect society's needs for mathematical literacy, past practice in mathematics education, and the values and expectations held by teachers, mathematics educators, mathematicians, and the general public. Finally, much of the content included here is based on the experiences and observations of the classroom teachers, teacher educators, educational researchers, and mathematicians in the Writing Group and on the input the Writing Group received throughout the drafting of the document.

Principles and Standards includes a number of classroom examples, instances of student work, and episodes that illustrate points made in the text. If drawn from another published source, the example or episode includes a citation to that source. If an episode does not have a citation and is written in the past tense, it is drawn from the experiences of a Writing Group member or a teacher colleague, with an indication of its source (such as unpublished observation notes) where appropriate. Episodes written in the present tense are hypothetical examples based on the experiences of the writers and are identified as such.

This document presents a vision of school mathematics—a set of goals toward which to strive. Throughout the document, this vision for mathematics education is expressed using words like "should, will, can, and must" to convey to readers the kind of mathematics teaching and learning that NCTM proposes. In no sense is this language meant to convey an assurance of some predetermined outcome; it is, rather, a means of describing the vision NCTM has constructed.

Principles and Standards is available in both print and electronic hypertext formats. *Principles and Standards for School Mathematics (Hypertext Edition)* includes tools to enhance navigation of the document, as well as a more extensive set of electronic examples (e-examples) to illuminate and enlarge the ideas in the text. It has also made possible the inclusion of links to resource and background material to enhance the messages of *Principles and Standards*. The e-examples are keyed to particular passages in the text and are signaled by an icon in the margin. The electronic version is available both on CD-ROM and on the World Wide Web at standards.nctm.org.

In the coming years, *Principles and Standards for School Mathematics* will provide focus and direction to the work of the Council. A number of initiatives related to *Principles and Standards* have already begun. The Council has established a task force to develop a series of materials, both print and electronic, with the working title of Navigations to assist and support teachers as they work in realizing the *Principles and Standards* in their classrooms, much as the Addenda series did following the release of the *Curriculum and Evaluation Standards for School Mathematics*. A series of institutes, organized by NCTM's new Academy for Professional Development, will give leaders a concentrated introduction to *Principles and Standards* and explore in depth various Standards or themes in the document.

Yet another task force is developing a plan and materials to help the Council effectively reach out to education administrators. The Standards Impact Research Group has been established to consider how the overall process of Standards-based education improvement set forth in the document can be better understood and subsequently refined in order to meet the goal of improving student learning. The E-Standards Task Force is considering ways to expand and improve future electronic versions (both Web and CD) of the *Principles and Standards*, and the Illuminations project is providing Web-based resources to "illuminate" the messages of the document, with funding provided by MCI WorldCom. These activities (and many others that will inevitably emerge in the coming years) build on the solid foundation of *Principles and Standards*, ensuring that the National Council of Teachers of Mathematics will continue to provide leadership toward its goal of improving the mathematics education of all students. For current information on these and other efforts and for other information surrounding about the document, visit www.nctm.org.

Letter of Appreciation
to the National Council of Teachers of Mathematics

Starting in 1989, the National Council of Teachers of Mathematics (NCTM) has developed and disseminated standards for curriculum, teaching, and assessment. These documents have guided many subsequent efforts to improve mathematics instruction in the U.S. While these first efforts predictably met with mixed interpretations and reactions, they have stimulated broad public and professional interest in the nature and formation of such standards.

As the NCTM has undertaken to update and refine the standards, producing the new *Principles and Standards for School Mathematics*, it recognized the necessity to enlist the broad critical participation of the diverse expert communities that bear some responsibility for mathematics education. In order to provide for this complex advisory function, the NCTM petitioned each of the professional organizations of the Conference Board of the Mathematical Sciences (CBMS) to form an Association Review Group (ARG) that would respond, in stages, to a series of substantial and focused questions framed by the *Principles and Standards* writing group in the course of its work.

This formidable undertaking has been, in the view of the participating organizations, dramatically successful, to the profit of both NCTM and the contributing organizations. It was a remarkable and unprecedented process that produced some of the most thoughtful and disciplined discussions of mathematics curriculum and instruction that we have seen in these professional communities. It contrasted with what was felt by some to be the inadequate participation by mathematics professionals in the formation of the original standards.

Of course, *Principles and Standards* addresses matters of educational goals and policies for which there is no simple right answer or formulation, and no clear or stable consensus. It represents the views of a team of writers assembled by the national professional organization of mathematics teachers, views deeply informed by the knowledge and dispositions of diverse professional communities. What can be objectively said is that the process of construction of the *Principles and Standards* has been open, rigorous, and well informed by the views of all professionals concerned with mathematics education, and that this has been achieved, in part, thanks to the innovative design of the ARG process. Indeed, the quality of that development process, including the ARGs, has been independently reviewed, at the invitation of NCTM itself, in a study by the National Research Council.

With this letter, representatives of the following member organizations of CBMS wish to register their appreciation to the NCTM for the design, and implementation with integrity, of this process. With this, the NCTM has established a model, heretofore all too rare, of how to stage civil, disciplined and probing discourse among diverse professionals on matters of mathematics education.

Conference Board of the Mathematical Sciences
Lynne Billard, *Chair*
Thomas R. Banchoff, *Chair-Elect*
Raymond L. Johnson, *Secretary-Treasurer*
Sadie C. Bragg, *Member-at-Large of Executive Committee*

American Mathematical Association of Two-Year Colleges
Susan S. Wood, *President*

American Mathematical Society
Felix E. Browder, *President*
Hyman Bass, *President Elect and Past Chair of the Committee on Education*

Association of Mathematics Teacher Educators
Susan Gay, *President*

American Statistical Association
W. Michael O'Fallon, *President*

Association for Symbolic Logic
C. Ward Henson, *Secretary-Treasurer*
Jean A. Larson, *Chair, Committee on Education*

Association of State Supervisors of Mathematics
Jacqueline P. Mitchell, *President*

Association for Women in Mathematics
Jean E. Taylor, *President*

Benjamin Banneker Association
Beatrice Moore-Harris, *President*

Institute for Operations Research and the Management Sciences
John R. Birge, *President*

Institute of Mathematical Statistics
Morris L. Eaton, *President*

Mathematical Association of America
Thomas R. Banchoff, *President*
Kenneth Ross, *Chair of the MAA Association Review Group*

National Association of Mathematicians
John W. Alexander, Jr., *President*

National Council of Supervisors of Mathematics
Jerry Cummins, *President*

Society for Industrial and Applied Mathematics
Gilbert Strang, *President*

Imagine a classroom, a school, or a school district where all students have access to high-quality, engaging mathematics instruction.

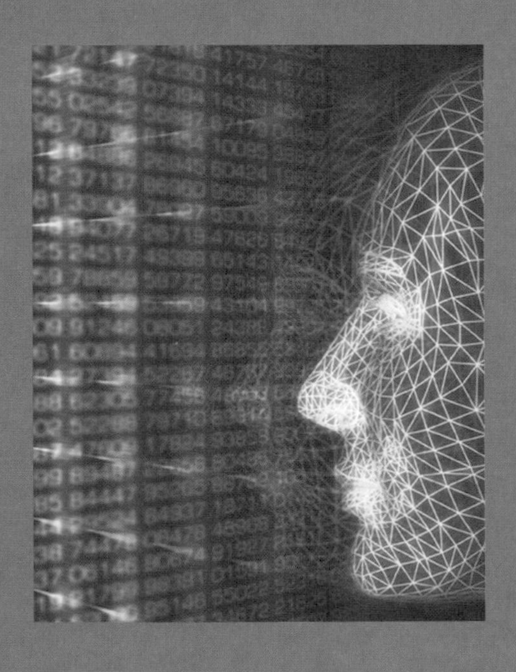

Chapter 1

A Vision for School Mathematics

Imagine a classroom, a school, or a school district where all students have access to high-quality, engaging mathematics instruction. There are ambitious expectations for all, with accommodation for those who need it. Knowledgeable teachers have adequate resources to support their work and are continually growing as professionals. The curriculum is mathematically rich, offering students opportunities to learn important mathematical concepts and procedures with understanding. Technology is an essential component of the environment. Students confidently engage in complex mathematical tasks chosen carefully by teachers. They draw on knowledge from a wide variety of mathematical topics, sometimes approaching the same problem from different mathematical perspectives or representing the mathematics in different ways until they find methods that enable them to make progress. Teachers help students make, refine, and explore conjectures on the basis of evidence and use a variety of reasoning and proof techniques to confirm or disprove those conjectures. Students are flexible and resourceful problem solvers. Alone or in groups and with access to technology, they work productively and reflectively, with the skilled guidance of their teachers. Orally and in writing, students communicate their ideas and results effectively. They value mathematics and engage actively in learning it.

The vision for mathematics education described in *Principles and Standards for School Mathematics* is highly ambitious. Achieving it requires solid mathematics curricula, competent and knowledgeable teachers who can integrate instruction with assessment, education policies that enhance and support learning, classrooms with ready access to technology, and a commitment to both equity and excellence. The challenge is enormous and meeting it is essential. Our students deserve and need the best mathematics education possible, one that enables them to fulfill personal ambitions and career goals in an ever-changing world.

Since the release in 1989 of the *Curriculum and Evaluation Standards for School Mathematics*—followed in 1991 by the *Professional Teaching Standards for School Mathematics* and in 1995 by the *Assessment Standards for School Mathematics*—the National Council of Teachers of Mathematics (NCTM) has remained committed to the view that standards can play a leading role in guiding the improvement of mathematics education. As an organization representing teachers of mathematics, NCTM shares with students, school leaders, and parents and other caregivers the responsibility to ensure that all students receive a high-quality mathematics education. All interested parties must work together to create mathematics classrooms where students of varied backgrounds and abilities work with expert teachers, learning important mathematical ideas with understanding, in environments that are equitable, challenging, supportive, and technologically equipped for the twenty-first century.

The Need for Mathematics in a Changing World

The need to understand and be able to use mathematics in everyday life and in the workplace has never been greater.

We live in a time of extraordinary and accelerating change. New knowledge, tools, and ways of doing and communicating mathematics continue to emerge and evolve. Calculators, too expensive for common use in the early eighties, now are not only commonplace and inexpensive but vastly more powerful. Quantitative information available to limited numbers of people a few years ago is now widely disseminated through popular media outlets.

The need to understand and be able to use mathematics in everyday life and in the workplace has never been greater and will continue to increase. For example:

- *Mathematics for life.* Knowing mathematics can be personally satisfying and empowering. The underpinnings of everyday life are increasingly mathematical and technological. For instance, making purchasing decisions, choosing insurance or health plans, and voting knowledgeably all call for quantitative sophistication.
- *Mathematics as a part of cultural heritage.* Mathematics is one of the greatest cultural and intellectual achievements of humankind, and citizens should develop an appreciation and understanding of that achievement, including its aesthetic and even recreational aspects.
- *Mathematics for the workplace.* Just as the level of mathematics needed for intelligent citizenship has increased dramatically, so too has the level of mathematical thinking and problem solving needed in the workplace, in professional areas ranging from health care to graphic design.
- *Mathematics for the scientific and technical community.* Although all careers require a foundation of mathematical knowledge, some are mathematics intensive. More students must pursue an educational path that will prepare them for lifelong work as mathematicians, statisticians, engineers, and scientists.

In this changing world, those who understand and can do mathematics will have significantly enhanced opportunities and options for shaping their futures. Mathematical competence opens doors to productive futures. A lack of mathematical competence keeps those doors closed. NCTM challenges the assumption that mathematics is only for the select few. On the contrary, everyone needs to understand mathematics. All students should have the opportunity and the support necessary to learn significant mathematics with depth and understanding. There is no conflict between equity and excellence.

Principles and Standards calls for a common foundation of mathematics to be learned by all students. This approach, however, does not imply that all students are alike. Students exhibit different talents, abilities, achievements, needs, and interests in mathematics. Nevertheless, all students must have access to the highest-quality mathematics instructional programs. Students with a deep interest in pursuing mathematical and scientific careers must have their talents and interests engaged. Likewise, students with special educational needs must have the opportunities and support they require to attain a substantial understanding of important mathematics. A society in which only a few have the mathematical knowledge needed to fill crucial economic, political, and scientific roles is not consistent with the values of a just democratic system or its economic needs.

There is no conflict between equity and excellence.

The Need for Continued Improvement of Mathematics Education

The vision described at the beginning of this chapter is idealized. Despite the concerted efforts of many classroom teachers, administrators, teacher-leaders, curriculum developers, teacher educators, mathematicians, and policymakers, the portrayal of mathematics teaching and learning in *Principles and Standards* is not the reality in the vast majority of classrooms, schools, and districts. Evidence from a variety of sources makes it clear that many students are not learning the mathematics they need or are expected to learn (Kenney and Silver 1997; Mullis et al. 1997, 1998; Beaton et al. 1996). The reasons for this deficiency are many: In some instances, students have not had the opportunity to learn important mathematics. In other instances, the curriculum offered to students does not engage them. Sometimes students lack a commitment to learning. The quality of mathematics teaching is highly variable. There is no question that the effectiveness of mathematics education in the United States and Canada can be improved substantially.

Standards can play a central role in the process of improvement. The previously released NCTM *Standards* (NCTM 1989, 1991, 1995) have influenced state standards and curriculum frameworks (Council of Chief State School Officers 1995; Raimi and Braden 1998), instructional materials (U.S. Department of Education 1999), teacher education (Mathematical Association of America 1991), and classroom practice (Ferrini-Mundy and Schram 1997). As with any educational innovation, however, the ideas of the *Standards* have been interpreted in many different ways and have been implemented with varying degrees of fidelity. Sometimes the changes made in the name of standards have been superficial or incomplete. For example, some of the pedagogical ideas from the NCTM *Standards*—such as the emphases on discourse, worthwhile mathematical

Sometimes the changes made in the name of standards have been superficial or incomplete.

tasks, or learning through problem solving—have been enacted without sufficient attention to students' understanding of mathematics content. Efforts to move in the directions of the original NCTM *Standards* are by no means fully developed or firmly in place.

The Role and Purpose of Standards

The introduction to the 1989 *Curriculum and Evaluation Standards* noted three reasons for a professional organization to formally adopt standards: to ensure quality, to indicate goals, and to promote change. One way in which standards documents help meet these goals is by shaping conversations about mathematics education. As with the previous NCTM *Standards*, *Principles and Standards* offers common language, examples, and recommendations to engage many groups of people in productive dialogue. Although there will never be complete consensus within the mathematics education profession or among the general public about the ideas advanced in any standards document, the Standards provide a guide for focused, sustained efforts to improve students' school mathematics education. *Principles and Standards* supplies guidance and vision while leaving specific curriculum decisions to the local level. This document is intended to—

- set forth a comprehensive and coherent set of goals for mathematics for all students from prekindergarten through grade 12 that will orient curricular, teaching, and assessment efforts during the next decades;
- serve as a resource for teachers, education leaders, and policymakers to use in examining and improving the quality of mathematics instructional programs;
- guide the development of curriculum frameworks, assessments, and instructional materials;
- stimulate ideas and ongoing conversations at the national, provincial or state, and local levels about how best to help students gain a deep understanding of important mathematics.

An Overview of *Principles and Standards*

Principles and Standards for School Mathematics builds on and consolidates messages from the previous *Standards* documents. The document is organized into four main parts:

- Principles for school mathematics (chapter 2)
- An overview of the Standards for mathematics education in prekindergarten through grade 12 (chapter 3)
- Standards for four separate grade bands: prekindergarten through grade 2 (chapter 4), grades 3–5 (chapter 5), grades 6–8 (chapter 6), and grades 9–12 (chapter 7)
- A discussion of the steps needed to move toward the vision embodied in *Principles and Standards* (chapter 8)

The principles are statements reflecting basic precepts that are fundamental to a high-quality mathematics education. The discussions in chapter 2 elaborate on the underlying assumptions, values, and evidence on which these Principles are founded. The Principles should be useful as perspectives on which educators can base decisions that affect school mathematics. NCTM's commitment to mathematics for all is reaffirmed in the Equity Principle. In the Curriculum Principle, a focused curriculum is shown to be an important aspect of what is needed to improve school mathematics. The Teaching Principle makes the case that students must have opportunities to learn important mathematics under the guidance of competent and committed teachers. The view of learning that is the basis for the document is taken up in the Learning Principle. The important roles of assessment and technology in school mathematics programs are discussed in the Assessment and Technology Principles.

Chapters 3–7 outline an ambitious and comprehensive set of curriculum standards for all students. Standards are descriptions of what mathematics instruction should enable students to know and do—statements of what is valued for school mathematics education. Each of the ten curriculum standards proposed in this document spans the entire range from prekindergarten through grade 12. Chapter 3 discusses each Standard in turn to convey its main ideas. In addition, these discussions give a sense of how the ideas encompassed in a Standard develop over all four grade bands, highlighting points at which certain levels of mastery or closure are appropriate. Chapters 4–7 present the Standards in detail for each grade band.

The first five Standards describe mathematical content goals in the areas of number and operations, algebra, geometry, measurement, and data analysis and probability. The next five Standards address the processes of problem solving, reasoning and proof, connections, communication, and representation. In each grade-band chapter, a set of "expectations" is identified and discussed for each Content Standard. The appendix displays the Content Standards and expectations in a chart that highlights the increasing sophistication of ideas across the grades. Each grade-band chapter discusses what each Process Standard should "look like" in that grade band and what the teacher's role is in supporting the development of that process.

One purpose of this document is to offer a way to focus curricula.

The mathematical Content and Process Standards discussed in chapters 3–7 are inextricably linked. One cannot solve problems without understanding and using mathematical content. Establishing geometric knowledge calls for reasoning. The concepts of algebra can be examined and communicated through representations.

One purpose of this document is to offer teachers, curriculum developers, and those responsible for establishing curriculum frameworks a way to focus curricula. Focus is promoted through attention to the idea of "moving on." School mathematics programs should not address every topic every year. Instead, students will reach certain levels of conceptual understanding and procedural fluency by certain points in the curriculum. Teachers should be able to assume that students possess these understandings and levels of fluency when they plan their mathematics instruction. Teachers and policymakers can then fashion instructional programs and curricular frameworks that develop progressively over the grades and that focus on important mathematical areas.

Chapter 8 discusses what it will take to move toward the vision described in the previous chapters. In particular, it discusses critical issues related to putting the Principles into action and outlines the key roles played by various groups and communities in realizing the vision of the *Principles and Standards*.

As We Move Forward

The task is enormous and essential.

Attaining the vision described at the beginning of this chapter will require the talents, energy, and attention of many individuals, including students, teachers, school administrators, teacher-leaders, policymakers, parents and other caregivers, mathematicians, mathematics educators, and the local community. It will require that the vision of this document be shared and understood and that all concerned be committed to improving the futures of our children. The task is enormous and essential. All students need an education in mathematics that will prepare them for a future of great and continual change.

The Principles describe particular features of high-quality mathematics education.

The power of these Principles as guides and tools for decision making derives from their interaction.

Principles for School Mathematics

Decisions made by teachers, school administrators, and other education professionals about the content and character of school mathematics have important consequences both for students and for society. These decisions should be based on sound professional guidance. *Principles and Standards for School Mathematics* is intended to provide such guidance. The Principles describe particular features of high-quality mathematics education. The Standards describe the mathematical content and processes that students should learn. Together, the Principles and Standards constitute a vision to guide educators as they strive for the continual improvement of mathematics education in classrooms, schools, and educational systems.

The six principles for school mathematics address overarching themes:

- ***Equity.*** Excellence in mathematics education requires equity—high expectations and strong support for all students.
- ***Curriculum.*** A curriculum is more than a collection of activities: it must be coherent, focused on important mathematics, and well articulated across the grades.
- ***Teaching.*** Effective mathematics teaching requires understanding what students know and need to learn and then challenging and supporting them to learn it well.
- ***Learning.*** Students must learn mathematics with understanding, actively building new knowledge from experience and prior knowledge.
- ***Assessment.*** Assessment should support the learning of important mathematics and furnish useful information to both teachers and students.
- ***Technology.*** Technology is essential in teaching and learning mathematics; it influences the mathematics that is taught and enhances students' learning.

These six Principles, which are discussed in depth below, do not refer to specific mathematics content or processes and thus are quite different from the Standards. They describe crucial issues that, although not unique to school mathematics, are deeply intertwined with school mathematics programs. They can influence the development of curriculum frameworks, the selection of curriculum materials, the planning of instructional units or lessons, the design of assessments, the assignment of teachers and students to classes, instructional decisions in the classroom, and the establishment of supportive professional development programs for teachers. The perspectives and assumptions underlying the Principles are compatible with, and foundational to, the Standards and expectations presented in chapters 3–7.

Each Principle is discussed separately, but the power of these Principles as guides and tools for decision making derives from their interaction in the thinking of educators. The Principles will come fully alive as they are used together to develop high-quality school mathematics programs.

The Equity Principle

Excellence in mathematics education requires equity—high expectations and strong support for all students.

The vision of equity in mathematics education challenges a pervasive societal belief in North America that only some students are capable of learning mathematics.

Making the vision of the *Principles and Standards for School Mathematics* a reality for all students, prekindergarten through grade 12, is both an essential goal and a significant challenge. Achieving this goal requires raising expectations for students' learning, developing effective methods of supporting the learning of mathematics by all students, and providing students and teachers with the resources they need.

Educational equity is a core element of this vision. All students, regardless of their personal characteristics, backgrounds, or physical challenges, must have opportunities to study—and support to learn—mathematics. Equity does not mean that every student should receive identical instruction; instead, it demands that reasonable and appropriate accommodations be made as needed to promote access and attainment for all students.

Equity is interwoven with the other Principles. All students need access each year to a coherent, challenging mathematics curriculum taught by competent and well-supported mathematics teachers. Moreover, students' learning and achievement should be assessed and reported in ways that point to areas requiring prompt additional attention. Technology can assist in achieving equity and must be accessible to all students.

Equity requires high expectations and worthwhile opportunities for all.

The vision of equity in mathematics education challenges a pervasive societal belief in North America that only some students are capable of learning mathematics. This belief, in contrast to the equally pervasive view that all students can and should learn to read and write in English, leads to low expectations for too many students. Low expectations are especially problematic because students who live in poverty, students who are not native speakers of English, students with disabilities, females, and many nonwhite students have traditionally been far more likely than their counterparts in other demographic groups to be the victims of low expectations. Expectations must be raised—mathematics can and must be learned by *all* students.

Mathematics can and must be learned by all students.

The Equity Principle demands that high expectations for mathematics learning be communicated in words and deeds to all students. Teachers communicate expectations in their interactions with students during classroom instruction, through their comments on students' papers, when assigning students to instructional groups, through the presence or absence of consistent support for students who are striving for high levels of attainment, and in their contacts with significant adults in a student's life. These actions, along with decisions and actions taken outside the classroom to assign students to different classes or curricula, also determine students' opportunities to learn and influence students' beliefs about their own abilities to succeed in mathematics. Schools have an obligation to ensure that all students participate in a strong instructional program that supports their mathematics learning. High expectations can be achieved in part with instructional programs that are interesting for students and help them see the importance and utility of continued mathematical study for their own futures.

Equity requires accommodating differences to help everyone learn mathematics.

Higher expectations are necessary, but they are not sufficient to accomplish the goal of an equitable school mathematics education for all students. All students should have access to an excellent and equitable mathematics program that provides solid support for their learning and is responsive to their prior knowledge, intellectual strengths, and personal interests.

Some students may need further assistance to meet high mathematics expectations. Students who are not native speakers of English, for instance, may need special attention to allow them to participate fully in classroom discussions. Some of these students may also need assessment accommodations. If their understanding is assessed only in English, their mathematical proficiency may not be accurately evaluated.

Some students may need further assistance to meet high mathematics expectations.

Students with disabilities may need increased time to complete assignments, or they may benefit from the use of oral rather than written assessments. Students who have difficulty in mathematics may need additional resources, such as after-school programs, peer mentoring, or cross-age tutoring. Likewise, students with special interests or exceptional talent in mathematics may need enrichment programs or additional resources to challenge and engage them. The talent and interest of these students must be nurtured and supported so that they have the opportunity and guidance to excel. Schools and school systems must take care to accommodate the special needs of some students without inhibiting the learning of others.

Technology can help achieve equity in the classroom. For example, technological tools and environments can give all students opportunities

Access to technology must not become yet another dimension of educational inequity.

to explore complex problems and mathematical ideas, can furnish structured tutorials to students needing additional instruction and practice on skills, or can link students in rural communities to instructional opportunities or intellectual resources not readily available in their locales. Computers with voice-recognition or voice-creation software can offer teachers and peers access to the mathematical ideas and arguments developed by students with disabilities who would otherwise be unable to share their thinking. Moreover, technology can be effective in attracting students who disengage from nontechnological approaches to mathematics. It is important that all students have opportunities to use technology in appropriate ways so that they have access to interesting and important mathematical ideas. Access to technology must not become yet another dimension of educational inequity.

Equity requires resources and support for all classrooms and all students.

Well-documented examples demonstrate that all children, including those who have been traditionally underserved, can learn mathematics when they have access to high-quality instructional programs that support their learning (Campbell 1995; Griffin, Case, and Siegler 1994; Knapp et al. 1995; Silver and Stein 1996). These examples should become the norm rather than the exception in school mathematics education.

Achieving equity requires a significant allocation of human and material resources in schools and classrooms. Instructional tools, curriculum materials, special supplemental programs, and the skillful use of community resources undoubtedly play important roles. An even more important component is the professional development of teachers. Teachers need help to understand the strengths and needs of students who come from diverse linguistic and cultural backgrounds, who have specific disabilities, or who possess a special talent and interest in mathematics. To accommodate differences among students effectively and sensitively, teachers also need to understand and confront their own beliefs and biases.

The Curriculum Principle

A curriculum is more than a collection of activities: it must be coherent, focused on important mathematics, and well articulated across the grades.

A school mathematics curriculum is a strong determinant of what students have an opportunity to learn and what they do learn. In a coherent curriculum, mathematical ideas are linked to and build on one another so that students' understanding and knowledge deepens and their ability to apply mathematics expands. An effective mathematics curriculum focuses on important mathematics—mathematics that will prepare students for continued study and for solving problems in a variety of school, home, and work settings. A well-articulated curriculum challenges students to learn increasingly more sophisticated mathematical ideas as they continue their studies.

A mathematics curriculum should be coherent.

Mathematics comprises different topical strands, such as algebra and geometry, but the strands are highly interconnected. The interconnections should be displayed prominently in the curriculum and in instructional materials and lessons. A coherent curriculum effectively organizes and integrates important mathematical ideas so that students can see how the ideas build on, or connect with, other ideas, thus enabling them to develop new understandings and skills.

Curricular coherence is also important at the classroom level. Researchers have analyzed lessons in the videotape study of eighth-grade mathematics classrooms that was part of the Third International Mathematics and Science Study (Stigler and Hiebert 1999). One important characteristic of the lessons had to do with the internal coherence of the mathematics. The researchers found that typical Japanese lessons were designed around one central idea, which was carefully developed and extended; in contrast, typical American lessons included several ideas or topics that were not closely related and not well developed.

A coherent curriculum effectively organizes and integrates important mathematical ideas.

In planning individual lessons, teachers should strive to organize the mathematics so that fundamental ideas form an integrated whole. Big ideas encountered in a variety of contexts should be established carefully, with important elements such as terminology, definitions, notation, concepts, and skills emerging in the process. Sequencing lessons coherently across units and school years is challenging. And teachers also need to be able to adjust and take advantage of opportunities to move lessons in unanticipated directions.

A mathematics curriculum should focus on important mathematics.

School mathematics curricula should focus on mathematics content and processes that are worth the time and attention of students. Mathematics topics can be considered important for different reasons, such as their utility in developing other mathematical ideas, in linking different areas of mathematics, or in deepening students' appreciation of mathematics as a discipline and as a human creation. Ideas may also merit curricular focus because they are useful in representing and solving problems within or outside mathematics.

Foundational ideas like place value, equivalence, proportionality, function, and rate of change should have a prominent place in the mathematics curriculum because they enable students to understand other mathematical ideas and connect ideas across different areas of mathematics. Mathematical thinking and reasoning skills, including making conjectures and developing sound deductive arguments, are important because they serve as a basis for developing new insights and promoting further study. Many concepts and processes, such as symmetry and generalization, can help students gain insights into the nature and beauty of mathematics. In addition, the curriculum should offer experiences that allow students to see that mathematics has powerful uses in modeling and predicting real-world phenomena. The curriculum also should emphasize the mathematics processes and skills that support the quantitative literacy of students. Members of an intelligent citizenry should be able to judge claims, find fallacies, evaluate risks, and weigh evidence (Price 1997).

Although any curriculum document is fixed at a point in time, the curriculum itself need not be fixed. Different configurations of important mathematical ideas are possible and to some extent inevitable. The relative importance of particular mathematics topics is likely to change over time in response to changing perceptions of their utility and to new demands and possibilities. For example, mathematics topics such as recursion, iteration, and the comparison of algorithms are receiving more attention in school mathematics because of their increasing relevance and utility in a technological world.

A mathematics curriculum should be well articulated across the grades.

A well-articulated curriculum gives guidance about when closure is expected for particular skills or concepts.

Learning mathematics involves accumulating ideas and building successively deeper and more refined understanding. A school mathematics curriculum should provide a road map that helps teachers guide students to increasing levels of sophistication and depths of knowledge. Such guidance requires a well-articulated curriculum so that teachers at each level understand the mathematics that has been studied by students at the previous level and what is to be the focus at successive levels. For example, in grades K–2 students typically explore similarities and differences among two-dimensional shapes. In grades 3–5 they can identify characteristics of various quadrilaterals. In grades 6–8 they may examine and make generalizations about properties of particular quadrilaterals. In grades 9–12 they may develop logical arguments to justify conjectures about particular polygons. As they reach higher levels, students should engage more deeply with mathematical ideas and their understanding and ability to use the knowledge is expected to grow.

Without a clear articulation of the curriculum across all grades, duplication of effort and unnecessary review are inevitable. A well-articulated curriculum gives teachers guidance regarding important ideas or major themes, which receive special attention at different points in time. It also gives guidance about the depth of study warranted at particular times and when closure is expected for particular skills or concepts.

The Teaching Principle

Effective mathematics teaching requires understanding what students know and need to learn and then challenging and supporting them to learn it well.

Students learn mathematics through the experiences that teachers provide. Thus, students' understanding of mathematics, their ability to use it to solve problems, and their confidence in, and disposition toward, mathematics are all shaped by the teaching they encounter in school. The improvement of mathematics education for all students requires effective mathematics teaching in all classrooms.

Teaching mathematics well is a complex endeavor, and there are no easy recipes for helping all students learn or for helping all teachers become effective. Nevertheless, much is known about effective mathematics teaching, and this knowledge should guide professional judgment and activity. To be effective, teachers must know and understand deeply the mathematics they are teaching and be able to draw on that knowledge with flexibility in their teaching tasks. They need to understand and be committed to their students as learners of mathematics and as human beings and be skillful in choosing from and using a variety of pedagogical and assessment strategies (National Commission on Teaching and America's Future 1996). In addition, effective teaching requires reflection and continual efforts to seek improvement. Teachers must have frequent and ample opportunities and resources to enhance and refresh their knowledge.

The *Professional Standards for Teaching Mathematics* (NCTM 1991) presented six standards for the teaching of mathematics. They address–

- worthwhile mathematical tasks;
- the teacher's role in discourse;
- the student's role in discourse;
- tools for enhancing discourse;
- the learning environment;
- the analysis of teaching and learning.

Effective teaching requires knowing and understanding mathematics, students as learners, and pedagogical strategies.

Teachers need several different kinds of mathematical knowledge—knowledge about the whole domain; deep, flexible knowledge about curriculum goals and about the important ideas that are central to their grade level; knowledge about the challenges students are likely to encounter in learning these ideas; knowledge about how the ideas can be represented to teach them effectively; and knowledge about how students' understanding can be assessed. This knowledge helps teachers make curricular judgments, respond to students' questions, and look ahead to where concepts are leading and plan accordingly. Pedagogical knowledge, much of which is acquired and shaped through the practice of teaching, helps teachers understand how students learn mathematics, become facile with a range of different teaching techniques and instructional materials, and organize and manage the classroom. Teachers need to understand the big ideas of mathematics and be able to represent mathematics as a coherent and connected enterprise (Schifter 1999; Ma 1999). Their decisions and their actions in the classroom—all of which affect how well their students learn mathematics—should be based on this knowledge.

This kind of knowledge is beyond what most teachers experience in standard preservice mathematics courses in the United States. For example, that fractions can be understood as parts of a whole, the quotient of two integers, or a number on a line is important for mathematics teachers (Ball and Bass forthcoming). Such understanding might be characterized as "profound understanding of fundamental mathematics" (Ma 1999). Teachers also need to understand the different representations of an idea, the relative strengths and weaknesses of each, and how they are related to one another (Wilson, Shulman, and Richert 1987). They need to know the ideas with which students often have difficulty and ways to help bridge common misunderstandings.

Teaching mathematics well is a complex endeavor, and there are no easy recipes.

There is no one "right way" to teach.

Effective mathematics teaching requires a serious commitment to the development of students' understanding of mathematics. Because students learn by connecting new ideas to prior knowledge, teachers must understand what their students already know. Effective teachers know how to ask questions and plan lessons that reveal students' prior knowledge; they can then design experiences and lessons that respond to, and build on, this knowledge.

Teachers have different styles and strategies for helping students learn particular mathematical ideas, and there is no one "right way" to teach. However, effective teachers recognize that the decisions they make shape students' mathematical dispositions and can create rich settings for learning. Selecting and using suitable curricular materials, using appropriate instructional tools and techniques, and engaging in reflective practice and continuous self-improvement are actions good teachers take every day.

One of the complexities of mathematics teaching is that it must balance purposeful, planned classroom lessons with the ongoing decision making that inevitably occurs as teachers and students encounter unanticipated discoveries or difficulties that lead them into uncharted territory. Teaching mathematics well involves creating, enriching, maintaining, and adapting instruction to move toward mathematical goals, capture and sustain interest, and engage students in building mathematical understanding.

Effective teaching requires a challenging and supportive classroom learning environment.

Teachers make many choices each day about how the learning environment will be structured and what mathematics will be emphasized. These decisions determine, to a large extent, what students learn. Effective teaching conveys a belief that each student can and is expected to understand mathematics and that each will be supported in his or her efforts to accomplish this goal.

Teachers establish and nurture an environment conducive to learning mathematics through the decisions they make, the conversations they orchestrate, and the physical setting they create. Teachers' actions are what encourage students to think, question, solve problems, and discuss their ideas, strategies, and solutions. The teacher is responsible for creating an intellectual environment where serious mathematical thinking is the norm. More than just a physical setting with desks, bulletin boards, and posters, the classroom environment communicates subtle messages about what is valued in learning and doing mathematics. Are students' discussion and collaboration encouraged? Are students expected to justify their thinking? If students are to learn to make conjectures, experiment with various approaches to solving problems, construct mathematical arguments and respond to others' arguments, then creating an environment that fosters these kinds of activities is essential.

In effective teaching, worthwhile mathematical tasks are used to introduce important mathematical ideas and to engage and challenge students intellectually. Well-chosen tasks can pique students' curiosity and draw them into mathematics. The tasks may be connected to the real-world experiences of students, or they may arise in contexts that are purely mathematical. Regardless of the context, worthwhile tasks should be intriguing, with a level of challenge that invites speculation and hard work. Such tasks often can be approached in more than one way, such as using an arithmetic counting approach, drawing a geometric diagram and enumerating possibilities, or using algebraic equations, which makes the tasks accessible to students with varied prior knowledge and experience.

Worthwhile tasks alone are not sufficient for effective teaching. Teachers must also decide what aspects of a task to highlight, how to organize and orchestrate the work of the students, what questions to ask to challenge those with varied levels of expertise, and how to support students without taking over the process of thinking for them and thus eliminating the challenge.

Opportunities to reflect on and refine instructional practice are crucial.

Effective teaching requires continually seeking improvement.

Effective teaching involves observing students, listening carefully to their ideas and explanations, having mathematical goals, and using the information to make instructional decisions. Teachers who employ such practices motivate students to engage in mathematical thinking and reasoning and provide learning opportunities that challenge students at all levels of understanding. Effective teaching requires continuing efforts to learn and improve. These efforts include learning about mathematics and pedagogy, benefiting from interactions with students and colleagues, and engaging in ongoing professional development and self-reflection.

Opportunities to reflect on and refine instructional practice—during class and outside class, alone and with others—are crucial in the vision of school mathematics outlined in *Principles and Standards.* To improve their mathematics instruction, teachers must be able to analyze what they and their students are doing and consider how those actions are affecting students' learning. Using a variety of strategies, teachers should monitor students' capacity and inclination to analyze situations, frame and solve problems, and make sense of mathematical concepts and procedures. They can use this information to assess their students' progress and to appraise how well the mathematical tasks, student discourse, and classroom environment are interacting to foster students' learning. They then use these appraisals to adapt their instruction.

Reflection and analysis are often individual activities, but they can be greatly enhanced by teaming with an experienced and respected colleague, a new teacher, or a community of teachers. Collaborating with colleagues regularly to observe, analyze, and discuss teaching and students' thinking or to do "lesson study" is a powerful, yet neglected, form of professional development in American schools (Stigler and Hiebert 1999). The work and time of teachers must be structured to allow and support professional development that will benefit them and their students.

The Learning Principle

Students must learn mathematics with understanding, actively building new knowledge from experience and prior knowledge.

The vision of school mathematics in *Principles and Standards* is based on students' learning mathematics with understanding. Unfortunately, learning mathematics *without* understanding has long been a common outcome of school mathematics instruction. In fact, learning without understanding has been a persistent problem since at least the 1930s, and it has been the subject of much discussion and research by psychologists and educators over the years (e.g., Brownell [1947]; Skemp [1976]; Hiebert and Carpenter [1992]). Learning the mathematics outlined in chapters 3–7 requires understanding and being able to apply procedures, concepts, and processes. In the twenty-first century, all students should be expected to understand and be able to apply mathematics.

Learning mathematics with understanding is essential.

In recent decades, psychological and educational research on the learning of complex subjects such as mathematics has solidly established the important role of conceptual understanding in the knowledge and activity of persons who are proficient. Being proficient in a complex domain such as mathematics entails the ability to use knowledge flexibly, applying what is learned in one setting appropriately in another. One of the most robust findings of research is that conceptual understanding is an important component of proficiency, along with factual knowledge and procedural facility (Bransford, Brown, and Cocking 1999).

Conceptual understanding is an important component of proficiency.

The alliance of factual knowledge, procedural proficiency, and conceptual understanding makes all three components usable in powerful ways. Students who memorize facts or procedures without understanding often are not sure when or how to use what they know, and such learning is often quite fragile (Bransford, Brown, and Cocking 1999). Learning with understanding also makes subsequent learning easier. Mathematics makes more sense and is easier to remember and to apply when students connect new knowledge to existing knowledge in meaningful ways (Schoenfeld 1988). Well-connected, conceptually grounded ideas are more readily accessed for use in new situations (Skemp 1976).

The requirements for the workplace and for civic participation in the contemporary world include flexibility in reasoning about and using quantitative information. Conceptual understanding is an essential component of the knowledge needed to deal with novel problems and settings. Moreover, as judgments change about the facts or procedures that are essential in an increasingly technological world, conceptual understanding becomes even more important. For example, most of the arithmetic and algebraic procedures long viewed as the heart of the school mathematics curriculum can now be performed with handheld calculators. Thus, more attention can be given to understanding the number concepts and the modeling procedures used in solving problems. Change is a ubiquitous feature of contemporary life, so learning with understanding is essential to enable students to use what they learn to solve the new kinds of problems they will inevitably face in the future.

A major goal of school mathematics programs is to create autonomous learners, and learning with understanding supports this goal. Students learn more and learn better when they can take control of their learning by defining their goals and monitoring their progress. When challenged with appropriately chosen tasks, students become confident in their ability to tackle difficult problems, eager to figure things out on their own, flexible in exploring mathematical ideas and trying alternative solution paths, and willing to persevere. Effective learners recognize the importance of reflecting on their thinking and learning from their mistakes. Students should view the difficulty of complex mathematical investigations as a worthwhile challenge rather than as an excuse to give up. Even when a mathematical task is difficult, it can be engaging and rewarding. When students work hard to solve a difficult problem or to understand a complex idea, they experience a very special feeling of accomplishment, which in turn leads to a willingness to continue and extend their engagement with mathematics.

Learning with understanding is essential to enable students to solve the new kinds of problems they will inevitably face in the future.

Students can learn mathematics with understanding.

Students will be served well by school mathematics programs that enhance their natural desire to understand what they are asked to learn. From a young age, children are interested in mathematical ideas. Through their experiences in everyday life, they gradually develop a rather complex set of informal ideas about numbers, patterns, shapes, quantities, data, and size, and many of these ideas are correct and robust. Thus children learn many mathematical ideas quite naturally even before they enter school (Gelman and Gallistel 1978; Resnick 1987). A pattern of building new learning on prior learning and experience is established early and repeated, albeit often in less obvious ways, throughout the school years (see, e.g., Steffe [1994]). Students of all ages have a considerable knowledge base on which to build, including ideas developed in prior school instruction and those acquired through everyday experience (Bransford, Brown, and Cocking 1999).

The kinds of experiences teachers provide clearly play a major role in determining the extent and quality of students' learning. Students' understanding of mathematical ideas can be built throughout their school years if they actively engage in tasks and experiences designed to deepen and connect their knowledge. Learning with understanding can be further enhanced by classroom interactions, as students propose mathematical ideas and conjectures, learn to evaluate their own thinking and that of others, and develop mathematical reasoning skills (Hanna and Yackel forthcoming). Classroom discourse and social interaction can be used to promote the recognition of connections among ideas and the reorganization of knowledge (Lampert 1986). By having students talk about their informal strategies, teachers can help them become aware of, and build on, their implicit informal knowledge (Lampert 1989; Mack 1990). Moreover, in such settings, procedural fluency and conceptual understanding can be developed through problem solving, reasoning, and argumentation.

The Assessment Principle

Assessment should support the learning of important mathematics and furnish useful information to both teachers and students.

Assessment should not merely be done to students; rather, it should also be done for students.

When assessment is an integral part of mathematics instruction, it contributes significantly to all students' mathematics learning. When assessment is discussed in connection with standards, the focus is sometimes on using tests to certify students' attainment, but there are other important purposes of assessment. Assessment should be more than merely a test at the end of instruction to see how students perform under special conditions; rather, it should be an integral part of instruction that informs and guides teachers as they make instructional decisions. Assessment should not merely be done *to* students; rather, it should also be done *for* students, to guide and enhance their learning.

Assessment should enhance students' learning.

The assertion that assessment should enhance students' learning may be surprising. After all, if assessment ascertains what students have learned and are able to do, how can it also have positive consequences for learning? Research indicates that making assessment an integral part of classroom practice is associated with improved student learning. Black and Wiliam (1998) reviewed about 250 research studies and concluded that the learning of students, including low achievers, is generally enhanced in classrooms where teachers include attention to formative assessment in making judgments about teaching and learning.

Good assessment can enhance students' learning in several ways. First, the tasks used in an assessment can convey a message to students about what kinds of mathematical knowledge and performance are valued. That message can in turn influence the decisions students make—for example, whether or where to apply effort in studying. Thus, it is important that assessment tasks be worthy of students' time and attention. Activities that are consistent with (and sometimes the same as) the activities used in instruction should be included. When teachers use assessment techniques such as observations, conversations and interviews with students, or interactive journals, students are likely to learn through the process of articulating their ideas and answering the teacher's questions.

Feedback from assessment tasks can also help students in setting goals, assuming responsibility for their own learning, and becoming more independent learners. For example, scoring guides, or rubrics, can help teachers analyze and describe students' responses to complex tasks and determine students' levels of proficiency. They can also help students understand the characteristics of a complete and correct response. Similarly, classroom discussions in which students present and evaluate different approaches to solving complex problems can hone their sense of the difference between an excellent response and one that is mediocre. Through the use of good tasks and the public discussion of

The *Assessment Standards for School Mathematics* (NCTM 1995) presented six standards about exemplary mathematics assessment. They address how assessment should–

- reflect the mathematics that students should know and be able to do;
- enhance mathematics learning;
- promote equity;
- be an open process;
- promote valid inference;
- be a coherent process.

criteria for good responses, teachers can cultivate in their students both the disposition and the capacity to engage in self-assessment and reflection on their own work and on ideas put forth by others. Such a focus on self-assessment and peer assessment has been found to have a positive impact on students' learning (Wilson and Kenney forthcoming).

Assessment is a valuable tool for making instructional decisions.

To ensure deep, high-quality learning for all students, assessment and instruction must be integrated so that assessment becomes a routine part of the ongoing classroom activity rather than an interruption. Such assessment also provides the information teachers need to make appropriate instructional decisions. In addition to formal assessments, such as tests and quizzes, teachers should be continually gathering information about their students' progress through informal means, such as asking questions during the course of a lesson, conducting interviews with individual students, and giving writing prompts.

Assessment should become a routine part of the ongoing classroom activity rather than an interruption.

When teachers have useful information about what students are learning, they can support their students' progress toward significant mathematical goals. The instructional decisions made by teachers—such as how and when to review prerequisite material, how to revisit a difficult concept, or how to adapt tasks for students who are struggling or for those who need enrichment—are based on inferences about what students know and what they need to learn. Assessment is a primary source of the evidence on which these inferences are based, and the decisions that teachers make will be only as good as that evidence.

Assessment should reflect the mathematics that all students need to know and be able to do, and it should focus on students' understanding as well as their procedural skills. Teachers need to have a clear sense of what is to be taught and learned, and assessment should be aligned with their instructional goals. By providing information about students' individual and collective progress toward the goals, assessment can help ensure that everyone moves productively in the right direction.

To make effective decisions, teachers should look for convergence of evidence from different sources. Formal assessments provide only one viewpoint on what students can do in a very particular situation—often working individually on paper-and-pencil tasks, with limited time to complete the tasks. Overreliance on such assessments may give an incomplete and perhaps distorted picture of students' performance. Because different students show what they know and can do in different ways, assessments should allow for multiple approaches, thus giving a well-rounded picture and allowing each student to show his or her best strengths.

Many assessment techniques can be used by mathematics teachers, including open-ended questions, constructed-response tasks, selected-response items, performance tasks, observations, conversations, journals, and portfolios. These methods can all be appropriate for classroom assessment, but some may apply more readily to particular goals. For example, quizzes using simple constructed-response or selected-response items may indicate whether students can apply procedures.

Constructed-response or performance tasks may better illuminate students' capacity to apply mathematics in complex or new situations. Observations and conversations in the classroom can provide insights into students' thinking, and teachers can monitor changes in students' thinking and reasoning over time with reflective journals and portfolios.

When teachers are selecting assessment methods, the age, experience, and special needs of students should be considered. Teachers must ensure that all students have an opportunity to demonstrate clearly and completely what they know and can do. For example, teachers should use English-enhancing and bilingual techniques to support students who are learning English.

When done well, assessment that helps teachers make decisions about the content or form of instruction (often called formative assessment) can also be used to judge students' attainment (summative assessment). The same sources of evidence can be assembled to build a picture of individual students' progress toward the goals of instruction. To maximize the instructional value of assessment, teachers need to move beyond a superficial "right or wrong" analysis of tasks to a focus on how students are thinking about the tasks. Efforts should be made to identify valuable student insights on which further progress can be based rather than to concentrate solely on errors or misconceptions. Although less straightforward than averaging scores on quizzes, assembling evidence from a variety of sources is more likely to yield an accurate picture of what each student knows and is able to do.

Assembling evidence from a variety of sources is more likely to yield an accurate picture.

Whether the focus is on formative assessment aimed at guiding instruction or on summative assessment of students' progress, teachers' knowledge is paramount in collecting useful information and drawing valid inferences. Teachers must understand their mathematical goals deeply, they must understand how their students may be thinking about mathematics, they must have a good grasp of possible means of assessing students' knowledge, and they must be skilled in interpreting assessment information from multiple sources. For teachers to attain the necessary knowledge, assessment must become a major focus in teacher preparation and professional development.

The Technology Principle

Technology is essential in teaching and learning mathematics; it influences the mathematics that is taught and enhances students' learning.

Electronic technologies—calculators and computers—are essential tools for teaching, learning, and doing mathematics. They furnish visual images of mathematical ideas, they facilitate organizing and analyzing data, and they compute efficiently and accurately. They can support investigation by students in every area of mathematics, including geometry, statistics, algebra, measurement, and number. When technological tools are available, students can focus on decision making, reflection, reasoning, and problem solving.

Students can learn more mathematics more deeply with the appropriate use of technology (Dunham and Dick 1994; Sheets 1993; Boers-van Oosterum 1990; Rojano 1996; Groves 1994). Technology should not be used as a replacement for basic understandings and intuitions; rather, it can and should be used to foster those understandings and intuitions. In mathematics-instruction programs, technology should be used widely and responsibly, with the goal of enriching students' learning of mathematics.

The existence, versatility, and power of technology make it possible and necessary to reexamine what mathematics students should learn as well as how they can best learn it. In the mathematics classrooms envisioned in *Principles and Standards*, every student has access to technology to facilitate his or her mathematics learning under the guidance of a skillful teacher.

Technology should not be used as a replacement for basic understandings and intuitions.

Technology enhances mathematics learning.

Technology can help students learn mathematics. For example, with calculators and computers students can examine more examples or representational forms than are feasible by hand, so they can make and explore conjectures easily. The graphic power of technological tools affords access to visual models that are powerful but that many students are unable or unwilling to generate independently. The computational capacity of technological tools extends the range of problems accessible to students and also enables them to execute routine procedures quickly and accurately, thus allowing more time for conceptualizing and modeling.

Students' engagement with, and ownership of, abstract mathematical ideas can be fostered through technology. Technology enriches the range and quality of investigations by providing a means of viewing mathematical ideas from multiple perspectives. Students' learning is assisted by feedback, which technology can supply: drag a node in a Dynamic Geometry® environment, and the shape on the screen changes; change the defining rules for a spreadsheet, and watch as dependent values are modified. Technology also provides a focus as students discuss with one another and with their teacher the objects on the screen and the effects of the various dynamic transformations that technology allows.

Technology offers teachers options for adapting instruction to special student needs. Students who are easily distracted may focus more intently on computer tasks, and those who have organizational difficulties may benefit from the constraints imposed by a computer environment. Students who have trouble with basic procedures can develop and demonstrate other mathematical understandings, which in turn can eventually help them learn the procedures. The possibilities for engaging students with physical challenges in mathematics are dramatically increased with special technologies.

The possibilities for engaging students with physical challenges in mathematics are dramatically increased with special technologies.

Technology supports effective mathematics teaching.

The effective use of technology in the mathematics classroom depends on the teacher. Technology is not a panacea. As with any teaching tool, it can be used well or poorly. Teachers should use technology to

Many ideas in the Technology principle are illustrated in the electronic examples in the hypertext edition; see the margin icons in chapters 4–7 referring to the CD-ROM that accompanies this book.

enhance their students' learning opportunities by selecting or creating mathematical tasks that take advantage of what technology can do efficiently and well—graphing, visualizing, and computing. For example, teachers can use simulations to give students experience with problem situations that are difficult to create without technology, or they can use data and resources from the Internet and the World Wide Web to design student tasks. Spreadsheets, dynamic geometry software, and computer microworlds are also useful tools for posing worthwhile problems.

Technology does not replace the mathematics teacher. When students are using technological tools, they often spend time working in ways that appear somewhat independent of the teacher, but this impression is misleading. The teacher plays several important roles in a technology-rich classroom, making decisions that affect students' learning in important ways. Initially, the teacher must decide if, when, and how technology will be used. As students use calculators or computers in the classroom, the teacher has an opportunity to observe the students and to focus on their thinking. As students work with technology, they may show ways of thinking about mathematics that are otherwise often difficult to observe. Thus, technology aids in assessment, allowing teachers to examine the processes used by students in their mathematical investigations as well as the results, thus enriching the information available for teachers to use in making instructional decisions.

Technology influences what mathematics is taught.

Technology not only influences how mathematics is taught and learned but also affects what is taught and when a topic appears in the curriculum. With technology at hand, young children can explore and solve problems involving large numbers, or they can investigate characteristics of shapes using dynamic geometry software. Elementary school students can organize and analyze large sets of data. Middle-grades students can study linear relationships and the ideas of slope and uniform change with computer representations and by performing physical experiments with calculator-based-laboratory systems. High school students can use simulations to study sample distributions, and they can work with computer algebra systems that efficiently perform most of the symbolic manipulation that was the focus of traditional high school mathematics programs. The study of algebra need not be limited to simple situations in which symbolic manipulation is relatively straightforward. Using technological tools, students can reason about more-general issues, such as parameter changes, and they can model and solve complex problems that were heretofore inaccessible to them. Technology also blurs some of the artificial separations among topics in algebra, geometry, and data analysis by allowing students to use ideas from one area of mathematics to better understand another area of mathematics.

Technology can help teachers connect the development of skills and procedures to the more general development of mathematical understanding. As some skills that were once considered essential are rendered less necessary by technological tools, students can be asked to work at higher levels of generalization or abstraction. Work with virtual manipulatives (computer simulations of physical manipulatives) or with Logo can allow young children to extend physical experience and to develop an initial understanding of sophisticated ideas like the use of algorithms. Dynamic geometry software can allow experimentation with families of geometric objects, with an explicit focus on geometric transformations. Similarly, graphing utilities facilitate the exploration of characteristics of classes of functions. Because of technology, many topics in discrete mathematics take on new importance in the contemporary mathematics classroom; the boundaries of the mathematical landscape are being transformed.

Standards are descriptions of what mathematics instruction should enable students to know and do.

Ambitious standards are required to achieve a society that has the capability to think and reason mathematically.

Standards for School Mathematics

Prekindergarten through Grade 12

What mathematical content and processes should students know and be able to use as they progress through school? *Principles and Standards for School Mathematics* presents NCTM's proposal for what should be valued in school mathematics education. Ambitious standards are required to achieve a society that has the capability to think and reason mathematically and a useful base of mathematical knowledge and skills.

The ten Standards presented in this chapter describe a connected body of mathematical understandings and competencies—a comprehensive foundation recommended for all students, rather than a menu from which to make curricular choices. Standards are descriptions of what mathematics instruction should enable students to know and do. They specify the understanding, knowledge, and skills that students should acquire from prekindergarten through grade 12. The Content Standards—Number and Operations, Algebra, Geometry, Measurement, and Data Analysis and Probability—explicitly describe the content that students should learn. The Process Standards—Problem Solving, Reasoning and Proof, Communication, Connections, and Representation—highlight ways of acquiring and using content knowledge.

Growth across the Grades: Aiming for Focus and Coherence

Each of these ten Standards applies across all grades, prekindergarten through grade 12. The set of Standards, which are discussed in detail in chapters 4 through 7, proposes the mathematics that all students should have the opportunity to learn. Each Standard comprises a small number of goals that apply across all grades—a commonality that promotes a focus on the growth in students' knowledge and sophistication as they progress through the curriculum. For each of the Content Standards, chapters 4 through 7 offer an additional set of expectations specific to each grade band.

The Table of Standards and expectations in the appendix highlights the growth of expectations across the grades. It is not expected that every topic will be addressed each year. Rather, students will reach a certain depth of understanding of the concepts and acquire certain levels of fluency with the procedures by prescribed points in the curriculum, so further instruction can assume and build on this understanding and fluency.

Even though each of these ten Standards applies to all grades, emphases will vary both within and between the grade bands. For instance, the emphasis on number is greatest in prekindergarten through grade 2, and by grades 9–12, number receives less instructional attention. And the total time for mathematical instruction will be divided differently according to particular needs in each grade band—for example, in the middle grades, the majority of instructional time would address algebra and geometry. Figure 3.1 shows roughly how the Content Standards might receive different emphases across the grade bands.

Fig. **3.1.**

The Content Standards should receive different emphases across the grade bands.

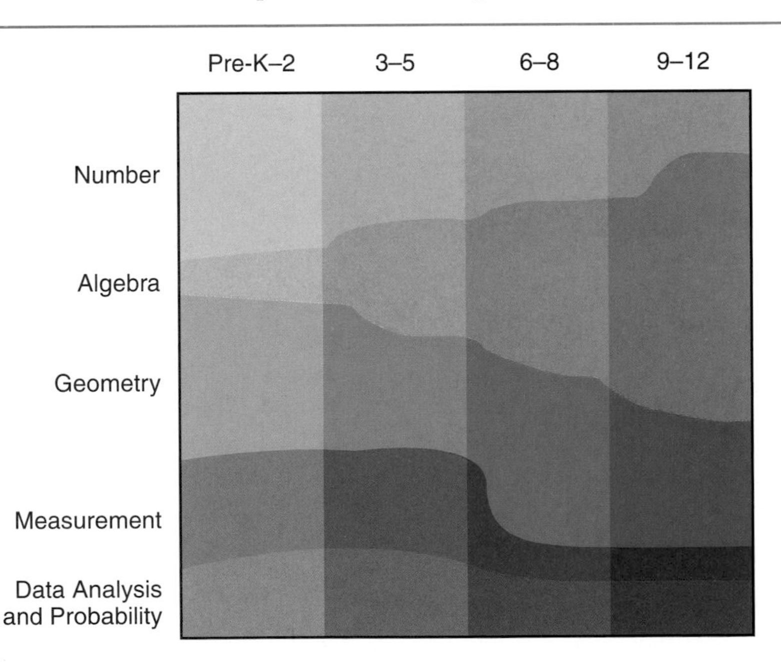

This set of ten Standards does not neatly separate the school mathematics curriculum into nonintersecting subsets. Because mathematics as a discipline is highly interconnected, the areas described by the Standards overlap and are integrated. Processes can be learned within the Content Standards, and content can be learned within the Process

Standards. Rich connections and intersections abound. Number, for example, pervades all areas of mathematics. Some topics in data analysis could be characterized as part of measurement. Patterns and functions appear throughout geometry. The processes of reasoning, proving, problem solving, and representing are used in all content areas.

The arrangement of the curriculum into these Standards is proposed as one coherent organization of significant mathematical content and processes. Those who design curriculum frameworks, assessments, instructional materials, and classroom instruction based on *Principles and Standards* will need to make their own decisions about emphasis and order; other labels and arrangements are certainly possible.

Where Is Discrete Mathematics?

The 1989 *Curriculum and Evaluation Standards for School Mathematics* introduced a Discrete Mathematics Standard in grades 9–12. In *Principles and Standards*, the main topics of discrete mathematics are included, but they are distributed across the Standards, instead of receiving separate treatment, and they span the years from prekindergarten through grade 12. As an active branch of contemporary mathematics that is widely used in business and industry, discrete mathematics should be an integral part of the school mathematics curriculum, and these topics naturally occur throughout the other strands of mathematics.

The areas described by the Standards overlap and are integrated.

Three important areas of discrete mathematics are integrated within these Standards: combinatorics, iteration and recursion, and vertex-edge graphs. These ideas can be systematically developed from prekindergarten through grade 12. In addition, matrices should be addressed in grades 9–12. Combinatorics is the mathematics of systematic counting. Iteration and recursion are used to model sequential, step-by-step change. Vertex-edge graphs are used to model and solve problems involving paths, networks, and relationships among a finite number of objects.

Number and Operations

Historically, number has been a cornerstone of the mathematics curriculum.

The Number and Operations Standard describes deep and fundamental understanding of, and proficiency with, counting, numbers, and arithmetic, as well as an understanding of number systems and their structures. The concepts and algorithms of elementary arithmetic are part of number and operations, as are the properties and characteristics of the classes of numbers that form the beginnings of number theory. Central to this Standard is the development of number sense—the ability to decompose numbers naturally, use particular numbers like 100 or 1/2 as referents, use the relationships among arithmetic operations to solve problems, understand the base-ten number system, estimate, make sense of numbers, and recognize the relative and absolute magnitude of numbers (Sowder 1992).

Historically, number has been a cornerstone of the entire mathematics curriculum internationally as well as in the United States and Canada (Reys and Nohda 1994). All the mathematics proposed for prekindergarten through grade 12 is strongly grounded in number. The principles that govern equation solving in algebra are the same as the structural properties of systems of numbers. In geometry and measurement, attributes are described with numbers. The entire area of data analysis involves making sense of numbers. Through problem solving, students can explore and solidify their understandings of number. Young children's earliest mathematical reasoning is likely to be about number situations, and their first mathematical representations will probably be of numbers. Research has shown that learning about number and operations is a complex process for children (e.g., Fuson [1992]).

In these Standards, understanding number and operations, developing number sense, and gaining fluency in arithmetic computation form the core of mathematics education for the elementary grades. As they progress from prekindergarten through grade 12, students should attain a rich understanding of numbers—what they are; how they are represented with objects, numerals, or on number lines; how they are related to one another; how numbers are embedded in systems that have structures and properties; and how to use numbers and operations to solve problems.

Knowing basic number combinations—the single-digit addition and multiplication pairs and their counterparts for subtraction and division—is essential. Equally essential is computational fluency—having and using efficient and accurate methods for computing. Fluency might be manifested in using a combination of mental strategies and jottings on paper or using an algorithm with paper and pencil, particularly when the numbers are large, to produce accurate results quickly. Regardless of the particular method used, students should be able to explain their method, understand that many methods exist, and see the usefulness of methods that are efficient, accurate, and general. Students also need to be able to estimate and judge the reasonableness of results. Computational fluency should develop in tandem with understanding of the role and meaning of arithmetic operations in number systems (Hiebert et.al., 1997; Thornton 1990).

Calculators should be available at appropriate times as computational tools, particularly when many or cumbersome computations are needed to solve problems. However, when teachers are working with students

Number and Operations Standard

Instructional programs from prekindergarten through grade 12 should enable all students to—

- understand numbers, ways of representing numbers, relationships among numbers, and number systems;
- understand meanings of operations and how they relate to one another;
- compute fluently and make reasonable estimates.

on developing computational algorithms, the calculator should be set aside to allow this focus. Today, the calculator is a commonly used computational tool outside the classroom, and the environment inside the classroom should reflect this reality.

Understand numbers, ways of representing numbers, relationships among numbers, and number systems

Understanding of number develops in prekindergarten through grade 2 as children count and learn to recognize "how many" in sets of objects. A key idea is that a number can be decomposed and thought about in many ways. For instance, 24 is 2 tens and 4 ones and also 2 sets of twelve. Making a transition from viewing "ten" as simply the accumulation of 10 ones to seeing it both as 10 ones *and* as 1 ten is an important first step for students toward understanding the structure of the base-ten number system (Cobb and Wheatley 1988). Throughout the elementary grades, students can learn about classes of numbers and their characteristics, such as which numbers are odd, even, prime, composite, or square.

Beyond understanding whole numbers, young children can be encouraged to understand and represent commonly used fractions in context, such as 1/2 of a cookie or 1/8 of a pizza, and to see fractions as part of a unit whole or of a collection. Teachers should help students develop an understanding of fractions as division of numbers. And in the middle grades, in part as a basis for their work with proportionality, students need to solidify their understanding of fractions as numbers. Students' knowledge about, and use of, decimals in the base-ten system should be very secure before high school. With a solid understanding of number, high school students can use variables that represent numbers to do meaningful symbolic manipulation.

Representing numbers with various physical materials should be a major part of mathematics instruction in the elementary school grades. By the middle grades, students should understand that numbers can be represented in various ways, so that they see that 1/4, 25%, and 0.25 are all different names for the same number. Students' understanding and ability to reason will grow as they represent fractions and decimals with physical materials and on number lines and as they learn to generate equivalent representations of fractions and decimals.

As students gain understanding of numbers and how to represent them, they have a foundation for understanding relationships among numbers. In grades 3 through 5, students can learn to compare fractions to familiar benchmarks such as 1/2. And, as their number sense develops, students should be able to reason about numbers by, for instance, explaining that 1/2 + 3/8 must be less than 1 because each addend is less than or equal to 1/2. In grades 6–8, it is important for students to be able to move flexibly among equivalent fractions, decimals, and percents and to order and compare rational numbers using a range of strategies. By extending from whole numbers to integers, middle-grades students' intuitions about order and magnitude will be more reliable, and they have a glimpse into the way that systems of numbers work. High school students can use variables and functions to represent relationships among sets of numbers, and to look at properties of classes of numbers.

Representing numbers with various physical materials should be a major part of mathematics instruction in the elementary grades.

Although other curricular areas are emphasized more than number in grades 9–12, in these grades students should see number systems from a more global perspective. They should learn about differences among number systems and about what properties are preserved and lost in moving from one system to another.

Understand meanings of operations and how they relate to one another

During the primary grades, students should encounter a variety of meanings for addition and subtraction of whole numbers. Researchers and teachers have learned about how children understand operations through their approaches to simple arithmetic problems like this:

> Bob got 2 cookies. Now he has 5 cookies. How many cookies did Bob have in the beginning?

To solve this problem, young children might use addition and count on from 2, keeping track with their fingers, to get to 5. Or they might recognize this problem as a subtraction situation and use the fact that $5 - 2 = 3$. Exploring thinking strategies like these or realizing that $7 + 8$ is the same as $7 + 7 + 1$ will help students see the meaning of the operations. Such explorations also help teachers learn what students are thinking. Multiplication and division can begin to have meaning for students in prekindergarten through grade 2 as they solve problems that arise in their environment, such as how to share a bag of raisins fairly among four people.

The same operation can be applied in problem situations that seem quite different from one another.

In grades 3–5, helping students develop meaning for whole-number multiplication and division should become a central focus. By creating and working with representations (such as diagrams or concrete objects) of multiplication and division situations, students can gain a sense of the relationships among the operations. Students should be able to decide whether to add, subtract, multiply, or divide for a particular problem. To do so, they must recognize that the same operation can be applied in problem situations that on the surface seem quite different from one another, know how operations relate to one another, and have an idea about what kind of result to expect.

In grades 6–8, operations with rational numbers should be emphasized. Students' intuitions about operations should be adapted as they work with an expanded system of numbers (Graeber and Campbell 1993). For example, multiplying a whole number by a fraction between 0 and 1 (e.g., $8 \times 1/2$) produces a result less than the whole number. This is counter to students' prior experience (with whole numbers) that multiplication always results in a greater number.

Working with proportions is a major focus proposed in these Standards for the middle grades. Students should become proficient in creating ratios to make comparisons in situations that involve pairs of numbers, as in the following problem:

> If three packages of cocoa make fifteen cups of hot chocolate, how many packages are needed to make sixty cups?

Students at this level also need to learn operations with integers. In grades 9–12, as students learn how to combine vectors and matrices arithmetically, they will experience other kinds of systems involving numbers in which new properties and patterns emerge.

Compute fluently and make reasonable estimates

Developing fluency requires a balance and connection between conceptual understanding and computational proficiency. On the one hand, computational methods that are over-practiced without understanding are often forgotten or remembered incorrectly (Hiebert 1999; Kamii, Lewis, and Livingston, 1993; Hiebert and Lindquist 1990). On the other hand, understanding without fluency can inhibit the problem-solving process (Thornton 1990). As children in prekindergarten through grade 2 develop an understanding of whole numbers and the operations of addition and subtraction, instructional attention should focus on strategies for computing with whole numbers so that students develop flexibility and computational fluency. Students will generate a range of interesting and useful strategies for solving computational problems, which should be shared and discussed. By the end of grade 2, students should know the basic addition and subtraction combinations, should be fluent in adding two-digit numbers, and should have methods for subtracting two-digit numbers. At the grades 3–5 level, as students develop the basic number combinations for multiplication and division, they should also develop reliable algorithms to solve arithmetic problems efficiently and accurately. These methods should be applied to larger numbers and practiced for fluency.

Developing fluency requires a balance and connection between conceptual understanding and computational proficiency.

Researchers and experienced teachers alike have found that when children in the elementary grades are encouraged to develop, record, explain, and critique one another's strategies for solving computational problems, a number of important kinds of learning can occur (see, e.g., Hiebert [1999]; Kamii, Lewis, and Livingston [1993]; Hiebert et al. [1997]). The efficiency of various strategies can be discussed. So can their generalizability: Will this work for any numbers or only the two involved here? And experience suggests that in classes focused on the development and discussion of strategies, various "standard" algorithms either arise naturally or can be introduced by the teacher as appropriate. The point is that students must become fluent in arithmetic computation—they must have efficient and accurate methods that are supported by an understanding of numbers and operations. "Standard" algorithms for arithmetic computation are one means of achieving this fluency.

The development of rational-number concepts is a major goal for grades 3–5, which should lead to informal methods for calculating with fractions. For example, a problem such as 1/4 + 1/2 should be solved mentally with ease because students can picture 1/2 and 1/4 or can use decomposition strategies, such as 1/4 + 1/2 = 1/4 + (1/4 + 1/4). In these grades, methods for computing with decimals should be developed and applied, and by grades 6–8, students should become fluent in computing with rational numbers in fraction and decimal form. When asked to estimate 12/13 + 7/8, only 24 percent of thirteen-year-old students in a national assessment said the answer was close to 2 (Carpenter et al. 1981). Most said it was close to 1, 19, or 21, all of which reflect common computational errors in adding fractions and suggest a lack of understanding of the operation being carried out. If students understand addition of fractions and have developed number sense, these errors should not occur. As they develop an understanding of the meaning and representation of integers, they should also develop methods for computing with

Part of being able to compute fluently means making smart choices about which tools to use and when.

integers. In grades 9–12, students should compute fluently with real numbers and have some basic proficiency with vectors and matrices in solving problems, using technology as appropriate.

Part of being able to compute fluently means making smart choices about which tools to use and when. Students should have experiences that help them learn to choose among mental computation, paper-and-pencil strategies, estimation, and calculator use. The particular context, the question, and the numbers involved all play roles in those choices. Do the numbers allow a mental strategy? Does the context call for an estimate? Does the problem require repeated and tedious computations? Students should evaluate problem situations to determine whether an estimate or an exact answer is needed, using their number sense to advantage, and be able to give a rationale for their decision.

Algebra

Algebra has its historical roots in the study of general methods for solving equations. The Algebra Standard emphasizes relationships among quantities, including functions, ways of representing mathematical relationships, and the analysis of change. Functional relationships can be expressed by using symbolic notation, which allows complex mathematical ideas to be expressed succinctly and change to be analyzed efficiently. Today, the methods and ideas of algebra support mathematical work in many areas. For example, distribution and communication networks, laws of physics, population models, and statistical results can all be represented in the symbolic language of algebra. In addition, algebra is about abstract structures and about using the principles of those structures in solving problems expressed with symbols.

Much of the symbolic and structural emphasis in algebra can build on students' extensive experiences with number. Algebra is also closely linked to geometry and to data analysis. The ideas included in the Algebra Standard constitute a major component of the school mathematics curriculum and help to unify it. Algebraic competence is important in adult life, both on the job and as preparation for postsecondary education. All students should learn algebra.

By viewing algebra as a strand in the curriculum from prekindergarten on, teachers can help students build a solid foundation of understanding and experience as a preparation for more-sophisticated work in algebra in the middle grades and high school. For example, systematic experience with patterns can build up to an understanding of the idea of function (Erick Smith forthcoming), and experience with numbers and their properties lays a foundation for later work with symbols and algebraic expressions. By learning that situations often can be described using mathematics, students can begin to form elementary notions of mathematical modeling.

Many adults equate school algebra with symbol manipulation—solving complicated equations and simplifying algebraic expressions. Indeed, the algebraic symbols and the procedures for working with them are a towering, historic mathematical accomplishment and are critical in mathematical work. But algebra is more than moving symbols around. Students need to understand the concepts of algebra, the structures and principles that govern the manipulation of the symbols, and how the symbols themselves can be used for recording ideas and gaining insights into situations. Computer technologies today can produce graphs of functions, perform operations on symbols, and instantaneously do calculations on columns of data. Students now need to learn how to interpret technological representations and how to use the technology effectively and wisely.

Often, algebra has not been treated explicitly in the school curriculum until the traditional algebra course offered in middle school or high school. By promoting algebra as a strand that is begun in the early grades, *Principles and Standards* supports other possibilities for configuring programs in the middle grades and secondary schools. The Standards for grades 6–8 include a significant emphasis on algebra, along with much more geometry than has normally been offered in the middle grades, and call for the integration of these two areas. The

All students should learn algebra.

Algebra Standard

Instructional programs from prekindergarten through grade 12 should enable all students to—

- understand patterns, relations, and functions;
- represent and analyze mathematical situations and structures using algebraic symbols;
- use mathematical models to represent and understand quantitative relationships;
- analyze change in various contexts.

Standards for grades 9–12, assuming that this strong foundation in algebra will be in place by the end of the eighth grade, describe an ambitious program in algebra, geometry, and data analysis and statistics and also call for integration and connections among ideas.

Understand patterns, relations, and functions

Early experiences with classifying and ordering objects are natural and interesting for young children. Teachers might help children notice that red-blue-blue-red-blue-blue can be extended with another red-blue-blue sequence or help them predict that the twelfth term is blue, assuming that the red-blue-blue pattern repeats indefinitely. Initially, students may describe the regularity in patterns verbally rather than with mathematical symbols (English and Warren 1998). In grades 3–5, they can begin to use variables and algebraic expressions as they describe and extend patterns. By the end of secondary school, they should be comfortable using the notation of functions to describe relationships.

Algebra is more than moving symbols around.

In the lower grades, students can describe patterns like 2, 4, 6, 8, … by focusing on how a term is obtained from the previous number—in this example, by adding 2. This is the beginning of recursive thinking. Later, students can study sequences that can best be defined and computed using recursion, such as the Fibonacci sequence, 1, 1, 2, 3, 5, 8, …, in which each term is the sum of the previous two terms. Recursive sequences appear naturally in many contexts and can be studied using technology.

As they progress from preschool through high school, students should develop a repertoire of many types of functions. In the middle grades, students should focus on understanding linear relationships. In high school, they should enlarge their repertoire of functions and learn about the characteristics of classes of functions.

Many college students understand the notion of function only as a rule or formula such as "given n, find 2^n for $n = 0, 1, 2$, and 3" (Vinner and Dreyfus 1989). By the middle grades, students should be able to understand the relationships among tables, graphs, and symbols and to judge the advantages and disadvantages of each way of representing relationships for particular purposes. As they work with multiple representations of functions—including numeric, graphic, and symbolic—they will develop a more comprehensive understanding of functions (see Leinhardt, Zaslavsky, and Stein 1990; Moschkovich, Schoenfeld and Arcavi 1993; NRC 1998).

Represent and analyze mathematical situations and structures using algebraic symbols

Students' understanding of properties of numbers develops gradually from preschool through high school. While young children are skip-counting by twos, they may notice that the numbers they are using end in 0, 2, 4, 6, and 8; they could then use this algebraic observation to extend the pattern. In grades 3–5, as students investigate properties of whole-number operations, they may find that they can multiply 18 by 14 mentally by computing 18×10 and adding it to 18×4; they are using the distributive property of multiplication over addition. Sometimes geometric arguments can be understood long before students can reasonably be expected to perform sophisticated manipulations of algebraic symbols. For example, the diagram in figure 3.2 might help lead upper elementary school students to the conjecture that the sum of the first n odd numbers is n^2. Middle school students should be able to understand how the diagram relates to the equation. Students in high school should be able to represent the relationship in general, with symbols, as $1 + 3 + \cdots + (2n - 1) = n^2$, and they should be able to prove the validity of their generalization.

Fig. **3.2.**

Demonstration that $1 + 3 + 5 + 7 = 4^2$

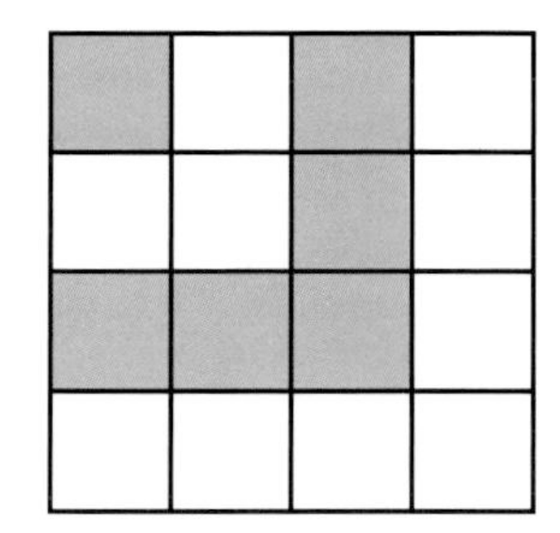

Research indicates a variety of student difficulties with the concept of variable (Küchemann 1978; Kieran 1983; Wagner and Parker 1993), so developing understanding of variable over the grades is important. In the elementary grades, students typically develop a notion of variable as a placeholder for a specific number, as in __ + 2 = 11. Later, they should learn that the variable x in the equation $3x + 2 = 11$ has a very different use from the variable x in the identity $0 \times x = 0$ and that both uses are quite different from the use of r in the formula $A = \pi r^2$. A thorough understanding of variable develops over a long time, and it needs to be grounded in extensive experience (Sfard 1991).

The notion of equality also should be developed throughout the curriculum. As a consequence of the instruction they have received, young students typically perceive the equals sign operationally, that is, as a signal to "do something" (Behr, Erlwanger, and Nichols 1976; Kieran 1981). They should come to view the equals sign as a symbol of equivalence and balance.

Students should begin to develop their skill in producing equivalent expressions and solving linear equations in the middle grades, both mentally and with paper and pencil. They should develop fluency in operating with symbols in their high school years, with by-hand or mental computation in simple cases and with computer algebra technology in all cases. In general, if students engage extensively in symbolic manipulation before they develop a solid conceptual foundation for their work, they will be unable to do more than mechanical manipulations (NRC 1998). The foundation for meaningful work with symbolic notation should be laid over a long time.

Use mathematical models to represent and understand quantitative relationships

One of the most powerful uses of mathematics is the mathematical modeling of phenomena. Students at all levels should have opportunities to model a wide variety of phenomena mathematically in ways that are appropriate to their level. In the lower elementary grades, students can use objects, pictures, and symbols to model situations that involve the addition and subtraction of whole numbers. When children demonstrate the situation "Gary has 4 apples, and Becky has 5 more" by arranging counters, they are doing beginning work with modeling.

In grades 3–5 students should use their models to make predictions, draw conclusions, or better understand quantitative situations. These uses of models will grow more sophisticated. For instance, in solving a problem about making punch, middle-grades students might describe the relationships in the problem with the formula $P = (8/3)J$, where P is the number of cups of punch and J is the number of cups of juice. This mathematical model that can be used to decide how much punch will be made from fifty cups of juice.

High school students should be able to develop models by drawing on their knowledge of many classes of functions—to decide, for instance, whether a situation would best be modeled with a linear function or a quadratic function—and be able to draw conclusions about the situation by analyzing the model. Using computer-based laboratories (devices that gather data, such as the speed or distance of an object, and transmit them directly to a computer so that graphs, tables, and equations can be generated), students can get reliable numerical data quickly from physical experiments. This technology allows them to build models in a wide range of interesting situations.

Analyze change in various contexts

Understanding change is fundamental to understanding functions and to understanding many ideas presented in the news.

Understanding change is fundamental to understanding functions and to understanding many ideas presented in the news. The study of mathematical change is formalized in calculus, when students study the concept of the derivative. Research indicates this is not an area that students typically understand with much depth, even after taking calculus (Smith forthcoming). If ideas of change receive a more explicit focus from the early grades on, perhaps students will eventually enter calculus with a stronger basis for understanding the ideas at that level. In prekindergarten through grades 2, students can, at first, describe qualitative change ("I grew taller over the summer") and then quantitative change ("I grew two inches in the last year"). Using graphs and tables, students in grades 3–5 can begin to notice and describe change, such as the changing nature of the growth of a plant—"It grows slowly, then grows faster, then slows down." And as they look at sequences, they can distinguish between arithmetic growth (2, 5, 8, 11, 14, …) and geometric growth (2, 4, 8, 16, …). With a strong middle-grades focus on linearity, students should learn about the idea that slope represents the constant rate of change in linear functions and be ready to learn in high school about classes of functions that have nonconstant rates of change.

Geometry

Through the study of geometry, students will learn about geometric shapes and structures and how to analyze their characteristics and relationships. Spatial visualization—building and manipulating mental representations of two- and three-dimensional objects and perceiving an object from different perspectives—is an important aspect of geometric thinking. Geometry is a natural place for the development of students' reasoning and justification skills, culminating in work with proof in the secondary grades. Geometric modeling and spatial reasoning offer ways to interpret and describe physical environments and can be important tools in problem solving.

Geometric ideas are useful in representing and solving problems in other areas of mathematics and in real-world situations, so geometry should be integrated when possible with other areas. Geometric representations can help students make sense of area and fractions, histograms and scatterplots can give insights about data, and coordinate graphs can serve to connect geometry and algebra. Spatial reasoning is helpful in using maps, planning routes, designing floor plans, and creating art. Students can learn to see the structure and symmetry around them. Using concrete models, drawings, and dynamic geometry software, students can engage actively with geometric ideas. With well-designed activities, appropriate tools, and teachers' support, students can make and explore conjectures about geometry and can learn to reason carefully about geometric ideas from the earliest years of schooling. Geometry is more than definitions; it is about describing relationships and reasoning. The notion of building understanding in geometry across the grades, from informal to more formal thinking, is consistent with the thinking of theorists and researchers (Burger and Shaughnessy 1986; Fuys, Geddes, and Tischler 1988; Senk 1989; van Hiele 1986).

Geometry has long been regarded as the place in the school mathematics curriculum where students learn to reason and to see the axiomatic structure of mathematics. The Geometry Standard includes a strong focus on the development of careful reasoning and proof, using definitions and established facts. Technology also has an important role in the teaching and learning of geometry. Tools such as dynamic geometry software enable students to model, and have an interactive experience with, a large variety of two-dimensional shapes. Using technology, students can generate many examples as a way of forming and exploring conjectures, but it is important for them to recognize that generating many examples of a particular phenomenon does not constitute a proof. Visualization and spatial reasoning are also improved by interaction with computer animations and in other technological settings (Clements et al. 1997; Yates 1988).

Analyze characteristics and properties of two- and three-dimensional geometric shapes and develop mathematical arguments about geometric relationships

Young students are inclined naturally to observe and describe a variety of shapes and to begin to notice their properties. Identifying shapes is

Geometric ideas are useful in representing and solving problems.

Geometry Standard

Instructional programs from prekindergarten through grade 12 should enable all students to—

- analyze characteristics and properties of two- and three-dimensional geometric shapes and develop mathematical arguments about geometric relationships;
- specify locations and describe spatial relationships using coordinate geometry and other representational systems;
- apply transformations and use symmetry to analyze mathematical situations;
- use visualization, spatial reasoning, and geometric modeling to solve problems.

important, too, but the focus on properties and their relationships should be strong. For example, students in prekindergarten through grade 2 may observe that rectangles work well for tiling because they have four right angles. At this level, students can learn about geometric shapes using objects that can be seen, held, and manipulated. Later, the study of the attributes of shapes and of their properties becomes more abstract. In higher grades, students can learn to focus on and discuss components of shapes, such as sides and angles, and the properties of classes of shapes. For example, using objects or dynamic geometric software to experiment with a variety of rectangles, students in grades 3–5 should be able to conjecture that rectangles always have congruent diagonals that bisect each other.

Through the middle grades and into high school, as they study such topics as similarity and congruence, students should learn to use deductive reasoning and more-formal proof techniques to solve problems and to prove conjectures. At all levels, students should learn to formulate convincing explanations for their conjectures and solutions. Eventually, they should be able to describe, represent, and investigate relationships within a geometric system and to express and justify them in logical chains. They should also be able to understand the role of definitions, axioms, and theorems and be able to construct their own proofs.

Specify locations and describe spatial relationships using coordinate geometry and other representational systems

At first, young children learn concepts of relative position, such as above, behind, near, and between. Later they can make and use rectangular grids to locate objects and measure the distance between points along vertical or horizontal lines. Experiences with the rectangular coordinate plane will be useful as they solve a wider array of problems in geometry and algebra. In the middle and secondary grades, the coordinate plane can be helpful as students work on discovering and analyzing properties of shapes. Finding distances between points in the plane by using scales on maps or the Pythagorean relationship is important in the middle grades. Geometric figures, such as lines in the middle grades or triangles and circles in high school, can be represented analytically, thus establishing a fundamental connection between algebra and geometry.

Students should gain experience in using a variety of visual and coordinate representations to analyze problems and study mathematics.

Students should gain experience in using a variety of visual and coordinate representations to analyze problems and study mathematics. In the elementary grades, for example, an interpretation of whole-number addition can be demonstrated on the number line. In later years, students can use the number line to represent operations on other types of numbers. In grades 3–5, grids and arrays can help students understand multiplication. Later, more-complex problems can be considered. For example, in trying to minimize the distance an ambulance would have to travel to reach a new hospital from any location in the community, students in the middle grades might use distances measured along streets. In high school, students can be asked to find the shortest airplane route between two cities and compare the results using a map to the results using a globe. If students were trying to minimize the distances of a car trip to several cities, they might use vertex-edge graphs. High school students should use Cartesian coordinates as a means both to solve problems and to prove their results.

Apply transformations and use symmetry to analyze mathematical situations

Young children come to school with intuitions about how shapes can be moved. Students can explore motions such as slides, flips, and turns by using mirrors, paper folding, and tracing. Later, their knowledge about transformations should become more formal and systematic. In grades 3–5 students can investigate the effects of transformations and begin to describe them in mathematical terms. Using dynamic geometry software, they can begin to learn the attributes needed to define a transformation. For example, to transform a figure using a rotation, students need to define the center of rotation, the direction of the rotation, and the angle of rotation, as illustrated in figure 3.3. In the middle grades, students should learn to understand what it means for a transformation to preserve distance, as translations, rotations, and reflections do. High school students should learn multiple ways of expressing transformations, including using matrices to show how figures are transformed on the coordinate plane, as well as function notation. They should also begin to understand the effects of compositions of transformations. At all grade levels, appropriate consideration of symmetry provides insights into mathematics and into art and aesthetics.

Fig. **3.3.**

A clockwise rotation of 120°

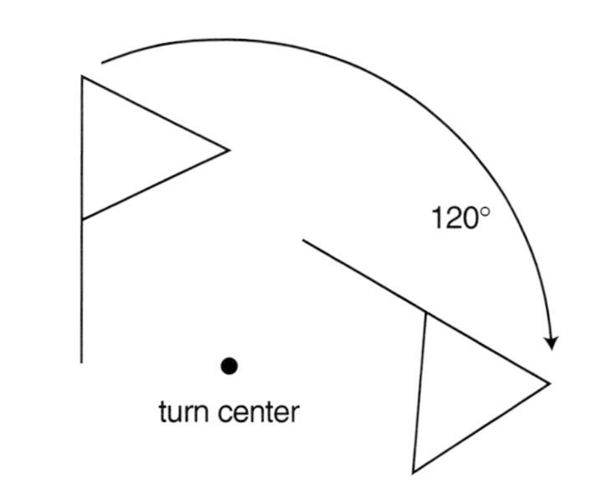

Use visualization, spatial reasoning, and geometric modeling to solve problems

Beginning in the early years of schooling, students should develop visualization skills through hands-on experiences with a variety of geometric objects and through the use of technology that allows them to turn, shrink, and deform two- and three-dimensional objects. Later, they should become comfortable analyzing and drawing perspective views, counting component parts, and describing attributes that cannot be seen but can be inferred. Students need to learn to physically and mentally change the position, orientation, and size of objects in systematic ways as they develop their understandings about congruence, similarity, and transformations.

One aspect of spatial visualization involves moving between two- and three-dimensional shapes and their representations. Elementary school students can wrap blocks in nets—two dimensional figures, usually made of paper, that can be folded to form three-dimensional objects—as a step toward learning to predict whether certain nets match certain solids. By the middle grades, they should be able to interpret and create top or side views of objects. This skill can be developed by challenging them to build a structure given only the side view and the front view, as in figure 3.4. In grades 3–5, students can determine if it is possible to build more than one structure satisfying both conditions. Middle-grades and secondary school students can be asked to find the minimum number of blocks needed to build the structure. High school students should be able to visualize and draw other cross-sections of the structures and of a range of geometric solids.

Fig. **3.4.**

A block structure
(from a presentation by J. de Lange)

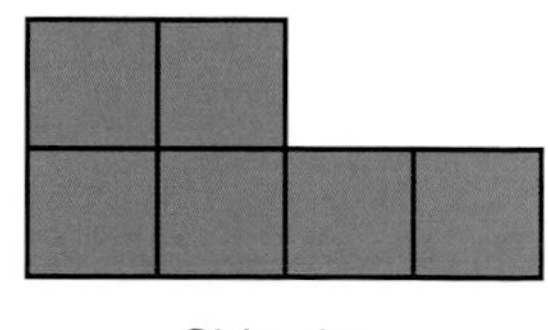

Side view

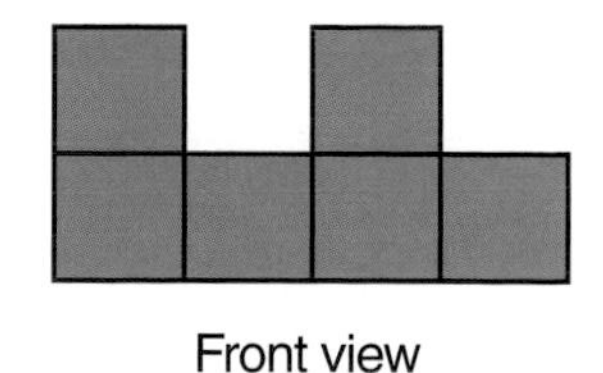

Front view

Measurement

Measurement is the assignment of a numerical value to an attribute of an object, such as the length of a pencil. At more-sophisticated levels, measurement involves assigning a number to a characteristic of a situation, as is done by the consumer price index. Understanding what a measurable attribute is and becoming familiar with the units and processes that are used in measuring attributes is a major emphasis in this Standard. Through their school experience, primarily in prekindergarten through grade 8, students should become proficient in using measurement tools, techniques, and formulas in a range of situations.

Instructional programs should not repeat the same measurement curriculum year after year.

The study of measurement is important in the mathematics curriculum from prekindergarten through high school because of the practicality and pervasiveness of measurement in so many aspects of everyday life. The study of measurement also offers an opportunity for learning and applying other mathematics, including number operations, geometric ideas, statistical concepts, and notions of function. It highlights connections within mathematics and between mathematics and areas outside of mathematics, such as social studies, science, art, and physical education.

Measurement lends itself especially well to the use of concrete materials. In fact, it is unlikely that children can gain a deep understanding of measurement without handling materials, making comparisons physically, and measuring with tools. Measurement concepts should grow in sophistication and breadth across the grades, and instructional programs should not repeat the same measurement curriculum year after year. However, it should be emphasized more in the elementary and middle grades than in high school.

Measurement Standard

Instructional programs from prekindergarten through grade 12 should enable all students to—

- understand measurable attributes of objects and the units, systems, and processes of measurement;
- apply appropriate techniques, tools, and formulas to determine measurements.

Understand measurable attributes of objects and the units, systems, and processes of measurement

A measurable attribute is a characteristic of an object that can be quantified. Line segments have length, plane regions have area, and physical objects have mass. As students progress through the curriculum from preschool through high school, the set of attributes they can measure should expand. Recognizing that objects have attributes that are measurable is the first step in the study of measurement. Children in prekindergarten through grade 2 begin by comparing and ordering objects using language such as *longer* and *shorter*. Length should be the focus in this grade band, but weight, time, area, and volume should also be explored. In grades 3–5, students should learn about area more thoroughly, as well as perimeter, volume, temperature, and angle measure. In these grades, they learn that measurements can be computed using formulas and need not always be taken directly with a measuring tool. Middle-grades students build on these earlier measurement experiences by continuing their study of perimeter, area, and volume and by beginning to explore derived measurements, such as speed. They should also become proficient in measuring angles and understanding angle relationships. In high school, students should understand how decisions about unit and scale can affect measurements. Whatever their grade level, students should have many informal experiences in understanding attributes before using tools to measure them or relying on formulas to compute measurements.

As they progress through school, not only should students' repertoire of measurable attributes expand, but their understanding of the relationships between attributes should also develop. Students in the elementary grades can explore how changing an object's attributes affects certain measurements. For example, cutting apart and rearranging the pieces of a shape may change the perimeter but will not affect the area. In the middle grades this idea can be extended to explorations of how the surface area of a rectangular prism can vary as the volume is held constant. Such observations can offer glimpses of sophisticated mathematical concepts such as invariance under certain transformations.

The types of units that students use for measuring and the ways they use them should expand and shift as students move through the prekindergarten through grade 2 curriculum. In preschool through grade 2, students should begin their study of measurement by using nonstandard units. They should be encouraged to use a wide variety of objects, such as paper clips to measure length, square tiles to measure area, and paper cups to measure volume. Young children should also have opportunities to use standard units like centimeters, pounds, and hours. The "standardization" of units should arise later in the lower grades, as students notice that using Joey's foot to measure the length of the classroom gives a different length from that found by using Aria's foot. Such experiences help students see the convenience and consistency of using standard units. As students progress through middle school and high school, they should learn how to use standard units to measure new abstract attributes, such as volume and density. By secondary school, as students are measuring abstract attributes, they should use more-complex units, such as pounds per square inch and person-days.

Understanding that different units are needed to measure different attributes is sometimes difficult for young children. Learning how to choose an appropriate unit is a major part of understanding measurement. For example, students in prekindergarten through grade 2 should learn that length can be measured using linear tools but area cannot be directly measured this way. Young children should see that to measure area they will need to use a unit of area such as a square region; middle-grades students should learn that square regions do not work for measuring volume and should explore the use of three-dimensional units. Students at all levels should learn to make wise choices of units or scales, depending on the problem situation. Choosing a convenient unit of measurement is also important. For example, although the length of a soccer field can be measured in centimeters, the result may be difficult to interpret and use. Students should have a reasonable understanding of the role of units in measurement by the end of their elementary school years.

The metric system has a simple and consistent internal organization. Each unit is always related to the previous unit by a power of 10: a centimeter is ten times larger than a millimeter, a decimeter is ten times larger than a centimeter, and so forth. Since the customary English system of measurement is still prevalent in the United States, students should learn both customary and metric systems and should know some rough equivalences between the metric and customary systems—for

Learning how to choose an appropriate unit is a major part of understanding measurement.

Students should learn both customary and metric systems.

Understanding that all measurements are approximations is a difficult but important concept for students.

example, that a two-liter bottle of soda is a little more than half a gallon. The study of these systems begins in elementary school, and students at this level should be able to carry out simple conversions within both systems. Students should develop proficiency in these conversions in the middle grades and should learn some useful benchmarks for converting between the two systems. The study of measurement systems can help students understand aspects of the base-ten system, such as place value. And in making conversions, students apply their knowledge of proportions.

Understanding that all measurements are approximations is a difficult but important concept for students. They should work with this notion in grades 3–5 through activities in which they measure certain objects, compare their measurements with those of the rest of the class, and note that many of the values do not agree. Class discussions of their observations can elicit the ideas of precision and accuracy. Middle-grades students should continue to develop an understanding of measurements as approximations. In high school, students should come to recognize the need to report an appropriate number of significant digits when computing with measurements.

Apply appropriate techniques, tools, and formulas to determine measurements

Measurement techniques are strategies used to determine a measurement, such as counting, estimating, and using formulas or tools. Measurement tools are the familiar devices that most people associate with taking measurements; they include rulers, measuring tapes, vessels, scales, clocks, and stopwatches. Formulas are general relationships that produce measurements when values are specified for the variables in the formula.

Students in prekindergarten through grade 2 should learn to use a variety of techniques, including counting and estimating, and such tools as rulers, scales, and analog clocks. Elementary and middle-grades students should continue to use these techniques and develop new ones. In addition, they ought to begin to adapt their current tools and invent new techniques to find more-complicated measurements. For example, they might use transparent grid paper to approximate the area of a leaf. Middle-grades students can use formulas for the areas of triangles and rectangles to find the area of a trapezoid. An important measurement technique in high school is successive approximation, a precursor to calculus concepts.

Students should begin to develop formulas for perimeter and area in the elementary grades. Middle-grades students should formalize these techniques, as well as develop formulas for the volume and surface area of objects like prisms and cylinders. Many elementary and middle-grades children have difficulty with understanding perimeter and area (Kenney and Kouba 1997; Lindquist and Kouba 1989). Often, these children are using formulas such as $P = 2l + 2w$ or $A = l \times w$ without understanding how these formulas relate to the attribute being measured or the unit of measurement being used. Teachers must help students see the connections between the formula and the actual object. In high school, as students use formulas in solving problems, they should recognize that the units in the measurements behave like variables under algebraic procedures, and they can use this observation to organize their conversions and computations using unit analysis.

Estimating is another measurement technique that should be developed throughout the school years. Estimation activities in prekindergarten through grade 2 should focus on helping children better understand the process of measuring and the role of the size of the unit. Elementary school and middle-grades students should have many opportunities to estimate measures by comparing them against some benchmark. For example, a student might estimate the teacher's height by noting that the teacher is about one and one-half times as tall as the student. Middle-grades students should also use benchmarks to estimate angle measures and should estimate derived measurements such as speed.

Finally, students in grades 3–5 should have opportunities to use maps and make simple scale drawings. Grades 6–8 students should extend their understanding of scaling to solve problems involving scale factors. These problems can help students make sense of proportional relationships and develop an understanding of similarity. High school students should study more-sophisticated aspects of scaling, including the effects of scale changes on a problem situation. They should also come to understand nonlinear scale changes such as logarithmic scaling and how such techniques are used in analyzing data and in modeling.

Many children have difficulty with understanding perimeter and area.

Data Analysis and Probability

The increased emphasis on data analysis is intended to span the grades.

The Data Analysis and Probability Standard recommends that students formulate questions that can be answered using data and addresses what is involved in gathering and using the data wisely. Students should learn how to collect data, organize their own or others' data, and display the data in graphs and charts that will be useful in answering their questions. This Standard also includes learning some methods for analyzing data and some ways of making inferences and conclusions from data. The basic concepts and applications of probability are also addressed, with an emphasis on the way that probability and statistics are related.

The amount of data available to help make decisions in business, politics, research, and everyday life is staggering: Consumer surveys guide the development and marketing of products. Polls help determine political-campaign strategies, and experiments are used to evaluate the safety and efficacy of new medical treatments. Statistics are often misused to sway public opinion on issues or to misrepresent the quality and effectiveness of commercial products. Students need to know about data analysis and related aspects of probability in order to reason statistically—skills necessary to becoming informed citizens and intelligent consumers.

The increased curricular emphasis on data analysis proposed in these Standards is intended to span the grades rather than to be reserved for the middle grades and secondary school, as is common in many countries. NCTM's 1989 *Curriculum and Evaluation Standards for School Mathematics* introduced standards in statistics and probability at all grade bands; a number of organizations have developed instructional materials and professional development programs to promote the teaching and learning of these topics. Building on this base, these Standards recommend a strong development of the strand, with concepts and procedures becoming increasingly sophisticated across the grades so that by the end of high school students have a sound knowledge of elementary statistics. To understand the fundamentals of statistical ideas, students must work directly with data. The emphasis on working with data entails students' meeting new ideas and procedures as they progress through the grades rather than revisiting the same activities and topics. The data and statistics strand allows teachers and students to make a number of important connections among ideas and procedures from number, algebra, measurement, and geometry. Work in data analysis and probability offers a natural way for students to connect mathematics with other school subjects and with experiences in their daily lives.

In addition, the processes used in reasoning about data and statistics will serve students well in work and in life. Some things children learn in school seem to them predetermined and rule bound. In studying data and statistics, they can also learn that solutions to some problems depend on assumptions and have some degree of uncertainty. The kind of reasoning used in probability and statistics is not always intuitive, and so students will not necessarily develop it if it is not included in the curriculum.

Data Analysis and Probability Standard

Instructional programs from prekindergarten through grade 12 should enable all students to—

- formulate questions that can be addressed with data and collect, organize, and display relevant data to answer them;
- select and use appropriate statistical methods to analyze data;
- develop and evaluate inferences and predictions that are based on data;
- understand and apply basic concepts of probability.

Formulate questions that can be addressed with data and collect, organize, and display relevant data to answer them

Because young children are naturally curious about their world, they often raise questions such as, How many? How much? What kind? or Which of these? Such questions often offer opportunities for beginning the study of data analysis and probability. Young children like to design questions about things close to their experience—What kind of pets do classmates have? What are children's favorite kinds of pizza? As students move to higher grades, the questions they generate for investigation can be based on current issues and interests. Students in grades 6–8, for example, may be interested in recycling, conservation, or manufacturers' claims. They may pose questions such as, Is it better to use paper or plastic plates in the cafeteria? or Which brand of batteries lasts longer? By grades 9–12, students will be ready to pose and investigate problems that explore complex issues.

Young children can devise simple data-gathering plans to attempt to answer their questions. In the primary grades, the teacher might help frame the question or provide a tally sheet, class roster, or chart on which data can be recorded as they are collected. The "data" might be real objects, such as children's shoes arranged in a bar graph or the children themselves arranged by interest areas. As students move through the elementary grades, they should spend more time planning the data collection and evaluating how well their methods worked in getting information about their questions. In the middle grades, students should work more with data that have been gathered by others or generated by simulations. By grades 9–12, students should understand the various purposes of surveys, observational studies, and experiments.

A fundamental idea in prekindergarten through grade 2 is that data can be organized or ordered and that this "picture" of the data provides information about the phenomenon or question. In grades 3–5, students should develop skill in representing their data, often using bar graphs, tables, or line plots. They should learn what different numbers, symbols, and points mean. Recognizing that some numbers represent the values of the data and others represent the frequency with which those values occur is a big step. As students begin to understand ways of representing data, they will be ready to compare two or more data sets. Books, newspapers, the World Wide Web, and other media are full of displays of data, and by the upper elementary grades, students ought to learn to read and understand these displays. Students in grades 6–8 should begin to compare the effectiveness of various types of displays in organizing the data for further analysis or in presenting the data clearly to an audience. As students deal with larger or more-complex data sets, they can reorder data and represent data in graphs quickly, using technology so that they can focus on analyzing the data and understanding what they mean.

Books, newspapers, the World Wide Web, and other media are full of displays of data.

Select and use appropriate statistical methods to analyze data

Although young children are often most interested in their own piece of data on a graph (*I* have five people in *my* family), putting all the students' information in one place draws attention to the *set* of data. Later, students should begin to describe the set of data as a whole. Although this transition is difficult (Konold forthcoming), students may, for example, note that "more students come to school by bus than by all the other ways combined." By grades 3–5, students should be developing an understanding of aggregated data. As older students begin to see a set of data as a whole, they need tools to describe this set. Statistics such as measures of center or location (e.g., mean, median, mode), measures of spread or dispersion (range, standard deviation), and attributes of the shape of the data become useful to students as descriptors. In the elementary grades, students' understandings can be grounded in informal ideas, such as middle, concentration, or balance point (Mokros and Russell 1995). With increasing sophistication in secondary school, students should choose particular summary statistics according to the questions to be answered.

Students should learn what it means to make valid statistical comparisons.

Throughout the school years, students should learn what it means to make valid statistical comparisons. In the elementary grades, students might say that one group has more or less of some attribute than another. By the middle grades, students should be quantifying these differences by comparing specific statistics. Beginning in grades 3–5 and continuing in the middle grades, the emphasis should shift from analyzing and describing one set of data to comparing two or more sets (Konold forthcoming). As they move through the middle grades into high school, students will need new tools, including histograms, stem-and-leaf plots, box plots, and scatterplots, to identify similarities and differences among data sets. Students also need tools to investigate association and trends in bivariate data, including scatterplots and fitted lines in grades 6–8 and residuals and correlation in grades 9–12.

Develop and evaluate inferences and predictions that are based on data

Central elements of statistical analysis—defining an appropriate sample, collecting data from that sample, describing the sample, and making reasonable inferences relating the sample and the population—should be understood as students move through the grades. In the early grades, students are most often working with census data, such as a survey of each child in the class about favorite kinds of ice cream. The notion that the class can be viewed as a sample from a larger population is not obvious at these grades. Upper elementary and early middle-grades students can begin to develop notions about statistical inference, but developing a deep understanding of the idea of sampling is difficult (Schwartz et al. 1998). Research has shown that students in grades 5–8 expect their own judgment to be more reliable than information obtained from data (Hancock, Kaput, and Goldsmith 1992). In the later middle grades and high school, students should address the ideas of sample selection and statistical inference and begin to understand that there are ways of quantifying how certain one can be about statistical results.

In addition, students in grades 9–12 should use simulations to learn about sampling distributions and make informal inferences. In particular, they should know that basic statistical techniques are used to monitor quality in the workplace. Students should leave secondary school with the ability to judge the validity of arguments that are based on data, such as those that appear in the press.

Understand and apply basic concepts of probability

A subject in its own right, probability is connected to other areas of mathematics, especially number and geometry. Ideas from probability serve as a foundation to the collection, description, and interpretation of data.

In prekindergarten through grade 2, the treatment of probability ideas should be informal. Teachers should build on children's developing vocabulary to introduce and highlight probability notions, for example, We'll *probably* have recess this afternoon, or It's *unlikely* to rain today. Young children can begin building an understanding of chance and randomness by doing experiments with concrete objects, such as choosing colored chips from a bag. In grades 3–5 students can consider ideas of chance through experiments—using coins, dice, or spinners—with known theoretical outcomes or through designating familiar events as impossible, unlikely, likely, or certain. Middle-grades students should learn and use appropriate terminology and should be able to compute probabilities for simple compound events, such as the number of expected occurrences of two heads when two coins are tossed 100 times. In high school, students should compute probabilities of compound events and understand conditional and independent events. Through the grades, students should be able to move from situations for which the probability of an event can readily be determined to situations in which sampling and simulations help them quantify the likelihood of an uncertain outcome.

Many of the phenomena that students encounter, especially in school, have predictable outcomes. When a fair coin is flipped, it is equally likely to come up heads or tails. Which outcome will result on a given flip is uncertain—even if ten flips in a row have resulted in heads, for many people it is counterintuitive that the eleventh flip has only a 50 percent likelihood of being tails. If an event is random and if it is repeated many, many times, then the distribution of outcomes forms a pattern. The idea that individual events are not predictable in such a situation but that a pattern of outcomes can be predicted is an important concept that serves as a foundation for the study of inferential statistics.

Probability is connected to other areas of mathematics.

Problem Solving

Problem solving is an integral part of all mathematics learning.

Problem solving means engaging in a task for which the solution method is not known in advance. In order to find a solution, students must draw on their knowledge, and through this process, they will often develop new mathematical understandings. Solving problems is not only a goal of learning mathematics but also a major means of doing so. Students should have frequent opportunities to formulate, grapple with, and solve complex problems that require a significant amount of effort and should then be encouraged to reflect on their thinking.

By learning problem solving in mathematics, students should acquire ways of thinking, habits of persistence and curiosity, and confidence in unfamiliar situations that will serve them well outside the mathematics classroom. In everyday life and in the workplace, being a good problem solver can lead to great advantages.

Problem solving is an integral part of all mathematics learning, and so it should not be an isolated part of the mathematics program. Problem solving in mathematics should involve all the five content areas described in these Standards. The contexts of the problems can vary from familiar experiences involving students' lives or the school day to applications involving the sciences or the world of work. Good problems will integrate multiple topics and will involve significant mathematics.

Problem Solving Standard

Instructional programs from prekindergarten through grade 12 should enable all students to—

- build new mathematical knowledge through problem solving;
- solve problems that arise in mathematics and in other contexts;
- apply and adapt a variety of appropriate strategies to solve problems;
- monitor and reflect on the process of mathematical problem solving.

Build new mathematical knowledge through problem solving

How can problem solving help students learn mathematics? Good problems give students the chance to solidify and extend what they know and, when well chosen, can stimulate mathematics learning. With young children, most mathematical concepts can be introduced through problems that come from their worlds. For example, suppose second graders wanted to find out whether there are more boys or girls in the four second-grade classes. To solve this problem, they would need to learn how to gather information, record data, and accurately add several numbers at a time. In the middle grades, the concept of proportion might be introduced through an investigation in which students are given recipes for punch that call for different amounts of water and juice and are asked to determine which is "fruitier." Since no two recipes yield the same amount of juice, this problem is difficult for students who do not have an understanding of proportion. As various ideas are tried, with good questioning and guidance by a teacher, students eventually converge on using proportions. In high school, many areas of the curriculum can be introduced through problems from mathematical or applications contexts.

Problem solving can and should be used to help students develop fluency with specific skills. For example, consider the following problem, which is adapted from the *Curriculum and Evaluation Standards for School Mathematics* (NCTM 1989, p. 24):

> I have pennies, dimes, and nickels in my pocket. If I take three coins out of my pocket, how much money could I have taken?

Knowledge is needed to solve this problem—knowledge of the value of pennies, dimes, and nickels and also some understanding of addition.

Working on this problem offers good practice in addition skills. But the important mathematical goal of this problem—helping students to think systematically about possibilities and to organize and record their thinking—need not wait until students can add fluently.

The teacher's role in choosing worthwhile problems and mathematical tasks is crucial. By analyzing and adapting a problem, anticipating the mathematical ideas that can be brought out by working on the problem, and anticipating students' questions, teachers can decide if particular problems will help to further their mathematical goals for the class. There are many, many problems that are interesting and fun but that may not lead to the development of the mathematical ideas that are important for a class at a particular time. Choosing problems wisely, and using and adapting problems from instructional materials, is a difficult part of teaching mathematics.

The teacher's role in choosing worthwhile problems and mathematical tasks is crucial.

Solve problems that arise in mathematics and in other contexts

People who see the world mathematically are said to have a "mathematical disposition." Good problem solvers tend naturally to analyze situations carefully in mathematical terms and to pose problems based on situations they see. They first consider simple cases before trying something more complicated, yet they will readily consider a more sophisticated analysis. For example, a task for middle-grades students presents data about two ambulance companies and asks which company is more reliable (Balanced Assessment for the Mathematics Curriculum 1999a). A quick answer found by looking at the average time customers had to wait for each company turns out to be misleading. A more careful mathematical analysis involving plotting response times versus time of day reveals a different solution. In this task, a disposition to analyze more deeply leads to a more complete understanding of the situation and a correct solution. Throughout the grades, teachers can help build this disposition by asking questions that help students find the mathematics in their worlds and experiences and by encouraging students to persist with interesting but challenging problems.

Posing problems comes naturally to young children: I wonder how long it would take to count to a million? How many soda cans would it take to fill the school building? Teachers and parents can foster this inclination by helping students make mathematical problems from their worlds. Teachers play an important role in the development of students' problem-solving dispositions by creating and maintaining classroom environments, from prekindergarten on, in which students are encouraged to explore, take risks, share failures and successes, and question one another. In such supportive environments, students develop confidence in their abilities and a willingness to engage in and explore problems, and they will be more likely to pose problems and to persist with challenging problems.

Apply and adapt a variety of appropriate strategies to solve problems

Of the many descriptions of problem-solving strategies, some of the best known can be found in the work of Pólya (1957). Frequently cited

Opportunities to use strategies must be embedded naturally in the curriculum across the content areas.

strategies include using diagrams, looking for patterns, listing all possibilities, trying special values or cases, working backward, guessing and checking, creating an equivalent problem, and creating a simpler problem. An obvious question is, How should these strategies be taught? Should they receive explicit attention, and how should they be integrated with the mathematics curriculum? As with any other component of the mathematical tool kit, strategies must receive instructional attention if students are expected to learn them. In the lower grades, teachers can help children express, categorize, and compare their strategies. Opportunities to use strategies must be embedded naturally in the curriculum across the content areas. By the time students reach the middle grades, they should be skilled at recognizing when various strategies are appropriate to use and should be capable of deciding when and how to use them. By high school, students should have access to a wide range of strategies, be able to decide which one to use, and be able to adapt and invent strategies.

Young children's earliest experiences with mathematics come through solving problems. Different strategies are necessary as students experience a wider variety of problems. Students must become aware of these strategies as the need for them arises, and as they are modeled during classroom activities, the teacher should encourage students to take note of them. For example, after a student has shared a solution and how it was obtained, the teacher may identify the strategy by saying, "It sounds like you made an organized list to find the solution. Did anyone solve the problem a different way?" This verbalization helps develop common language and representations and helps other students understand what the first student was doing. Such discussion also suggests that no strategy is learned once and for all; strategies are learned over time, are applied in particular contexts, and become more refined, elaborate, and flexible as they are used in increasingly complex problem situations.

Monitor and reflect on the process of mathematical problem solving

Effective problem solvers constantly monitor and adjust what they are doing. They make sure they understand the problem. If a problem is written down, they read it carefully; if it is told to them orally, they ask questions until they understand it. Effective problem solvers plan frequently. They periodically take stock of their progress to see whether they seem to be on the right track. If they decide they are not making progress, they stop to consider alternatives and do not hesitate to take a completely different approach. Research (Garofalo and Lester 1985; Schoenfeld 1987) indicates that students' problem-solving failures are often due not to a lack of mathematical knowledge but to the ineffective use of what they do know.

Good problem solvers become aware of what they are doing and frequently monitor, or self-assess, their progress or adjust their strategies as they encounter and solve problems (Bransford et al. 1999). Such reflective skills (called *metacognition*) are much more likely to develop in a classroom environment that supports them. Teachers play an important role in helping to enable the development of these reflective habits of mind by asking questions such as “Before we go on, are we sure we understand this?” “What are our options?” “Do we have a plan?” “Are we making progress or should we reconsider what we are doing?” “Why do we think this is true?” Such questions help students get in the habit of checking their understanding as they go along. This habit should begin in the lowest grades. As teachers maintain an environment in which the development of understanding is consistently monitored through reflection, students are more likely to learn to take responsibility for reflecting on their work and make the adjustments necessary when solving problems.

Reasoning and Proof

A mathematical proof is a formal way of expressing particular kinds of reasoning and justification.

Mathematical reasoning and proof offer powerful ways of developing and expressing insights about a wide range of phenomena. People who reason and think analytically tend to note patterns, structure, or regularities in both real-world situations and symbolic objects; they ask if those patterns are accidental or if they occur for a reason; and they conjecture and prove. Ultimately, a mathematical proof is a formal way of expressing particular kinds of reasoning and justification.

Being able to reason is essential to understanding mathematics. By developing ideas, exploring phenomena, justifying results, and using mathematical conjectures in all content areas and—with different expectations of sophistication—at all grade levels, students should see and expect that mathematics makes sense. Building on the considerable reasoning skills that children bring to school, teachers can help students learn what mathematical reasoning entails. By the end of secondary school, students should be able to understand and produce mathematical proofs—arguments consisting of logically rigorous deductions of conclusions from hypotheses—and should appreciate the value of such arguments.

Reasoning and proof cannot simply be taught in a single unit on logic, for example, or by "doing proofs" in geometry. Proof is a very difficult area for undergraduate mathematics students. Perhaps students at the postsecondary level find proof so difficult because their only experience in writing proofs has been in a high school geometry course, so they have a limited perspective (Moore 1994). Reasoning and proof should be a consistent part of students' mathematical experience in prekindergarten through grade 12. Reasoning mathematically is a habit of mind, and like all habits, it must be developed through consistent use in many contexts.

Reasoning and Proof Standard

Instructional programs from prekindergarten through grade 12 should enable all students to—

- recognize reasoning and proof as fundamental aspects of mathematics;
- make and investigate mathematical conjectures;
- develop and evaluate mathematical arguments and proofs;
- select and use various types of reasoning and methods of proof.

Recognize reasoning and proof as fundamental aspects of mathematics

From children's earliest experiences with mathematics, it is important to help them understand that assertions should always have reasons. Questions such as "Why do you think it is true?" and "Does anyone think the answer is different, and why do you think so?" help students see that statements need to be supported or refuted by evidence. Young children may wish to appeal to others as sources for their reasons ("My sister told me so") or even to vote to determine the best explanation, but students need to learn and agree on what is acceptable as an adequate argument in the *mathematics* classroom. These are the first steps toward realizing that mathematical reasoning is based on specific assumptions and rules.

Part of the beauty of mathematics is that when interesting things happen, it is usually for good reason. Mathematics students should understand this. Consider, for example, the following "magic trick" one might find in a book of mathematical recreations:

> Write down your age. Add 5. Multiply the number you just got by 2. Add 10 to this number. Multiply this number by 5. Tell me the result. I can tell you your age.

The procedure given to find the answer is, Drop the final zero from the number you are given and subtract 10. The result is the person's age. Why does it work? Students at all grade levels can explore and explain problems such as this one.

Systematic reasoning is a defining feature of mathematics. It is found in all content areas and, with different requirements of rigor, at all grade levels. For example, first graders can note that even and odd numbers alternate; third graders can conjecture and justify—informally, perhaps, by paper folding—that the diagonals of a square are perpendicular. Middle-grades students can determine the likelihood of an even or odd product when two number cubes are rolled and the numbers that come up are multiplied. And high school students could be asked to consider what happens to a correlation coefficient under linear transformation of the variables.

Make and investigate mathematical conjectures

Doing mathematics involves discovery. Conjecture—that is, informed guessing—is a major pathway to discovery. Teachers and researchers agree that students can learn to make, refine, and test conjectures in elementary school. Beginning in the earliest years, teachers can help students learn to make conjectures by asking questions: What do you think will happen next? What is the pattern? Is this true always? Sometimes? Simple shifts in how tasks are posed can help students learn to conjecture. Instead of saying, "Show that the mean of a set of data doubles when all the values in the data set are doubled," a teacher might ask, "Suppose all the values of a sample are doubled. What change, if any, is there in the mean of the sample? Why?" High school students using dynamic geometry software could be asked to make observations about the figure formed by joining the midpoints of successive sides of a parallelogram and attempt to prove them. To make conjectures, students need multiple opportunities and rich, engaging contexts for learning.

Young children will express their conjectures and describe their thinking in their own words and often explore them using concrete materials and examples. Students at all grade levels should learn to investigate their conjectures using concrete materials, calculators and other tools, and increasingly through the grades, mathematical representations and symbols. They also need to learn to work with other students to formulate and explore their conjectures and to listen to and understand conjectures and explanations offered by classmates.

Conjecture is a major pathway to discovery.

Teachers can help students revisit conjectures that hold in one context to check to see whether they still hold in a new setting. For instance, the common notion that "multiplication makes bigger" is quite appropriate for young children working with whole numbers larger than 1. As they move to fractions, this conjecture needs to be revisited. Students may not always have the mathematical knowledge and tools they need to find a justification for a conjecture or a counterexample to refute it. For example, on the basis of their work with graphing calculators, high school students might be quite convinced that if a polynomial function has a value that is greater than 0 and a value that is less than 0 then it will cross the x-axis somewhere. Teachers can point out that a rigorous proof requires more knowledge than most high school students have.

Develop and evaluate mathematical arguments and proofs

Along with making and investigating conjectures, students should learn to answer the question, Why does this work? Children in the lower grades will tend to justify general claims using specific cases. For instance, students might represent the odd number 9 as in figure 3.5 and note that "an odd number is something that has one number left over" (Ball and Bass forthcoming, p. 33). Students might then reason that any odd number will have an "extra" unit in it, and so when two odd numbers are added, the two "extra" units will become a pair, giving an even number, with no "extras." By the upper elementary grades, justifications should be more general and can draw on other mathematical results. Using the fact that congruent shapes have equal area, a fifth grader might claim that a particular triangle and rectangle have the same area because each was formed by dividing one of two congruent rectangles in half. In high school, students should be expected to construct relatively complex chains of reasoning and provide mathematical reasons. To help students develop and justify more-general conjectures and also to refute conjectures, teachers can ask, "Does this always work? Sometimes? Never? Why?" This extension to general cases draws on more-sophisticated mathematical knowledge that should build up over the grades.

Fig. **3.5.**

A representation of 9 as an odd number

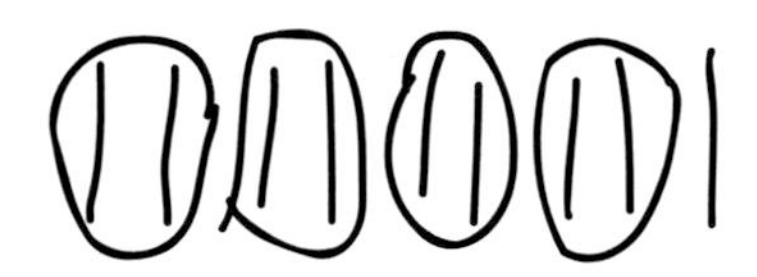

Students can learn about reasoning through class discussion of claims that other students make. The statement, If a number is divisible by 6 and by 4, then it is divisible by 24, could be examined in various ways. Middle-grades students could find a counterexample—the number 12 is divisible by 6 and by 4 but not by 24. High school students might find a related conjecture involving prime numbers that they could verify. Or students could explore the converse. In any event, both plausible and flawed arguments that are offered by students create an opportunity for discussion. As students move through the grades, they should compare their ideas with others' ideas, which may cause them to modify, consolidate, or strengthen their arguments or reasoning. Classrooms in which students are encouraged to present their thinking and in which everyone contributes by evaluating one another's thinking provide rich environments for learning mathematical reasoning.

Young children's explanations will be in their own language and often will be represented verbally or with objects. Students can learn to articulate their reasoning by presenting their thinking to their groups, their classmates, and to others outside the classroom. High school students should be able to present mathematical arguments in written forms that would be acceptable to professional mathematicians. The particular format of a mathematical justification or proof, be it narrative argument, "two-column proof," or a visual argument, is less important than a clear and correct communication of mathematical ideas appropriate to the students' grade level.

Select and use various types of reasoning and methods of proof

In the lower grades, the reasoning that children learn and use in mathematics class is informal compared to the logical deduction used by the mathematician. Over the years of schooling, as teachers help students learn the norms for mathematical justification and proof, the repertoire of the types of reasoning available to students—algebraic and geometric reasoning, proportional reasoning, probabilistic reasoning, statistical reasoning, and so forth—should expand. Students need to encounter and build proficiency in all these forms with increasing sophistication as they move through the curriculum.

Young children should be encouraged to reason from what they know. A child who solves the problem 6 + 7 by calculating 6 + 6 and then adding 1 is drawing on her knowledge of adding pairs, of adding 1, and of associativity. Students can be taught how to make explicit the knowledge they are using as they create arguments and justifications.

Early efforts at justification by young children will involve trial-and-error strategies or the unsystematic trying of many cases. With guidance and many opportunities to explore, students can learn by the upper elementary grades how to be systematic in their explorations, to know that they have tried all cases, and to create arguments using cases. One research study (Maher and Martino 1996, p. 195) reported a fifth grader's elegant proof by cases in response to the problem in figure 3.6.

Proof by contradiction is also possible with young children. A first grader argued from his knowledge of whole-number patterns that the number 0 is even: "If 0 were odd, then 0 and 1 would be two odd numbers in a row. Even and odd numbers alternate. So 0 must be even." Beginning in the elementary grades, children can learn to disprove conjectures by finding counterexamples. At all levels, students will reason inductively from patterns and specific cases. Increasingly over the grades, they should also learn to make effective deductive arguments based on the mathematical truths they are establishing in class.

Fig. **3.6.**

Stephanie's elegant "proof by cases" produced in grade 5 (from Maher and Martino [1996])

Name Stephanie Date

Please send a letter to a student who is ill and unable come to school. Describe all the different towers you have built that are three cubes tall, when you have two colors available to work with. Why were you sure that you had made every possible tower and had not left any out?

Dear Laura,

Today we made towers 3 high and with 2 colors. We have to be sure to make every posible pateren. There are 8 patterns total. I know because all you have to do is multyply 2 x the number you would get for towers of two. so It is 2x 4. I will prove it. If I put the towers in color order The colors are red & white R stands for Red & W stands for White.

no Red	1 Red top	1 Red middle	1 Red bottom	2 Red top	2 Red bottom	2 Red top bottom	3 Red
W	R	W	W	R	W	R	R
W	W	R	W	R	R	W	R
W	W	W	R	W	R	R	R

If this doesent convince you I'll tell you more → over →

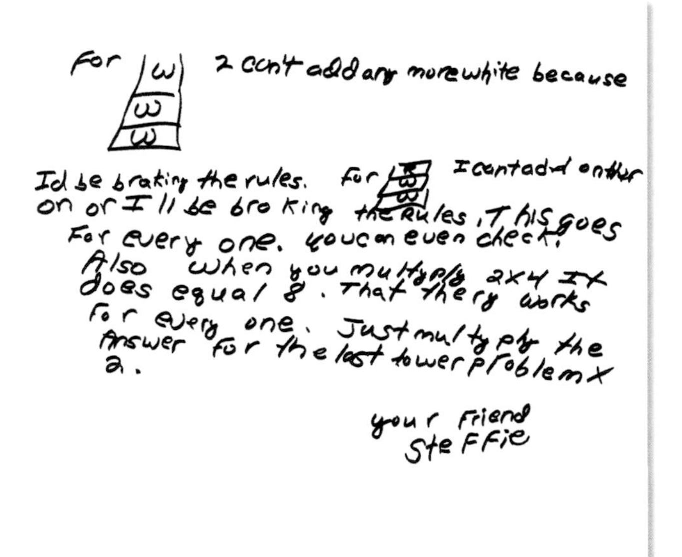
For [W/W/W] I can't add any more white because I'd be braking the rules. For [W/W] I can't add enther one or I'll be broking the rules, This goes for every one. you can even check. Also when you multyply 2x4 It does equal 8. That they works for every one. Just multyply the answer for the last tower problem x 2.

your friend
Steffie

Communication

Communication is an essential part of mathematics and mathematics education.

Communication Standard

Instructional programs from prekindergarten through grade 12 should enable all students to—

- organize and consolidate their mathematical thinking though communication;
- communicate their mathematical thinking coherently and clearly to peers, teachers, and others;
- analyze and evaluate the mathematical thinking and strategies of others;
- use the language of mathematics to express mathematical ideas precisely.

Communication is an essential part of mathematics and mathematics education. It is a way of sharing ideas and clarifying understanding. Through communication, ideas become objects of reflection, refinement, discussion, and amendment. The communication process also helps build meaning and permanence for ideas and makes them public. When students are challenged to think and reason about mathematics and to communicate the results of their thinking to others orally or in writing, they learn to be clear and convincing. Listening to others' explanations gives students opportunities to develop their own understandings. Conversations in which mathematical ideas are explored from multiple perspectives help the participants sharpen their thinking and make connections. Students who are involved in discussions in which they justify solutions—especially in the face of disagreement—will gain better mathematical understanding as they work to convince their peers about differing points of view (Hatano and Inagaki 1991). Such activity also helps students develop a language for expressing mathematical ideas and an appreciation of the need for precision in that language. Students who have opportunities, encouragement, and support for speaking, writing, reading, and listening in mathematics classes reap dual benefits: they communicate to learn mathematics, and they learn to communicate mathematically.

Because mathematics is so often conveyed in symbols, oral and written communication about mathematical ideas is not always recognized as an important part of mathematics education. Students do not necessarily talk about mathematics naturally; teachers need to help them learn how to do so (Cobb, Wood, and Yackel 1994). As students progress through the grades, the mathematics about which they communicate should become more complex and abstract. Students' repertoire of tools and ways of communicating, as well as the mathematical reasoning that supports their communication, should become increasingly sophisticated. Support for students is vital. Students whose primary language is not English may need some additional support in order to benefit from communication-rich mathematics classes, but they can participate fully if classroom activities are appropriately structured (Silver, Smith, and Nelson 1995).

Students need to work with mathematical tasks that are worthwhile topics of discussion. Procedural tasks for which students are expected to have well-developed algorithmic approaches are usually not good candidates for such discourse. Interesting problems that "go somewhere" mathematically can often be catalysts for rich conversations. Technology is another good basis for communication. As students generate and examine numbers or objects on the calculator or computer screen, they have a common (and often easily modifiable) referent for their discussion of mathematical ideas.

Organize and consolidate their mathematical thinking through communication

Students gain insights into their thinking when they present their methods for solving problems, when they justify their reasoning to a classmate or teacher, or when they formulate a question about something that is puzzling to them. Communication can support students' learning of new mathematical concepts as they act out a situation, draw, use objects, give verbal accounts and explanations, use diagrams, write, and use mathematical symbols. Misconceptions can be identified and addressed. A side benefit is that it reminds students that they share responsibility with the teacher for the learning that occurs in the lesson (Silver, Kilpatrick, and Schlesinger 1990).

Reflection and communication are intertwined processes in mathematics learning. With explicit attention and planning by teachers, communication for the purposes of reflection can become a natural part of mathematics learning. Children in the early grades, for example, can learn to explain their answers and describe their strategies. Young students can be asked to "think out loud," and thoughtful questions posed by a teacher or classmate can provoke them to reexamine their reasoning. With experience, students will gain proficiency in organizing and recording their thinking.

Reflection and communication are intertwined processes in mathematics learning.

Writing in mathematics can also help students consolidate their thinking because it requires them to reflect on their work and clarify their thoughts about the ideas developed in the lesson. Later, they may find it helpful to reread the record of their own thoughts.

Communicate their mathematical thinking coherently and clearly to peers, teachers, and others

In order for a mathematical result to be recognized as correct, the proposed proof must be accepted by the community of professional mathematicians. Students need opportunities to test their ideas on the basis of shared knowledge in the mathematical community of the classroom to see whether they can be understood and if they are sufficiently convincing. When such ideas are worked out in public, students can profit from being part of the discussion, and the teacher can monitor their learning (Lampert 1990). Learning what is acceptable as evidence in mathematics should be an instructional goal from prekindergarten through grade 12.

To support classroom discourse effectively, teachers must build a community in which students will feel free to express their ideas. Students in the lower grades need help from teachers in order to share mathematical ideas with one another in ways that are clear enough for other students to understand. In these grades, learning to see things from other people's perspectives is a challenge for students. Starting in grades 3–5, students should gradually take more responsibility for participating in whole-class discussions and responding to one another directly. They should become better at listening, paraphrasing, questioning, and interpreting others' ideas. For some students, participation in class discussions is a challenge. For example, students in the middle grades are often reluctant to stand out in any way during group interactions. Despite this fact, teachers can succeed in creating communication-rich environments in middle-grades mathematics classrooms. By the time students graduate from high school, they should have internalized standards of dialogue and argument so that they always aim to present clear and complete arguments and work to clarify and complete them

Written communication should be nurtured.

when they fall short. Modeling and carefully posed questions can help clarify age-appropriate expectations for student work.

Written communication should be nurtured in a similar fashion. Students begin school with few writing skills. In the primary grades, they may rely on other means, such as drawing pictures, to communicate. Gradually they will also write words and sentences. In grades 3–5, students can work on sequencing ideas and adding details, and their writing should become more elaborate. In the middle grades, they should become more explicit about basing their writing on a sense of audience and purpose. For some purposes it will be appropriate for students to describe their thinking informally, using ordinary language and sketches, but they should also learn to communicate in more-formal mathematical ways, using conventional mathematical terminology, through the middle grades and into high school. By the end of the high school years, students should be able to write well-constructed mathematical arguments using formal vocabulary.

Examining and discussing both exemplary and problematic pieces of mathematical writing can be beneficial at all levels. Since written assessments of students' mathematical knowledge are becoming increasingly prevalent, students will need practice responding to typical assessment prompts. The process of learning to write mathematically is similar to that of learning to write in any genre. Practice, with guidance, is important. So is attention to the specifics of mathematical argument, including the use and special meanings of mathematical language and the representations and standards of explanation and proof.

As students practice communication, they should express themselves increasingly clearly and coherently. They should also acquire and recognize conventional mathematical styles of dialogue and argument. Through the grades, their arguments should become more complete and should draw directly on the shared knowledge in the classroom. Over time, students should become more aware of, and responsive to, their audience as they explain their ideas in mathematics class. They should learn to be aware of whether they are convincing and whether others can understand them. As students mature, their communication should reflect an increasing array of ways to justify their procedures and results. In the lower grades, providing empirical evidence or a few examples may be enough. Later, short deductive chains of reasoning based on previously accepted facts should become expected. In the middle grades and high school, explanations should become more mathematically rigorous and students should increasingly state in their supporting arguments the mathematical properties they used.

Analyze and evaluate the mathematical thinking and strategies of others

In the process of working on problems with other students, learners gain several benefits. Often, a student who has one way of seeing a problem can profit from another student's view, which may reveal a different aspect of the problem. For example, students who try to solve the following problem (Krutetskii 1976, p. 121) algebraically often have difficulty setting up the equations, and they benefit from the insights provided by students who approach the problem using visual representations.

There are some rabbits and some hutches. If one rabbit is put in each hutch, one rabbit will be left without a place. If two rabbits are put in each hutch, one hutch will remain empty. How many rabbits and how many hutches are there?

It is difficult for students to learn to consider, evaluate, and build on the thinking of others, especially when their peers are still developing their own mathematical understandings. A good setting in which young students can share and analyze one another's strategies is in solving arithmetic problems, where students' invented strategies can become objects of discussion and critique. Students must also learn to question and probe one another's thinking in order to clarify underdeveloped ideas. Moreover, since not all methods have equal merit, students must learn to examine the methods and ideas of others in order to determine their strengths and limitations. By carefully listening to, and thinking about, the claims made by others, students learn to become critical thinkers about mathematics.

Use the language of mathematics to express mathematical ideas precisely

As students articulate their mathematical understanding in the lower grades, they begin by using everyday, familiar language. This provides a base on which to build a connection to formal mathematical language. Teachers can help students see that some words that are used in everyday language, such as *similar*, *factor*, *area*, or *function*, are used in mathematics with different or more-precise meanings. This observation is the foundation for understanding the concept of mathematical definitions. It is important to give students experiences that help them appreciate the power and precision of mathematical language. Beginning in the middle grades, students should understand the role of mathematical definitions and should use them in mathematical work. Doing so should become pervasive in high school. However, it is important to avoid a premature rush to impose formal mathematical language; students need to develop an appreciation of the need for precise definitions and for the communicative power of conventional mathematical terms by first communicating in their own words. Allowing students to grapple with their ideas and develop their own informal means of expressing them can be an effective way to foster engagement and ownership.

Technology affords other opportunities and challenges for the development and analysis of language. The symbols used in a spreadsheet may be related to, but are not the same as, the algebraic symbols used generally by mathematicians. Students will profit from experiences that require comparisons of standard mathematical expressions with those used with popular tools like spreadsheets or calculators.

It is important to avoid a premature rush to impose formal mathematical language.

Connections

When students can connect mathematical ideas, their understanding is deeper and more lasting. They can see mathematical connections in the rich interplay among mathematical topics, in contexts that relate mathematics to other subjects, and in their own interests and experience. Through instruction that emphasizes the interrelatedness of mathematical ideas, students not only learn mathematics, they also learn about the utility of mathematics.

Mathematics is not a collection of separate strands or standards, even though it is often partitioned and presented in this manner. Rather, mathematics is an integrated field of study. Viewing mathematics as a whole highlights the need for studying and thinking about the connections within the discipline, as reflected both within the curriculum of a particular grade and between grade levels. To emphasize the connections, teachers must know the needs of their students as well as the mathematics that the students studied in the preceding grades and what they will study in the following grades. As the Learning Principle emphasizes, understanding involves making connections. Teachers should build on students' previous experiences and not repeat what students have already done. This approach requires students to be responsible for what they have learned and for using that knowledge to understand and make sense of new ideas.

Connections Standard

Instructional programs from prekindergarten through grade 12 should enable all students to—

- recognize and use connections among mathematical ideas;
- understand how mathematical ideas interconnect and build on one another to produce a coherent whole;
- recognize and apply mathematics in contexts outside of mathematics.

Recognize and use connections among mathematical ideas

By emphasizing mathematical connections, teachers can help students build a disposition to use connections in solving mathematical problems, rather than see mathematics as a set of disconnected, isolated concepts and skills. This disposition can be fostered through the guiding questions that teachers ask, for instance, "How is our work today with similar triangles related to the discussion we had last week about scale drawings?" Students need to be made explicitly aware of the mathematical connections.

The notion that mathematical ideas are connected should permeate the school mathematics experience at all levels. The mathematical experiences of children first entering school have not been separated into categories, and this integration of mathematics in many contexts should continue in school. Children can learn to recognize mathematical patterns in the rhythms of the songs they sing, identify the hexagonal shape in a honeycomb, and count the number of times they can jump rope successfully. As students move into grades 3–5, their mathematical activity should expand into more-abstract contexts. They can begin to see the connections among arithmetic operations, understanding, for example, how multiplication can be thought of as repeated addition. As they see how mathematical operations can be used in different contexts, they can develop an appreciation for the abstraction of mathematics. In grades 6–8, students should see mathematics as a discipline of connected ideas. The key mathematical ideas in the middle grades are themselves closely connected, and ideas about rational numbers, proportionality, and linear relationships will pervade much of their mathematical and everyday activity. In grades 9–12, students not only learn to expect connections but they learn to take advantage of them, using insights gained in one context to solve problems in another.

Throughout the pre-K–12 span, students should routinely ask themselves, "How is this problem or mathematical topic like things I have studied before?" From the perspective of connections, new ideas are seen as extensions of previously learned mathematics. Students learn to use what they already know to address new situations. Elementary school students link their knowledge of the subtraction of whole numbers to the subtraction of decimals or fractions. Middle-grades students recognize and connect multiple representations of the same mathematical idea, such as the ratio that represents rate of change and the tilt or slope of a line. High school students connect ideas in algebra and geometry.

Some activities can be especially productive for featuring mathematical connections. For instance, the relationship between the diameter and the circumference of a circle can be studied empirically by collecting a variety of circular objects and measuring their circumferences and diameters. Middle-grades students might collect and graph data for the two variables—circumference (C) and diameter (d). By doing so, they can see that all the points lie close to a straight line through (0, 0), which suggests that the ratio of C/d is constant. This activity usually leads to an average value for C/d that lies between 3.1 and 3.2—a rough approximation of π. The problem involves ideas from measurement, data analysis, geometry, algebra, and number.

Understand how mathematical ideas interconnect and build on one another to produce a coherent whole

As students progress through their school mathematics experience, their ability to see the same mathematical structure in seemingly different settings should increase. Prekindergarten through grade 2 students recognize instances of counting, number, and shape; upper elementary school students look for instances of arithmetic operations, and middle-grades students look for examples of rational numbers, proportionality, and linear relationships. High school students are ready to look for connections among the many mathematical ideas they are encountering. For instance, a method for finding the volume of the truncated square pyramid shown at the top of figure 3.7, is suggested by the method for finding the area of the trapezoid that follows in the figure (Banchoff 1990, pp. 20–22).

As students develop a view of mathematics as a connected and integrated whole, they will have less of a tendency to view mathematical skills and concepts separately. If conceptual understandings are linked to procedures, students will not perceive mathematics as an arbitrary set of rules. This integration of procedures and concepts should be central in school mathematics.

Recognize and apply mathematics in contexts outside of mathematics

School mathematics experiences at all levels should include opportunities to learn about mathematics by working on problems arising in

Fig. **3.7.**

Connections between methods for finding the volume of a truncated pyramid and for finding the area of a trapezoid

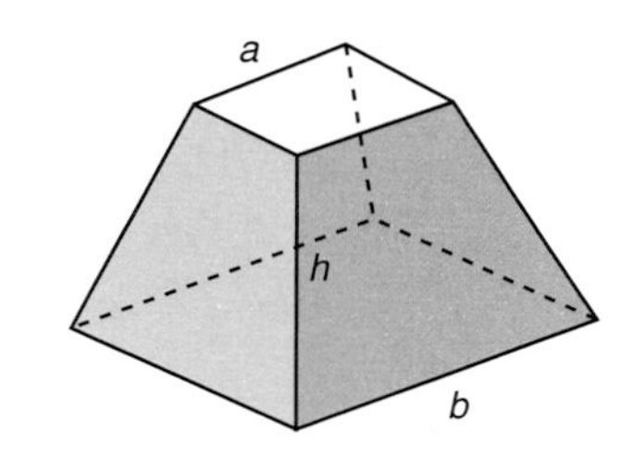

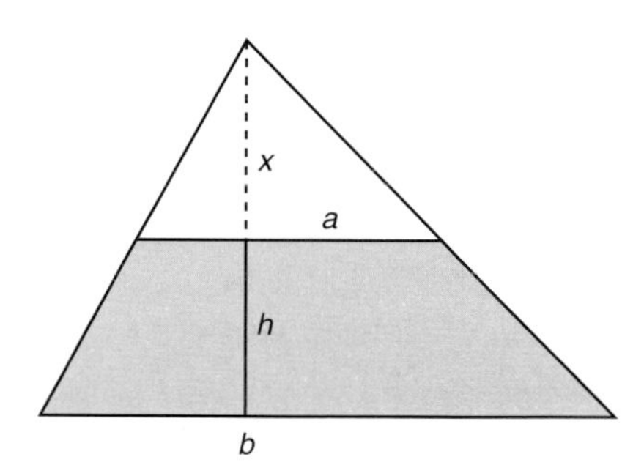

$$\text{Area of small triangle} = \frac{1}{2}ax.$$

$$\text{Area of large triangle} = \frac{1}{2}b(x+h).$$

$$\text{Area of trapezoid} = \frac{1}{2}b(x+h) - \frac{1}{2}ax$$

$$= \frac{1}{2}bx + \frac{1}{2}bh - \frac{1}{2}ax.$$

By similarity of triangles,

$$\frac{x}{(x+h)} = \frac{a}{b}$$

$$bx = a(x+h)$$

$$\text{Area of trapezoid} = a(x+h) + \frac{1}{2}bh - \frac{1}{2}ax$$

$$= \frac{1}{2}h(a+b).$$

contexts outside of mathematics. These connections can be to other subject areas and disciplines as well as to students' daily lives. Prekindergarten through grade 2 students can learn about mathematics primarily through connections with the real world. Students in grades 3–5 should learn to apply important mathematical ideas in other subject areas. This set of ideas expands in grades 6–8, and in grades 9–12 students should be confidently using mathematics to explain complex applications in the outside world.

The opportunity for students to experience mathematics in a context is important.

The opportunity for students to experience mathematics in a context is important. Mathematics is used in science, the social sciences, medicine, and commerce. The link between mathematics and science is not only through content but also through process. The processes and content of science can inspire an approach to solving problems that applies to the study of mathematics. In the *National Science Education Standards*, a yearlong elementary school science activity about weather is described (National Research Council 1996, pp. 131–33). The connections to mathematics in this activity are substantial: students design instruments for measuring weather conditions and plan for how to organize and communicate their data.

Steinberg (1998, p. 97) reports the following incident in which eleventh-grade students at a high school worked with the CVS Corporation to locate a new pharmacy in a Boston neighborhood:

> Although fully aware that the company would probably not rely only on their calculations in making a monetary decision as to where to locate a store, the students still felt involved in a real problem.... Organized into small work teams supported by experts from various departments within the CVS organization, students analyzed demographic and economic data to determine market demand for a CVS pharmacy in different neighborhoods. Students also worked with CVS staff to identify and evaluate several possible locations for the new store.... Students worked with architects on design options for the new store and worked with accountants on financing plans.

This project was incorporated into the students' mathematics and humanities classes. The students saw the connections of mathematics to the world of commerce and to other disciplines, and they also saw the connections within mathematics as they applied knowledge from several different areas.

Data analysis and statistics are useful in helping students clarify issues related to their personal lives. Students in prekindergarten through grade 2 who are working on calendar activities can collect data on the weather by recording rainy, cloudy, or sunny days. They can record the data, count days, generalize about conditions, and make predictions for the future. Students in grades 3–5 can use the Internet to collaborate with students in other classrooms to collect and analyze data about acid rain, deforestation, and other phenomena. By grades 9–12, students should be able to use their knowledge of data analysis and mathematical modeling to understand societal issues and workplace problems in reasonable depth.

Representation

The ways in which mathematical ideas are represented is fundamental to how people can understand and use those ideas. Consider how much more difficult multiplication is using Roman numerals (for those who have not worked extensively with them) than using Arabic base-ten notation. Many of the representations we now take for granted—such as numbers expressed in base-ten or binary form, fractions, algebraic expressions and equations, graphs, and spreadsheet displays—are the result of a process of cultural refinement that took place over many years. When students gain access to mathematical representations and the ideas they represent, they have a set of tools that significantly expand their capacity to think mathematically.

The term *representation* refers both to process and to product—in other words, to the act of capturing a mathematical concept or relationship in some form and to the form itself. The child who wrote her age as shown in figure 3.8 used a representation. The graph of $f(x) = x^3$ is a representation. Moreover, the term applies to processes and products that are observable externally as well as to those that occur "internally," in the minds of people doing mathematics. All these meanings of representation are important to consider in school mathematics.

Fig. **3.8.**

A child's representation of five and one-half

Some forms of representation—such as diagrams, graphical displays, and symbolic expressions—have long been part of school mathematics. Unfortunately, these representations and others have often been taught and learned as if they were ends in themselves. Representations should be treated as essential elements in supporting students' understanding of mathematical concepts and relationships; in communicating mathematical approaches, arguments, and understandings to one's self and to others; in recognizing connections among related mathematical concepts; and in applying mathematics to realistic problem situations through modeling. New forms of representation associated with electronic technology create a need for even greater instructional attention to representation.

Representation Standard

Instructional programs from prekindergarten through grade 12 should enable all students to—

- create and use representations to organize, record, and communicate mathematical ideas;
- select, apply, and translate among mathematical representations to solve problems;
- use representations to model and interpret physical, social, and mathematical phenomena.

Create and use representations to organize, record, and communicate mathematical ideas

Students should understand that written representations of mathematical ideas are an essential part of learning and doing mathematics. It is important to encourage students to represent their ideas in ways that make sense to them, even if their first representations are not conventional ones. It is also important that they learn conventional forms of representation to facilitate both their learning of mathematics and their communication with others about mathematical ideas.

The fact that representations are such effective tools may obscure how difficult it was to develop them and, more important, how much work it takes to understand them. For instance, base-ten notation is difficult for young children, and the curriculum should allow many opportunities for making connections between students' emerging understanding of the counting numbers and the structure of base-ten representation. But as students move through the curriculum, the focus tends to be increasingly on presenting the mathematics itself, perhaps under the assumption that students who are old enough to think in formal terms do not, like their younger counterparts, need to negotiate between their naive conceptions and the mathematical formalisms. Research indicates, however, that students at all levels need to work at developing their understandings of the complex ideas captured in conventional representations. A representation as seemingly clear as the variable x can be difficult for students to comprehend.

The idiosyncratic representations constructed by students as they solve problems and investigate mathematical ideas can play an important role in helping students understand and solve problems and providing meaningful ways to record a solution method and to describe the method to others. Teachers can gain valuable insights into students' ways of interpreting and thinking about mathematics by looking at their representations. They can build bridges from students' personal representations to more-conventional ones, when appropriate. It is important that students have opportunities not only to learn conventional forms of representation but also to construct, refine, and use their own representations as tools to support learning and doing mathematics.

Through the middle grades, children's mathematical representations usually are about objects and actions from their direct experience. Primary school students might use objects to represent the number of wheels on four bicycles or the number of fireflies in a story. They may represent larger numbers of objects using place-value mats or base-ten blocks. In the middle grades, students can begin to create and use mathematical representations for more-abstract objects, such as rational numbers, rates, or linear relationships. High school students should use conventional representations as a primary means for expressing and understanding more-abstract mathematical concepts. Through their representations, they should be ready to see a common structure in mathematical phenomena that come from very different contexts.

Representations can help students organize their thinking. Students' use of representations can help make mathematical ideas more concrete and available for reflection. In the lower grades, for example, children can use representations to provide a record for their teachers and their peers of their efforts to understand mathematics. In the middle grades, they should use representations more to solve problems or to portray, clarify, or extend a mathematical idea. They might, for example, focus on collecting a large amount of data in a weather experiment over an extended time and use a spreadsheet and related graphs to organize and represent the data. They might develop an algebraic representation for a real-world relationship (e.g., the number of unit tiles around a rectangular pool with dimensions m units by n units where m and n are integers; see fig 3.9) and begin to recognize that symbolic representations that appear different can describe the same phenomenon. For instance, the number of tiles in the border of the $m \times n$ pool can be expressed as $2n + 2m + 4$ or as $2(m + 2) + 2n$.

Fig. **3.9.**

Rectangular pool and border

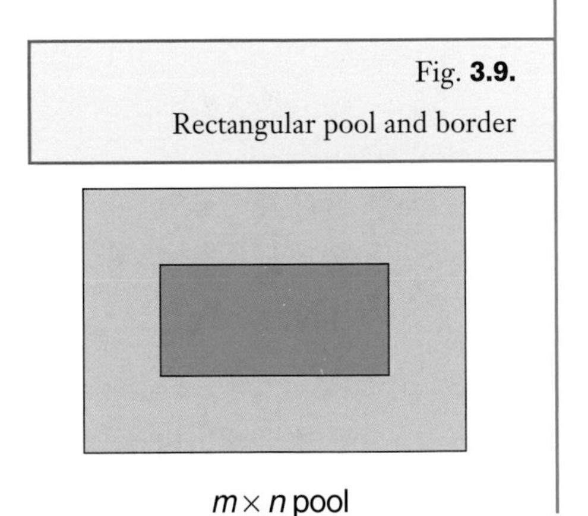

$m \times n$ pool

Computers and calculators change what students can do with conventional representations and expand the set of representations with which they can work. For example, students can flip, invert, stretch, and zoom in on graphs using graphing utilities or dynamic geometry software. They can use computer algebra systems to manipulate expressions, and they can investigate complex data sets using spreadsheets. As students learn to use these new, versatile tools, they also can consider ways in which some representations used in electronic technology differ from conventional representations. For example, numbers in scientific notation are represented differently in calculators and in textbooks. The algebraic expressions on computer algebra systems may look different from those in textbooks.

Select, apply, and translate among mathematical representations to solve problems

Different representations often illuminate different aspects of a complex concept or relationship. For example, students usually learn to represent fractions as sectors of a circle or as pieces of a rectangle or some other figure. Sometimes they use physical displays of pattern blocks or fraction bars that convey the part-whole interpretation of fractions. Such displays can help students see fraction equivalence and the meaning of the addition of fractions, especially when the fractions have the same denominator and when their sum is less than 1. Yet this form of representation does not convey other interpretations of fraction, such as ratio, indicated division, or fraction as number. Other common representations for fractions, such as points on a number line or ratios of discrete elements in a set, convey some but not all aspects of the complex fraction concept. Thus, in order to become deeply knowledgeable about fractions—and many other concepts in school mathematics—students will need a variety of representations that support their understanding.

The importance of using multiple representations should be emphasized throughout students' mathematical education. For example, a prekindergarten through grade 2 student should know how to represent three groups of four through repeated addition, skip-counting, or an array of objects. Primary-grades students begin to see how some representations make it easier to understand some properties. Using the arrays in figure 3.10, a teacher could make commutativity visible.

In grades 3–5, students' repertoires of representations should expand to include more-complex pictures, tables, graphs, and words to model problems and situations. For middle-grades students, representations are useful in developing ideas about algebra. As students become mathematically sophisticated, they develop an increasingly large repertoire of mathematical representations as well as a knowledge of how to use them productively. Such knowledge includes choosing and moving between representations and learning to ask such questions as, Would a graph give me more insight than a symbolic expression to solve this problem?

One of the powerful aspects of mathematics is its use of *abstraction*—the stripping away by symbolization of some features of a problem that are not necessary for analysis, allowing the "naked symbols" to be operated on easily. In many ways, this fact lies behind the power of mathematical applications and modeling. Consider this problem:

> From a ship on the sea at night, the captain can see three lighthouses and can measure the angles between them. If the captain knows the position of the lighthouses from a map, can the captain determine the position of the ship?

When the problem is translated into a mathematical representation, the ship and the lighthouses become points in the plane, and the problem

Fig. **3.10.**

These arrays can help teachers explain commutativity.

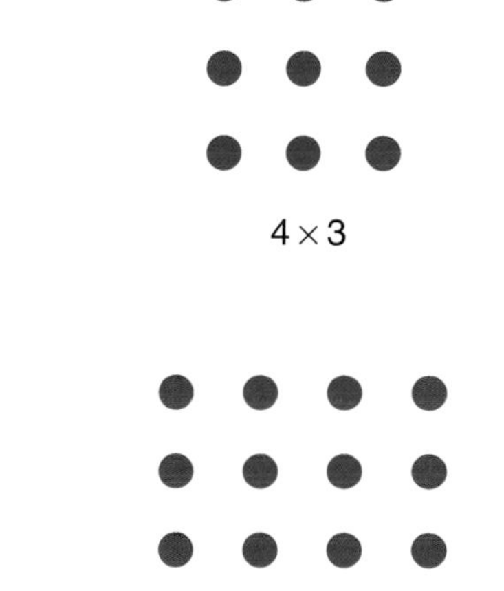

can be solved without knowing that it is about ships. Many other problems from different contexts may have similar representations. As soon as the problem is represented in some form, the classic solution methods for that mathematical form may be used to solve the problem.

Technological tools now offer opportunities for students to have more and different experiences with the use of multiple representations. For instance, in prekindergarten through grade 2, teachers and students can work with on-screen versions of concrete manipulatives, getting accuracy and immediate feedback. Later on, dynamic geometry tools can be used for generating conjectures. Several software packages allow students to view a function simultaneously in tabular, graphical, and equation form. Such software can allow students to examine how certain changes in one representation, such as varying a parameter in the equation $ax^2 + bx + c = 0$, simultaneously affect the other representations. Computer and calculator simulations can be used to investigate physical phenomena, such as motion.

As students' representational repertoire expands, it is important for students to reflect on their use of representations to develop an understanding of the relative strengths and weaknesses of various representations for different purposes. For example, when students learn different representational forms for displaying statistical data, they need opportunities to consider the kinds of data and questions for which a circle graph might be more appropriate than a line graph or a box-and-whiskers plot more appropriate than a histogram.

Use representations to model and interpret physical, social, and mathematical phenomena

The term *model* has many different meanings. So it is not surprising that the word is used in many different ways in discussions about mathematics education. For example, *model* is used to refer to physical materials with which students work in school—manipulative models. The term is also used to suggest exemplification or simulation, as in when a teacher models the problem-solving process for her students. Yet another usage treats the term as if it were roughly synonymous with *representation*. The term *mathematical model*, which is the focus in this context, means a mathematical representation of the elements and relationships in an idealized version of a complex phenomenon. Mathematical models can be used to clarify and interpret the phenomenon and to solve problems.

Models allow a view of a real-world phenomenon.

In some activities, models allow a view of a real-world phenomenon, such as the flow of traffic, through an analytic structure imposed on it. An example of a general question to be explored might be, How long should a traffic light stay green to let a reasonable number of cars flow through the intersection? Students can gather data about how long (on average) it takes the first car to go through, the second car, and so on. They can represent these data statistically, or they can construct analytic functions to work on the problem in the abstract, considering the wait time before a car starts moving, how long it takes a car to get up to regular traffic speed, and so on.

Technological tools now allow students to explore iterative models for situations that were once studied in much more advanced courses. For example, it is now possible for students in grades 9–12 to model predator-prey relations. The initial set-up might be that a particular habitat houses so many wolves and so many rabbits, which are the wolves' primary food source. When the wolves are well fed, they reproduce well (and more wolves eat more rabbits); when the wolves are starved, they die off. The rabbits multiply readily when wolves are scarce but lose numbers rapidly when the wolf population is large. Modeling software that uses difference equations allows students to enter initial conditions and the rules for change and then see what happens to the system dynamically.

Students' use of representations to model physical, social, and mathematical phenomena should grow through the years. In prekindergarten through grade 2, students can model distributing 24 cookies among 8 children, using tiles or blocks in a variety of ways. As students continue to encounter representations in grades 3–5, they begin to use them to model phenomena in the world around them and to aid them in noticing quantitative patterns. As middle-grades students model and solve problems that arise in the real and mathematical worlds, they learn to use variables to represent unknowns and also learn to employ equations, tables, and graphs to represent and analyze relationships. High school students create and interpret models of phenomena drawn from a wider range of contexts—including physical and social environments—by identifying essential elements of the context and by devising representations that capture mathematical relationships among those elements. With electronic technologies, students can use representations for problems and methods that until recently were difficult to explore meaningfully in high school. Iterative numerical methods, for example, can be used to develop an intuitive concept of limit and its applications. The asymptotic behavior of functions is more easily understood graphically, as are the effects of transformations on functions. These tools and understandings give students access to models that can be used to analyze a greatly expanded range of realistic and interesting situations.

Students' use of representations to model physical, social, and mathematical phenomena should grow through the years.

Throughout the early years, the Standards recommended here can help parents and educators give children a solid affective and cognitive foundation in mathematics.

Chapter 4

Standards for Grades Pre-K–2

During the years from birth to age four, much important mathematical development occurs in young children. Whether they are cared for by family members during their preschool years or receive care from persons outside their families, all children need their innate desire for learning to be nurtured and supported. In kindergarten through grade 2, considerations such as high-quality educational settings and experiences become paramount. Throughout the early years, the Standards and specific expectations for mathematics learning recommended here can help parents and educators give children a solid affective and cognitive foundation in mathematics.

Mathematics for the Youngest Learner

The foundation for children's mathematical development is established in the earliest years. Mathematics learning builds on the curiosity and enthusiasm of children and grows naturally from their experiences. Mathematics at this age, if appropriately connected to a child's world, is more than "getting ready" for school or accelerating them into elementary arithmetic. Appropriate mathematical experiences challenge young children to explore ideas related to patterns, shapes, numbers, and space with increasing sophistication.

The principle that all children can learn mathematics applies to all ages. Many mathematics concepts, at least in their intuitive beginnings, develop before school. For instance, infants spontaneously recognize and discriminate small numbers of objects (Starkey and Cooper 1980). Before they enter school, many children possess a substantive informal knowledge of mathematics. They use mathematical ideas in everyday life and develop mathematical knowledge that can be quite complex and sophisticated (Baroody 1992;

Clements et al. 1999; Gelman 1994; Ginsburg, Klein, and Starkey 1998). Children's long-term success in learning and development requires high-quality experiences during the "years of promise" (Carnegie Corporation 1999). Adults can foster children's mathematical development by providing environments rich in language, where thinking is encouraged, uniqueness is valued, and exploration is supported. Play is children's work. Adults support young children's diligence and mathematical development when they direct attention to the mathematics children use in their play, challenge them to solve problems, and encourage their persistence.

Children must learn to trust their own abilities to make sense of mathematics.

Children learn through exploring their world; thus, interests and everyday activities are natural vehicles for developing mathematical thinking. When a parent places crackers in a toddler's hands and says, "Here are two crackers—one, two," or when a three-year-old chooses how she wants her sandwich cut—into pieces shaped like triangles, rectangles, or small squares—mathematical thinking is occurring. As a child arranges stuffed animals by size, an adult might ask, "Which animal is the smallest?" When children recognize a stop sign by focusing on the octagonal shape, adults have an opportunity to talk about different shapes in the environment. Through careful observation, conversations, and guidance, adults can help children make connections between the mathematics in familiar situations and new ones.

Because young children develop a disposition for mathematics from their early experiences, opportunities for learning should be positive and supportive. Children must learn to trust their own abilities to make sense of mathematics. Mathematical foundations are laid as playmates create streets and buildings in the sand or make playhouses with empty boxes. Mathematical ideas grow as children count steps across the room or sort collections of rocks and other treasures. They learn mathematical concepts through everyday activities: sorting (putting toys or groceries away), reasoning (comparing and building with blocks), representing (drawing to record ideas), recognizing patterns (talking about daily routines, repeating nursery rhymes, and reading predictable books), following directions (singing motion songs such as "Hokey Pokey"), and using spatial visualization (working puzzles). Using objects, role-playing, drawing, and counting, children show what they know.

High-quality learning results from formal and informal experiences during the preschool years. "Informal" does not mean unplanned or haphazard. Since the most powerful mathematics learning for preschoolers often results from their explorations with problems and materials that interest them, adults should take advantage of opportunities to monitor and influence how children spend their time. Adults can provide access to books and stories with numbers and patterns; to music with actions and directions such as up, down, in, and out; or to games that involve rules and taking turns. All these activities help children understand a range of mathematical ideas. Children need things to count, sort, compare, match, put together, and take apart.

"Informal" does not mean unplanned or haphazard.

Children need introductions to the language and conventions of mathematics, at the same time maintaining a connection to their informal knowledge and language. They should hear mathematical language being used in meaningful contexts. For example, a parent may ask a child to get the *same* number of forks *as* spoons; or a sibling may be *taller than* they are, but the same sibling may be *shorter than* the girl next door. Young children need to learn words for comparing and for indicating position and direction at the same time they are developing an understanding of counting and number words.

Children are likely to enter formal school settings with various levels of mathematics understanding. However, "not knowing" more often reflects a lack of opportunity to learn than an inability to learn. Some children will need additional support so that they do not start school at a disadvantage. They should be identified with appropriate assessments that are adapted to the needs and characteristics of young children. Interviews and observations, for example, are more appropriate assessment techniques than group tests, which often do not yield complete data. Early assessments should be used to gain information for teaching and for potential early interventions rather than for sorting children. Pediatricians and other health-care providers often recognize indicators of early learning difficulties and can suggest community resources to address these challenges.

Mathematics Education in Prekindergarten through Grade 2

Like the years from birth to formal schooling, prekindergarten through grade 2 (pre-K–2) is a time of profound developmental change for young students. At no other time in schooling is cognitive growth so remarkable. Because young students are served in many different educational settings and begin educational programs at various ages, we refer to children at this level as *students* to denote their enrollment in formal educational programs. Teachers of young students—including parents and other caregivers—need to be knowledgeable about the many ways students learn mathematics, and they need to have high expectations for what can be learned during these early years.

Most students enter school confident in their own abilities, and they are curious and eager to learn more about numbers and mathematical objects. They make sense of the world by reasoning and problem solving, and teachers must recognize that young students can think in sophisticated ways. Young students are active, resourceful individuals who construct, modify, and integrate ideas by interacting with the physical world and with peers and adults. They make connections that clarify and extend their knowledge, thus adding new meaning to past experiences. They learn by talking about what they are thinking and doing and by collaborating and sharing their ideas. Students' abilities to communicate through language, pictures, and other symbolic means develop rapidly during these years. Although students' ways of knowing, representing, and communicating may be different from those of adults, by the end of grade 2, students should be using many conventional mathematical representations with understanding.

It is especially important in the early years for every child to develop a solid mathematical foundation. Children's efforts and their confidence that mathematics learning is within their reach must be supported. Young students are building beliefs about what mathematics is, about what it means to know and do mathematics, and about themselves as mathematics learners. These beliefs influence their thinking, performance, attitudes, and decisions about studying mathematics in later years (Kamii 2000). Therefore, it is imperative to provide all students with high-quality programs that include significant mathematics presented in a manner that respects both the mathematics and the nature of young children. These programs should build on and extend students' intuitive and informal mathematics knowledge. They should be grounded in a knowledge of child development and take place in environments that encourage students to be active learners and accept new challenges. They should develop a strong conceptual framework while encouraging and developing students' skills and their natural inclination to solve problems. Number activities oriented toward problem solving can be successful even with very young children and can develop not only counting and number abilities but also such reasoning abilities as classifying and ordering (Clements 1984). Recent research has confirmed that an appropriate curriculum strengthens the development of young students' knowledge of number and geometry (Griffin and Case 1997; Klein, Starkey, and Wakeley 1999; Razel and Eylon 1991).

Mathematics teaching in the lower grades should encourage students' strategies and build on them as ways of developing more-general ideas and systematic approaches. By asking questions that lead to clarifications, extensions, and the development of new understandings, teachers can facilitate students' mathematics learning. Teachers should ensure that interesting problems and stimulating mathematical conversations are a part of each day. They should honor individual students' thinking and reasoning and use formative assessment to plan instruction that enables students to connect new mathematics learning with what they know. Schools should furnish materials that allow students to continue to learn mathematics through counting, measuring, constructing with blocks and clay, playing games and doing puzzles, listening to stories, and engaging in dramatic play, music, and art.

Early education must build on the principle that all students can learn significant mathematics.

In prekindergarten through grade 2, mathematical concepts develop at different times and rates for each child. If students are to attain the mathematical goals described in *Principles and Standards for School Mathematics*, their mathematics education must include much more than short-term learning of rote procedures. All students need adequate time and opportunity to develop, construct, test, and reflect on their increasing understanding of mathematics. Early education must build on the principle that all students can learn significant mathematics. Along with their expectations for students, teachers should also set equally high standards for themselves, seeking, if necessary, the new knowledge and skills they need to guide and nurture all students. School leaders and teachers must take the responsibility for supporting learning so that all students leave grade 2 confident and competent in the mathematics described for this grade band.

The ten Standards presented in this document are not separate topics for study but are carefully interwoven strands designed to support the learning of connected mathematical ideas. At the core of mathematics in the early years are the Number and Geometry Standards. Numbers and their relationships, operations, place value, and attributes of shapes are examples of important ideas from these standards. Each of the other mathematical Content Standards, including Algebra, Measurement, and Data Analysis and Probability, contribute to, and is learned in conjunction with, the Number and Geometry Standards. The Process Standards of Problem Solving, Reasoning and Proof, Communication, Connections, and Representation support the learning of, and are developed through, the Content Standards. And learning content involves learning and using mathematics processes.

Students should recognize when using a calculator is appropriate.

The mathematics program in prekindergarten through grade 2 should take advantage of technology. Guided work with calculators can enable students to explore number and pattern, focus on problem-solving processes, and investigate realistic applications. Through their experiences and with the teacher's guidance, students should recognize when using a calculator is appropriate and when it is more efficient to compute mentally. Computers also can make powerful and unique contributions to students' learning by providing feedback and connections between representations. They benefit all students and are especially helpful for learners with physical limitations or those who interact more comfortably with technology than with classmates (Clements 1999a; Wright and Shade 1994).

Young students frequently possess greater knowledge than they are able to express in writing. Teachers need to determine what students already know and what they still have to learn. Information from a wide variety of classroom assessments—classroom routines, conversations, written work (including pictures), and observations—helps teachers plan meaningful tasks that offer support for students whose understandings are not yet complete and helps teachers challenge students who are ready to grapple with new problems and ideas. Teachers must maintain a balance, helping students develop both conceptual understanding and procedural facility (skill). Students' development of number sense should move through increasingly sophisticated levels of constructing ideas and skills, of recognizing and using relationships to solve problems, and of connecting new learning with old. As discussed in the Learning Principle (chapter 2), skills are most effectively acquired when understanding is the foundation for learning.

Mathematics learning for students at this level must be active, rich in natural and mathematical language, and filled with thought-provoking opportunities. Students respond to the challenge of high expectations, and mathematics should be taught for understanding rather than around preconceptions about children's limitations. This does not mean abandoning children's ways of knowing and representing; rather, it is a clear call to create opportunities for young students to learn new, important mathematics in ways that make sense to them.

Number and Operations Standard for Grades Pre-K–2

Instructional programs from prekindergarten through grade 12 should enable all students to—

Expectations

In prekindergarten through grade 2 all students should–

Standard	Expectations
Understand numbers, ways of representing numbers, relationships among numbers, and number systems	• count with understanding and recognize "how many" in sets of objects; • use multiple models to develop initial understandings of place value and the base-ten number system; • develop understanding of the relative position and magnitude of whole numbers and of ordinal and cardinal numbers and their connections; • develop a sense of whole numbers and represent and use them in flexible ways, including relating, composing, and decomposing numbers; • connect number words and numerals to the quantities they represent, using various physical models and representations; • understand and represent commonly used fractions, such as 1/4, 1/3, and 1/2.
Understand meanings of operations and how they relate to one another	• understand various meanings of addition and subtraction of whole numbers and the relationship between the two operations; • understand the effects of adding and subtracting whole numbers; • understand situations that entail multiplication and division, such as equal groupings of objects and sharing equally.
Compute fluently and make reasonable estimates	• develop and use strategies for whole-number computations, with a focus on addition and subtraction; • develop fluency with basic number combinations for addition and subtraction; • use a variety of methods and tools to compute, including objects, mental computation, estimation, paper and pencil, and calculators.

Number and Operations

The concepts and skills related to number and operations are a major emphasis of mathematics instruction in prekindergarten through grade 2. Over this span, the small child who holds up two fingers in response to the question "How many is two?" grows to become the second grader who solves sophisticated problems using multidigit computation strategies. In these years, children's understanding of number develops significantly. Children come to school with rich and varied informal knowledge of number (Baroody 1992; Fuson 1988; Gelman 1994). During the early years teachers must help students strengthen their sense of number, moving from the initial development of basic counting techniques to more-sophisticated understandings of the size of numbers, number relationships, patterns, operations, and place value.

Teachers should not underestimate what young students can learn.

Students' work with numbers should be connected to their work with other mathematics topics. For example, computational fluency (having and using efficient and accurate methods for computing) can both enable and be enabled by students' investigations of data; a knowledge of patterns supports the development of skip-counting and algebraic thinking; and experiences with shape, space, and number help students develop estimation skills related to quantity and size.

As they work with numbers, students should develop efficient and accurate strategies that they understand, whether they are learning the basic addition and subtraction number combinations or computing with multidigit numbers. They should explore numbers into the hundreds and solve problems with a particular focus on two-digit numbers. Although good judgment must be used about which numbers are important for students of a certain age to work with, teachers should be careful not to underestimate what young students can learn about number. Students are often surprisingly adept when they encounter numbers, even large numbers, in problem contexts. Therefore, teachers should regularly encourage students to demonstrate and deepen their understanding of numbers and operations by solving interesting, contextualized problems and by discussing the representations and strategies they use.

Understand numbers, ways of representing numbers, relationships among numbers, and number systems

Counting is a foundation for students' early work with number. Young children are motivated to count everything from the treats they eat to the stairs they climb, and through their repeated experience with the counting process, they learn many fundamental number concepts. They can associate number words with small collections of objects and gradually learn to count and keep track of objects in larger groups. They can establish one-to-one correspondence by moving, touching, or pointing to objects as they say the number words. They should learn that counting objects in a different order does not alter the result, and they may notice that the next whole number in the counting sequence is one more than the number just named. Children should learn that the last number named represents the last object as well as the total number of objects in the collection. They often solve addition and subtraction problems by counting concrete objects, and many children invent problem-solving

Flexibility in thinking about numbers ... is a hallmark of number sense.

strategies based on counting strategies (Ginsburg, Klein, and Starkey 1998; Siegler 1996).

Throughout the early years, teachers should regularly give students varied opportunities to continue to develop, use, and practice counting as they quantify collections of objects, measure attributes of shapes, identify locations, and solve problems. Preschool and kindergarten teachers, for example, should use naturally occurring opportunities to help students develop number concepts by posing questions such as, How many pencils do we need at this table? Shall we count how many steps to the playground? Who is third in line? Students often use different approaches when dealing with smaller numbers versus larger numbers. They may look at a small group of objects (about six items or fewer) and recognize "how many," but they may need to count a group of ten or twelve objects to find a total. The ability to recognize at a glance small groups within a larger group supports the development of visually grouping objects as a strategy for estimating quantities.

In these early years, students develop the ability to deal with numbers mentally and to think about numbers without having a physical model (Steffe and Cobb 1988). Some students will develop this capacity before entering school, and others will acquire it during their early school years. Thus, in a first-grade class where students are asked to tell how many blocks are hidden when the total number, say, seven, is known but some, say, three, are covered, students will vary in how they deal with the covered blocks. Some may be able to note that there are four visible blocks and then count up from there, saying, "Five, six, seven. There are three hidden!" But others may not be able to answer the question unless they see all the objects; they may need to uncover and point at or touch the blocks as they count them.

As students work with numbers, they gradually develop flexibility in thinking about numbers, which is a hallmark of number sense. Students may model twenty-five with beans and bean sticks or with two dimes and a nickel, or they may say that it is 2 tens and 5 ones, five more than twenty, or halfway between twenty and thirty. Number sense develops as students understand the size of numbers, develop multiple ways of thinking about and representing numbers, use numbers as referents, and develop accurate perceptions about the effects of operations on numbers (Sowder 1992). Young students can use number sense to reason with numbers in complex ways. For example, they may estimate the number of cubes they can hold in one hand by referring to the number of cubes that their teacher can hold in one hand. Or if asked whether four plus three is more or less than ten, they may recognize that the sum is less than ten because both numbers are less than five and five plus five makes ten.

Concrete models can help students represent numbers and develop number sense; they can also help bring meaning to students' use of written symbols and can be useful in building place-value concepts. But using materials, especially in a rote manner, does not ensure understanding. Teachers should try to uncover students' thinking as they work with concrete materials by asking questions that elicit students' thinking and reasoning. In this way, teachers can watch for students' misconceptions, such as interpreting the 2 tens and 3 ones in figure 4.1c merely as five objects. Teachers should also choose interesting tasks that engage students in mathematical thinking and reasoning, which builds their understanding of numbers and relationships among numbers.

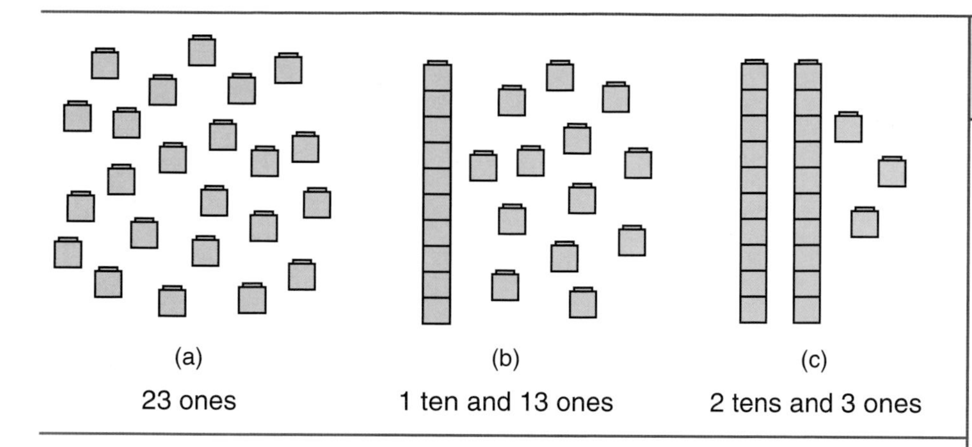

Fig. **4.1.**
Different ways of representing 23

It is absolutely essential that students develop a solid understanding of the base-ten numeration system and place-value concepts by the end of grade 2. Students need many instructional experiences to develop their understanding of the system, including how numbers are written. They should understand, for example, that multiples of 10 provide bridges when counting (e.g., 38, 39, **40**, 41) and that "10" is a special unit within the base-ten system. They should recognize that the word *ten* may represent a single entity (1 ten) and, at the same time, ten separate units (10 ones) and that these representations are interchangeable (Cobb and Wheatley 1988). Using concrete materials can help students learn to group and ungroup by tens. For example, such materials can help students express "23" as 23 ones (units), 1 ten and 13 ones, or 2 tens and 3 ones (see fig. 4.1). Of course, students should also note the ways in which using concrete materials to represent a number differs from using conventional notation. For example, when the numeral for the collection shown in figure 4.1 is written, the arrangement of digits matters—the digit for the tens must be written to the left of the digit for the units. In contrast, when base-ten blocks or connecting cubes are used, the value is not affected by the arrangement of the blocks.

Using materials, especially in a rote manner, does not ensure understanding.

Technology can help students develop number sense, and it may be especially helpful for those with special needs. For example, students who may be uncomfortable interacting with groups or who may not be physically able to represent numbers and display corresponding symbols can use computer manipulatives. The computer simultaneously links the student's actions with symbols. When the block arrangement is changed, the number displayed is automatically changed. As with connecting cubes, students can break computer base-ten blocks into ones or join ones together to form tens.

Place-value concepts can be developed and reinforced using calculators. For example, students can observe values displayed on a calculator and focus on which digits are changing. If students add 1 repeatedly on a calculator, they can observe that the units digit changes every time, but the tens digit changes less frequently. Through classroom conversations about such activities and patterns, teachers can help focus students' attention on important place-value ideas. Figure 4.2 shows another example—a challenging calculator activity for second graders—that could be used to strengthen students' understanding of place value. In this activity, students begin at one number and add or subtract to reach a target number. Since they are not limited in what they can add or subtract, activities like this allow them to use various approaches to reach the target numbers. They can decide whether to add or subtract ones or multiples of 10 or how they might use multiple steps to arrive at the target. By having students share and discuss the different strategies employed by members of a class, a teacher can highlight the ways in which students use place-value concepts in their strategies.

Fig. **4.2.**

A calculator activity to help develop understanding of place value

Make a New Number

Use a calculator. Start with 78. Do not press clear. Make the next number:	What numbers did you add or subtract?
98	
Next make → 48	
18	
118	
119	

Students also develop understanding of place value through the strategies they invent to compute.

Students also develop understanding of place value through the strategies they invent to compute (Fuson et al. 1997). Thus, it is not necessary to wait for students to fully develop place-value understandings before giving them opportunities to solve problems with two- and three-digit numbers. When such problems arise in interesting contexts, students can often invent ways to solve them that incorporate and deepen their understanding of place value, especially when students have opportunities to discuss and explain their invented strategies and approaches. Teachers emphasize place value by asking appropriate questions and choosing problems such as finding ten more than or ten less than a number and helping them contrast the answers with the initial number. As a result of regular experiences with problems that develop place-value concepts, second-grade students should be counting into the hundreds, discovering patterns in the numeration system related to place value, and composing (creating through different combinations) and decomposing (breaking apart in different ways) two- and three-digit numbers.

In addition to work with whole numbers, young students should also have some experience with simple fractions through connections to everyday situations and meaningful problems, starting with the common fractions expressed in the language they bring to the classroom, such as "half." At this level, it is more important for students to recognize when things are divided into equal parts than to focus on fraction notation. Second graders should be able to identify three parts out of four equal parts, or three-fourths of a folded paper that has been shaded, and to understand that "fourths" means four equal parts of a whole. Although fractions are not a topic for major emphasis for pre-K–2 students, informal experiences at this age will help develop a foundation for deeper learning in higher grades.

Understand meanings of operations and how they relate to one another

As students in the early grades work with complex tasks in a variety of contexts, they also build an understanding of operations on numbers. Appropriate contexts can arise through student-initiated activities, teacher-created stories, and in many other ways. As students explain their written work, solutions, and mental processes, teachers gain insight into their students' thinking. See the "Communication" section of this chapter for a more general discussion of these issues and more examples related to the development of students' understanding of number and operations.

An understanding of addition and subtraction can be generated when young students solve "joining" and take-away problems by directly modeling the situation or by using counting strategies, such as counting on or counting back (Carpenter and Moser 1984). Students develop further understandings of addition when they solve missing-addend problems that arise from stories or real situations. Further understandings of subtraction are conveyed by situations in which two collections need to be made equal or one collection needs to be made a desired size. Some problems, such as "Carlos had three cookies. María gave him some more, and now he has eight. How many did she give him?" can help students see the relationship between addition and subtraction. As they build an understanding of addition and subtraction of whole numbers, students also develop a repertoire of representations. For more discussion, see the section on "Representation" in this chapter.

In developing the meaning of operations, teachers should ensure that students repeatedly encounter situations in which the same numbers appear in different contexts. For example, the numbers 3, 4, and 7 may appear in problem-solving situations that could be represented by $4 + 3$, or $3 + 4$, or $7 - 3$, or $7 - 4$. Although different students may initially use quite different ways of thinking to solve problems, teachers should help students recognize that solving one kind of problem is related to solving another kind. Recognizing the inverse relationship between addition and subtraction can allow students to be flexible in using strategies to solve problems. For example, suppose a student solves the problem $27 + \square = 36$ by starting at 27 and counting up to 36, keeping track of the 9 counts. Then, if the student is asked to solve $36 - 9 = \square$, he may say immediately, "27." If asked how he knows, he might respond, "Because we just did it." This student understands that 27 and 9 are numbers in their own right, as well as two parts that make up the whole, 36. He also understands that subtraction is the inverse of addition (Steffe and Cobb 1988). Another student, one who does not use the relationship between addition and subtraction, might try to solve the problem by counting back 9 units from 36, which is a much more difficult strategy to apply correctly.

In developing the meaning of addition and subtraction with whole numbers, students should also encounter the properties of operations, such as the commutativity and associativity of addition. Although some students discover and use properties of operations naturally, teachers can

Teachers should help students recognize that solving one kind of problem is related to solving another kind.

bring these properties to the forefront through class discussions. For example, 6 + 9 + 4 may be easier to solve as 6 + 4 + 9, allowing students to add 6 and 4 to get 10 and 10 and 9 to get 19. Students notice that adding and subtracting the same number in a computation is equivalent to adding 0. For instance, 40 – 10 + 10 = 40 + 0 = 40. Some students recognize that equivalent quantities can be substituted: 8 + 7 = 8 + 2 + 5 because 7 = 2 + 5. They may realize that adding the same number (e.g., 100) to both terms of a difference (e.g., 50 – 10 = 40) does not change the result (150 – 110 = 40). The use of these properties is a sign of young students' growing number sense. Different students, however, need different amounts of time to make these properties their own. What some students learn in one year may take two or more years for others.

Different students need different amounts of time.

In prekindergarten through grade 2, students should also begin to develop an understanding of the concepts of multiplication and division. Through work in situations involving equal subgroups within a collection, students can associate multiplication with the repeated joining (addition) of groups of equal size. Similarly, they can investigate division with real objects and through story problems, usually ones involving the distribution of equal shares. The strategies used to solve such problems—the repeated joining of, and partitioning into, equal subgroups—thus become closely associated with the meaning of multiplication and division, respectively.

Compute fluently and make reasonable estimates

Young children often initially compute by using objects and counting; however, prekindergarten through grade 2 teachers need to encourage them to shift, over time, to solving many computation problems mentally or with paper and pencil to record their thinking. Students should develop strategies for knowing basic number combinations (the single-digit addition pairs and their counterparts for subtraction) that build on their thinking about, and understanding of, numbers. Fluency with basic addition and subtraction number combinations is a goal for the pre-K–2 years. By *fluency* we mean that students are able to compute efficiently and accurately with single-digit numbers. Teachers can help students increase their understanding and skill in single-digit addition and subtraction by providing tasks that (*a*) help them develop the relationships within subtraction and addition combinations and (*b*) elicit counting on for addition and counting up for subtraction and unknown-addend situations.

Teachers should also encourage students to share the strategies they develop in class discussions. Students can develop and refine strategies as they hear other students' descriptions of their thinking about number combinations. For example, a student might compute 8 + 7 by counting on from 8: "…, 9, 10, 11, 12, 13, 14, 15." But during a class discussion of solutions for this problem, she might hear another student's strategy, in which he uses knowledge about 10; namely, 8 and 2 make 10, and 5 more is 15. She may then be able to adapt and apply this strategy later when she computes 28 + 7 by saying, "28 and 2 make 30, and 5 more is 35."

Students learn basic number combinations and develop strategies for computing that make sense to them when they solve problems with interesting and challenging contexts. Through class discussions, they can compare the ease of use and ease of explanation of various strategies. In some cases, their strategies for computing will be close to conventional algorithms; in other cases, they will be quite different. Many times, students' invented approaches are based on a sound understanding of numbers and operations, and they can often be used efficiently and accurately. Some sense of the diversity of approaches students use can be seen in figure 4.3, which shows the ways several students in the same second-grade class computed 25 + 37. Students 1 and 2 have represented their thinking fairly completely, the first with words and the second with tallies. Both demonstrate an understanding of the meaning of the numbers involved. Students 3 and 4 have each used a process that resulted in an accurate answer, but the thinking that underlies the process is not as apparent in their recordings. Students 5 and 6 both illustrate a common source of error—treating the digits in ways that do not reflect their place value and thus generating an unreasonable result.

During a class discussion, other students reported strategies based on composing and decomposing numbers. One student started with 37 and used the fact that 25 could be decomposed into 20 plus 3 plus 2 to solve the problem as follows: 37 + 20 = 57, 57 + 3 = 60, and 60 + 2 = 62. Another student used flexible composing and decomposing in other ways to create an equivalent, easier problem: Take 3 from 25 and use it to turn 37 into 40. Then add 40 and 22 to get 62.

As students work with larger numbers, their strategies for computing play an important role in linking less formal knowledge with more-

Fig. **4.3.**

Six students' solutions to 25 + 37

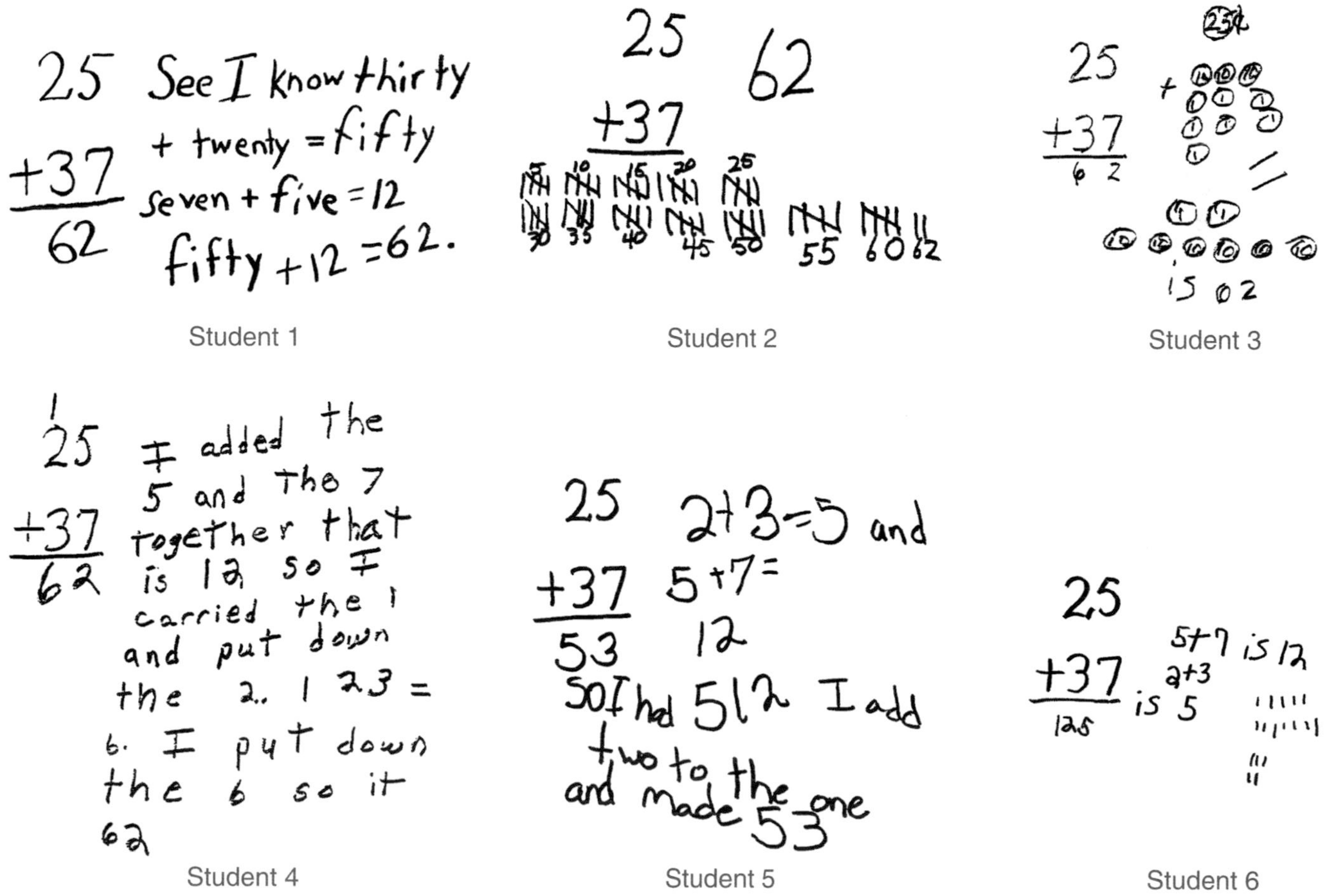

25 + 37 = 62 — See I know thirty + twenty = fifty, seven + five = 12, fifty + 12 = 62.

Student 1

25 + 37 = 62 (tallies counted 5, 10, 15, 20, 25, 30, 35, 40, 45, 50, 55, 60, 62)

Student 2

25 + 37 = 62

Student 3

25 + 37 = 62 — I added the 5 and the 7 together that is 12 so I carried the 1 and put down the 2. 1 2 3 = 6. I put down the 6 so it 62

Student 4

25 + 37 = 53 — 2+3=5 and 5+7= 12 So I had 512 I add two to the one and made 53

Student 5

25 + 37 = 125 — 5+7 is 12, 2+3 is 5

Student 6

Research provides evidence that students will rely on their own computational strategies.

sophisticated mathematical thinking. Research provides evidence that students will rely on their own computational strategies (Cobb et al. 1991). Such inventions contribute to their mathematical development (Gravemeijer 1994; Steffe 1994). Moreover, students who used invented strategies before they learned standard algorithms demonstrated a better knowledge of base-ten concepts and could better extend their knowledge to new situations, such as finding how much of $4.00 would be left after a purchase of $1.86 (Carpenter et al. 1998, p. 9). Thus, when students compute with strategies they invent or choose because they are meaningful, their learning tends to be robust—they are able to remember and apply their knowledge. Children with specific learning disabilities can actively invent and transfer strategies if given well-designed tasks that are developmentally appropriate (Baroody 1999).

Teachers have a very important role to play in helping students develop facility with computation. By allowing students to work in ways that have meaning for them, teachers can gain insight into students' developing understanding and give them guidance. To do this well, teachers need to become familiar with the range of ways that students might think about numbers and work with them to solve problems. Consider the following hypothetical story, in which a teacher poses this problem to a class of second graders:

> We have 153 students at our school. There are 273 students at the school down the street. How many students are in both schools?

As would be expected in most classrooms, the students give a variety of responses that illustrate a range of understandings. For example, Randy models the problem with bean sticks that the class has made earlier in the year, using hundreds rafts, tens sticks, and loose beans. He models the numbers and combines the bean sticks, but he is not certain how to record the results. He draws a picture of the bean sticks and labels the parts, "3 rafts," "12 tens," "6 beans" (fig. 4.4).

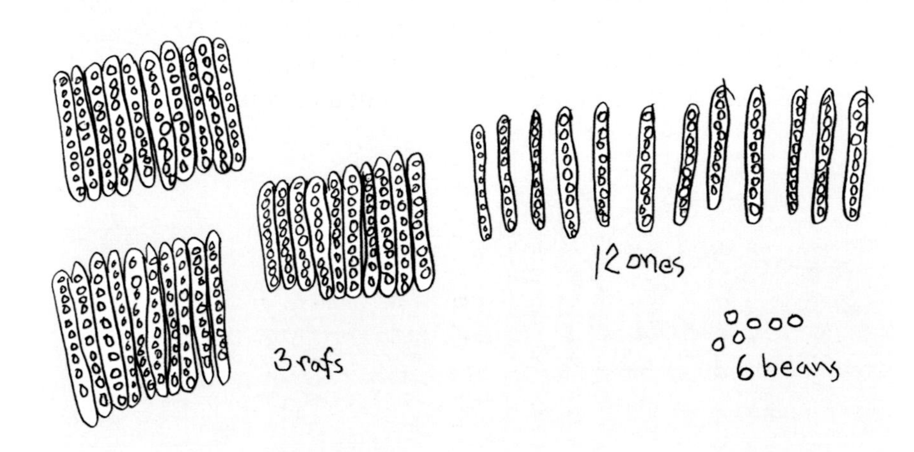

Fig. **4.4.**

Randy models 153 and 273 using beans, bean sticks, and rafts of bean sticks.

Ana first adds the hundreds, recording 300 as an intermediate result; then she adds the tens, keeping the answer in her head; she then adds the ones; and finally she adds the partial results and writes down the answer. Her written record is shown in figure 4.5.

Some students use the conventional algorithm (stacking the addends and then adding the ones, adding the tens and renaming them as hundreds and tens, and finally adding the hundreds) accurately, but others write 3126 as their answer, demonstrating a lack of understanding that the teacher needs to address. Becky finds the answer using mental computation and writes nothing down except her answer. When asked to explain, she says, “Well, 2 hundreds and 1 hundred are 3 hundreds, and 5 tens and 5 tens are 10 tens, or another hundred, so that’s 4 hundreds. There’s still 2 tens left over, and 3 and 3 is 6, so it’s 426.”

Fig. **4.5.**

A written record of the computation of 153 + 273 with intermediate results

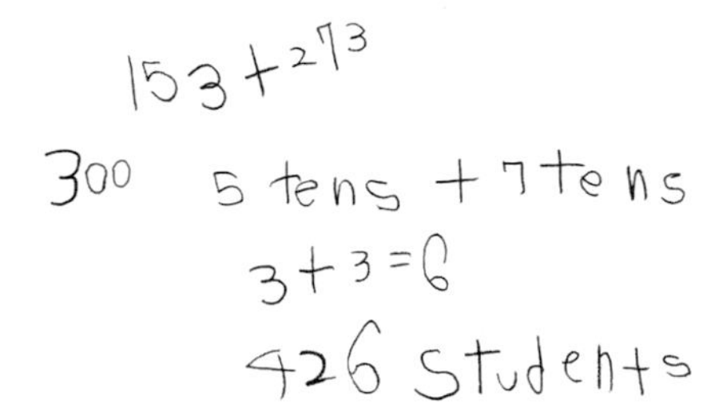

153 + 273
300 5 tens + 7 tens
3 + 3 = 6
426 students

Meaningful practice is necessary to develop fluency with basic number combinations and strategies with multidigit numbers. The example above illustrates that teachers can learn about students’ understanding and at the same time get information to gauge the need for additional attention and work. Practice needs to be motivating and systematic if students are to develop computational fluency, whether mentally, with manipulative materials, or with paper and pencil. Practice can be conducted in the context of other activities, including games that require computation as part of score keeping, questions that emerge from children’s literature, situations in the classroom, or focused activities that are part of another mathematical investigation. Practice should be purposeful and should focus on developing thinking strategies and a knowledge of number relationships rather than drill isolated facts.

The teacher’s responsibility is to gain insight into how students are thinking about various problems by encouraging them to explain what they did with the numbers (Carpenter et al. 1989). Teachers also must decide what new tasks will challenge students and encourage them to construct strategies that are efficient and accurate and that can be generalized. Class discussions and interesting tasks help students build directly on their knowledge and skills while providing opportunities for invention, practice, and the development of deeper understanding. Students’ explanations of solutions permit teachers to assess their development of number sense. As in the previous example, different levels of sophistication in understanding number relationships can be seen in second graders’ responses (fig. 4.6) to the following problem. Notice that all the students used their understanding of counting by fives or of five as a unit in their solutions.

Meaningful practice is necessary to develop fluency.

> There are 93 students going to the circus. Five students can ride in each car. How many cars will be needed?

Students can learn to compute accurately and efficiently through regular experience with meaningful procedures. They benefit from instruction that blends procedural fluency and conceptual understanding (Ginsburg, Klein, and Starkey 1998; Hiebert 1999). This is true for all students, including those with special educational needs. Many children with learning disabilities can learn when they receive high-quality, conceptually oriented instruction. Special instructional interventions for those who need them often focus narrowly on skills instead of offering balanced and comprehensive instruction that uses the child’s abilities to offset weaknesses and provides better long-term results (Baroody 1996). As students encounter problem situations in which computations are more cumbersome or tedious, they should be

93 childrern
5 10 15 20 25 30 35
40 45 50 55 60 65 70
75 80 85 90 3 extra!
19 cars

I knew 100 was 20 5's so
I said that 90 was 18, but I Knew their
was 3 more so I said it must be 19.

Fives

5	1
10	2
15	3
20	4
25	5
30	6
35	7
40	8
45	9
50	10
55	11
60	12
65	13
70	14
75	15
80	16
85	17
90	18
93	19

Olny 18 cars will be fall
and 3 will ride in the
19th car
answer
re.3 18 cars
19 cars

Fig. **4.6.**

Students' computation strategies exhibit different levels of sophistication.

encouraged to use calculators to aid in problem solving. In this way, even students who are slow to gain fluency with computation skills will not be deprived of worthwhile opportunities to solve complex mathematics problems and to develop and deepen their understanding of other aspects of number.

Algebra Standard for Grades Pre-K–2

Instructional programs from prekindergarten through grade 12 should enable all students to—

Expectations

In prekindergarten through grade 2 all students should–

Understand patterns, relations, and functions	• sort, classify, and order objects by size, number, and other properties; • recognize, describe, and extend patterns such as sequences of sounds and shapes or simple numeric patterns and translate from one representation to another; • analyze how both repeating and growing patterns are generated.
Represent and analyze mathematical situations and structures using algebraic symbols	• illustrate general principles and properties of operations, such as commutativity, using specific numbers; • use concrete, pictorial, and verbal representations to develop an understanding of invented and conventional symbolic notations.
Use mathematical models to represent and understand quantitative relationships	• model situations that involve the addition and subtraction of whole numbers, using objects, pictures, and symbols.
Analyze change in various contexts	• describe qualitative change, such as a student's growing taller; • describe quantitative change, such as a student's growing two inches in one year.

Algebra

Algebraic concepts can evolve and continue to develop during prekindergarten through grade 2. They will be manifested through work with classification, patterns and relations, operations with whole numbers, explorations of function, and step-by-step processes. Although the concepts discussed in this Standard are algebraic, this does not mean that students in the early grades are going to deal with the symbolism often taught in a traditional high school algebra course.

Even before formal schooling, children develop beginning concepts related to patterns, functions, and algebra. They learn repetitive songs, rhythmic chants, and predictive poems that are based on repeating and growing patterns. The recognition, comparison, and analysis of patterns are important components of a student's intellectual development. When students notice that operations seem to have particular properties, they are beginning to think algebraically. For example, they realize that changing the order in which two numbers are added does not change the result or that adding zero to a number leaves that number unchanged. Students' observations and discussions of how quantities relate to one another lead to initial experiences with function relationships, and their representations of mathematical situations using concrete objects, pictures, and symbols are the beginnings of mathematical modeling. Many of the step-by-step processes that students use form the basis of understanding iteration and recursion.

Patterns are a way for young students to recognize order and to organize their world.

Understand patterns, relations, and functions

Sorting, classifying, and ordering facilitate work with patterns, geometric shapes, and data. Given a package of assorted stickers, children quickly notice many differences among the items. They can sort the stickers into groups having similar traits such as color, size, or design and order them from smallest to largest. Caregivers and teachers should elicit from children the criteria they are using as they sort and group objects. Patterns are a way for young students to recognize order and to organize their world and are important in all aspects of mathematics at this level. Preschoolers recognize patterns in their environment and, through experiences in school, should become more skilled in noticing patterns in arrangements of objects, shapes, and numbers and in using patterns to predict what comes next in an arrangement. Students know, for example, that "first comes breakfast, then school," and "Monday we go to art, Tuesday we go to music." Students who see the digits "0, 1, 2, 3, 4, 5, 6, 7, 8, 9" repeated over and over will see a pattern that helps them learn to count to 100—a formidable task for students who do not recognize the pattern.

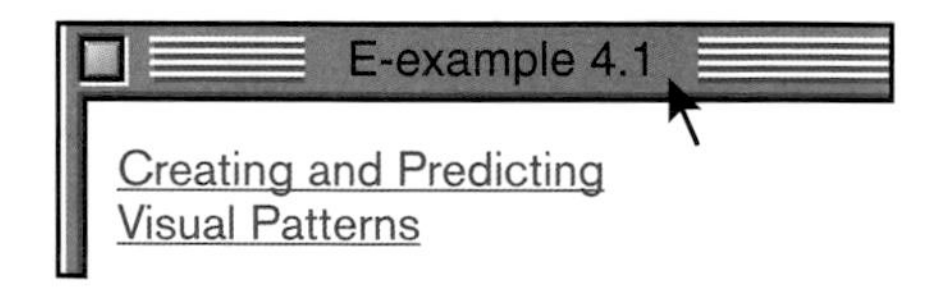

Teachers should help students develop the ability to form generalizations by asking such questions as "How could you describe this pattern?" or "How can it be repeated or extended?" or "How are these patterns alike?" For example, students should recognize that the color pattern "blue, blue, red, blue, blue, red" is the same in form as "clap, clap, step, clap, clap, step." This recognition lays the foundation for the idea that two very different situations can have the same mathematical features and thus are the same in some important ways. Knowing that each pattern above could be described as having the form AABAAB is for students an early introduction to the power of algebra.

By encouraging students to explore and model relationships using language and notation that is meaningful for them, teachers can help students see different relationships and make conjectures and generalizations from their experiences with numbers. Teachers can, for instance, deepen students' understanding of numbers by asking them to model the same quantity in many ways—for example, eighteen is nine groups of two, 1 ten and 8 ones, three groups of six, or six groups of three. Pairing counting numbers with a repeating pattern of objects can create a function (see fig. 4.7) that teachers can explore with students: What is the second shape? To continue the pattern, what shape comes next? What number comes next when you are counting? What do you notice about the numbers that are beneath the triangles? What shape would 14 be?

Fig. **4.7.**

Pairing counting numbers with a repeating pattern

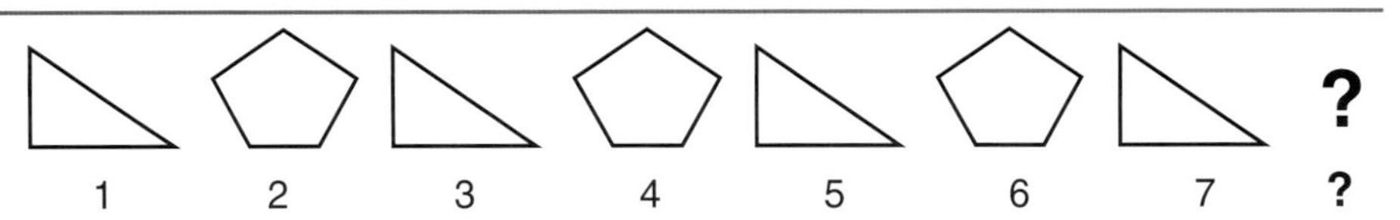

Students should learn to solve problems by identifying specific processes. For example, when students are skip-counting three, six, nine, twelve, ..., one way to obtain the next term is to add three to the previous number. Students can use a similar process to compute how much to pay for seven balloons if one balloon costs 20¢. If they recognize the sequence 20, 40, 60, ... and continue to add 20, they can find the cost for seven balloons. Alternatively, students can realize that the total amount to be paid is determined by the number of balloons bought and find a way to compute the total directly. Teachers in grades 1 and 2 should provide experiences for students to learn to use charts and tables for recording and organizing information in varying formats (see figs. 4.8 and 4.9). They also should discuss the different notations for showing amounts of money. (One balloon costs 20¢, or \$0.20, and seven balloons cost \$1.40.)

Fig. **4.8.**

A vertical chart for recording and organizing information

Cost of Balloons

Number of Balloons	Cost of Balloons in Cents
1	20
2	40
3	60
4	80
5	?
6	?
7	?

Skip-counting by different numbers can create a variety of patterns on a hundred chart that students can easily recognize and describe (see fig. 4.10). Teachers can simultaneously use hundred charts to help students learn about number patterns and to assess students' understanding of counting patterns. By asking questions such as "If you count by tens beginning at 36, what number would you color next?" and "If you continued counting by tens, would you color 87?" teachers can observe whether students understand the correspondence between the visual pattern formed by the shaded numbers and the counting pattern. Using a calculator and a hundred chart enables the students to see the same pattern in two different formats.

Fig. **4.9.**

A horizontal chart for recording and organizing information

Cost of Balloons

Number of balloons	1	2	3	4	5	6	7
Cost of balloons in cents	20	40	60	80	?	?	?

1	2	**3**	4	5	**6**	7	8	**9**	10
11	**12**	13	14	**15**	16	17	**18**	19	20
21	22	23	**24**	25	26	**27**	28	29	**30**
31	32	**33**	34	35	**36**	37	38	**39**	40
41	**42**	43	44	**45**	46	47	**48**	49	50
51	52	53	**54**	55	56	**57**	58	59	**60**
61	62	**63**	64	65	**66**	67	68	**69**	70
71	**72**	73	74	**75**	76	77	**78**	79	80
81	82	83	**84**	85	86	**87**	88	89	**90**
91	92	**93**	94	95	**96**	97	98	**99**	100

Counting by threes

1	2	3	4	5	**6**	7	8	9	10
11	**12**	13	14	15	16	17	**18**	19	20
21	22	23	**24**	25	26	27	28	29	**30**
31	32	33	34	35	**36**	37	38	39	40
41	**42**	43	44	45	46	47	**48**	49	50
51	52	53	**54**	55	56	57	58	59	**60**
61	62	63	64	65	**66**	67	68	69	70
71	**72**	73	74	75	76	77	**78**	79	80
81	82	83	**84**	85	86	87	88	89	**90**
91	92	93	94	95	**96**	97	98	99	100

Counting by sixes

Fig. **4.10.**
Skip-counting on a hundred chart

Represent and analyze mathematical situations and structures using algebraic symbols

Two central themes of algebraic thinking are appropriate for young students. The first involves making generalizations and using symbols to represent mathematical ideas, and the second is representing and solving problems (Carpenter and Levi 1999). For example, adding pairs of numbers in different orders such as 3 + 5 and 5 + 3 can lead students to infer that when two numbers are added, the order does not matter. As students generalize from observations about number and operations, they are forming the basis of algebraic thinking.

Similarly, when students decompose numbers in order to compute, they often use the associative property for the computation. For instance, they may compute 8 + 5, saying, "8 + 2 is 10, and 3 more is 13." Students often discover and make generalizations about other properties. Although it is not necessary to introduce vocabulary such as *commutativity* or *associativity*, teachers must be aware of the algebraic properties used by students at this age. They should build students' understanding of the importance of their observations about mathematical situations and challenge them to investigate whether specific observations and conjectures hold for all cases.

Teachers should take advantage of their observations of students, as illustrated in this story drawn from an experience in a kindergarten class.

> The teacher had prepared two groups of cards for her students. In the first group, the number on the front and back of each card differed by 1. In the second group, these numbers differed by 2.
>
> The teacher showed the students a card with 12 written on it and explained, "On the back of this card, I've written another number." She turned the card over to show the number 13. Then she showed the students a second card with 15 on the front and 16 on the back.

> As she continued showing the students the cards, each time she asked the students, “What do you think will be on the back?” Soon the students figured out that she was adding 1 to the number on the front to get the number on the back of the card.
>
> Then the teacher brought out a second set of cards. These were also numbered front and back, but the numbers differed by 2, for example, 33 and 35, 46 and 48, 22 and 24. Again, the teacher showed the students a sample card and continued with other cards, encouraging them to predict what number was on the back of each card. Soon the students figured out that the numbers on the backs of the cards were 2 more than the numbers on the fronts.
>
> When the set of cards was exhausted, the students wanted to play again. “But,” said the teacher, “we can’t do that until I make another set of cards.” One student spoke up, “You don’t have to do that, we can just flip the cards over. The cards will all be minus 2.”

As a follow-up to the discussion, this teacher could have described what was on each group of cards in a more algebraic manner. The numbers on the backs of the cards in the first group could be named as “front number plus 1” and the second as “front number plus 2.” Following the student’s suggestion, if the cards in the second group were flipped over, the numbers on the backs could then be described as “front number minus 2.” Such activities, together with the discussions and analysis that follow them, build a foundation for understanding the inverse relationship.

Through classroom discussions of different representations during the pre-K–2 years, students should develop an increased ability to use symbols as a means of recording their thinking. In the earliest years, teachers may provide scaffolding for students by writing for them until they have the ability to record their ideas. Original representations remain important throughout the students’ mathematical study and should be encouraged. Symbolic representation and manipulation should be embedded in instructional experiences as another vehicle for understanding and making sense of mathematics.

Equality is an important algebraic concept that students must encounter and begin to understand in the lower grades. A common explanation of the equals sign given by students is that “the answer is coming,” but they need to recognize that the equals sign indicates a relationship—that the quantities on each side are equivalent, for example, 10 = 4 + 6 or 4 + 6 = 5 + 5. In the later years of this grade band, teachers should provide opportunities for students to make connections from symbolic notation to the representation of the equation. For example, if a student records the addition of four 7s as shown on the left in figure 4.11, the teacher could show a series of additions correctly, as shown on the right, and use a balance and cubes to demonstrate the equalities.

Fig. **4.11.**

A student’s representation of adding four 7s (left) and a teacher’s correct representation of the same addition

7+7=14+7=21+7=28

7 + 7 = 14
14 + 7 = 21
21 + 7 = 28

Use mathematical models to represent and understand quantitative relationships

Students should learn to make models to represent and solve problems. For example, a teacher may pose the following problem:

> There are six chairs and stools. The chairs have four legs and the stools have three legs. Altogether there are twenty legs. How many chairs and how many stools are there?

One student may represent the situation by drawing six circles and then putting tallies inside to represent the number of legs. Another student may represent the situation by using symbols, making a first guess that the number of stools and chairs is the same and adding 3 + 3 + 3 + 4 + 4 + 4. Realizing that the sum is too large, the student might adjust the number of chairs and stools so that the sum of their legs is 20.

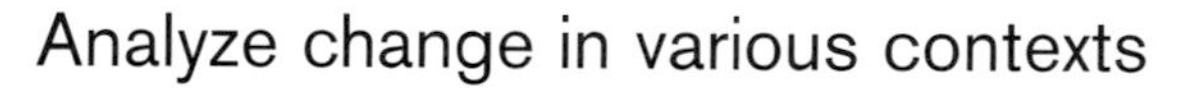

Analyze change in various contexts

Change is an important idea that students encounter early on. When students measure something over time, they can describe change both qualitatively (e.g., "Today is colder than yesterday") and quantitatively (e.g., "I am two inches taller than I was a year ago"). Some changes are predictable. For instance, students grow taller, not shorter, as they get older. The understanding that most things change over time, that many such changes can be described mathematically, and that many changes are predictable helps lay a foundation for applying mathematics to other fields and for understanding the world.

Geometry STANDARD for Grades Pre-K–2

Instructional programs from prekindergarten through grade 12 should enable all students to—

Expectations

In prekindergarten through grade 2 all students should–

Instructional programs from prekindergarten through grade 12 should enable all students to—	In prekindergarten through grade 2 all students should–
Analyze characteristics and properties of two- and three-dimensional geometric shapes and develop mathematical arguments about geometric relationships	• recognize, name, build, draw, compare, and sort two- and three-dimensional shapes; • describe attributes and parts of two- and three-dimensional shapes; • investigate and predict the results of putting together and taking apart two- and three-dimensional shapes.
Specify locations and describe spatial relationships using coordinate geometry and other representational systems	• describe, name, and interpret relative positions in space and apply ideas about relative position; • describe, name, and interpret direction and distance in navigating space and apply ideas about direction and distance; • find and name locations with simple relationships such as "near to" and in coordinate systems such as maps.
Apply transformations and use symmetry to analyze mathematical situations	• recognize and apply slides, flips, and turns; • recognize and create shapes that have symmetry.
Use visualization, spatial reasoning, and geometric modeling to solve problems	• create mental images of geometric shapes using spatial memory and spatial visualization; • recognize and represent shapes from different perspectives; • relate ideas in geometry to ideas in number and measurement; • recognize geometric shapes and structures in the environment and specify their location.

Geometry

The geometric and spatial knowledge children bring to school should be expanded by explorations, investigations, and discussions of shapes and structures in the classroom. Students should use their notions of geometric ideas to become more proficient in describing, representing, and navigating their environment. They should learn to represent two- and three-dimensional shapes through drawings, block constructions, dramatizations, and words. They should explore shapes by decomposing them and creating new ones. Their knowledge of direction and position should be refined through the use of spoken language to locate objects by giving and following multistep directions.

Geometry offers students an aspect of mathematical thinking that is different from, but connected to, the world of numbers. As students become familiar with shape, structure, location, and transformations and as they develop spatial reasoning, they lay the foundation for understanding not only their spatial world but also other topics in mathematics and in art, science, and social studies. Some students' capabilities with geometric and spatial concepts exceed their numerical skills. Building on these strengths fosters enthusiasm for mathematics and provides a context in which to develop number and other mathematical concepts (Razel and Eylon 1991).

Geometry offers an aspect of mathematical thinking that is different from, but connected to, the world of numbers.

Analyze characteristics and properties of two- and three-dimensional geometric shapes and develop mathematical arguments about geometric relationships

Children begin forming concepts of shape long before formal schooling. The primary grades are an ideal time to help them refine and extend their understandings. Students first learn to recognize a shape by its appearance as a whole (van Hiele 1986) or through qualities such as "pointiness" (Lehrer, Jenkins, and Osana 1998). They may believe that a given figure is a rectangle because "it looks like a door."

Pre-K–2 geometry begins with describing and naming shapes. Young students begin by using their own vocabulary to describe objects, talking about how they are alike and how they are different. Teachers must help students gradually incorporate conventional terminology into their descriptions of two- and three-dimensional shapes. However, terminology itself should not be the focus of the pre-K–2 geometry program. The goal is that early experiences with geometry lay the foundation for more-formal geometry in later grades. Using terminology to focus attention and to clarify ideas during discussions can help students build that foundation.

Teachers must provide materials and structure the environment appropriately to encourage students to explore shapes and their attributes. For example, young students can compare and sort building blocks as they put them away on shelves, identifying their similarities and differences. They can use commonly available materials such as cereal boxes to explore attributes of shapes or folded paper to investigate symmetry and congruence. Students can create shapes on geoboards or dot paper and represent them in drawings, block constructions, and dramatizations.

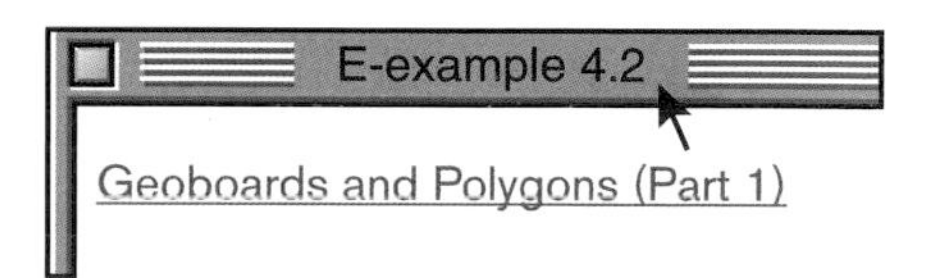

Students need to see many examples of shapes that correspond to the same geometrical concept as well as a variety of shapes that are nonexamples of the concept. For example, teachers must ensure that students see collections of triangles in different positions and with different sizes of angles (see fig. 4.12) and shapes that have a resemblance to triangles (see fig. 4.13) but are not triangles. Through class discussions of such examples and nonexamples, geometric concepts are developed and refined.

Fig. **4.12.**

Examples of triangles

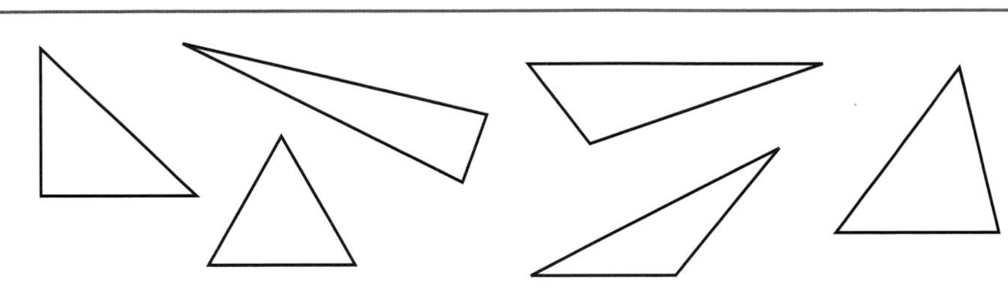

Fig. **4.13.**

Examples of nontriangles

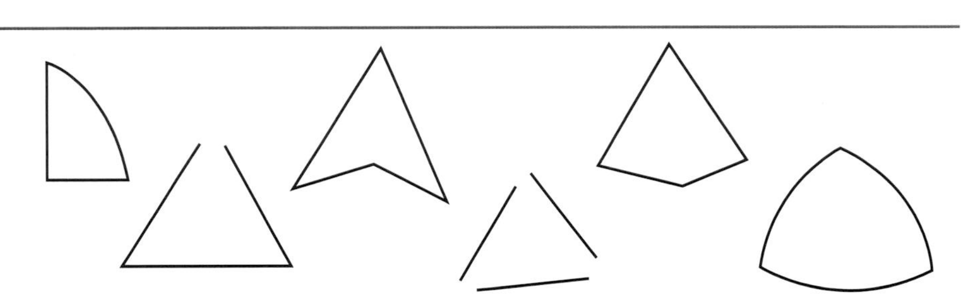

Students also learn about geometric properties by combining or cutting apart shapes to form new shapes. For example, second-grade students can be challenged to find and record all the different shapes that can be created with the two triangles shown in figure 4.14. Interactive computer programs provide a rich environment for activities in which students put together or take apart (compose and decompose) shapes. Technology can help all students understand mathematics, and interactive computer programs may give students with special instructional needs access to mathematics they might not otherwise experience.

Fig. **4.14.**

Two triangles can be combined to make different shapes.

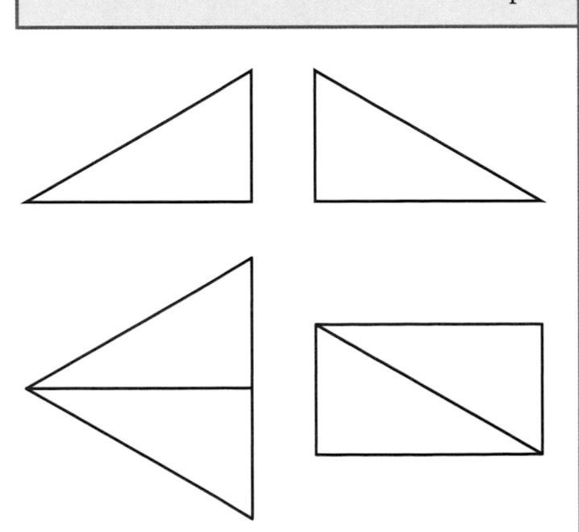

Specify locations and describe spatial relationships using coordinate geometry and other representational systems

Four types of mathematical questions regarding navigation and maps can help students develop a variety of spatial understandings: direction (which way?), distance (how far?), location (where?), and representation (what objects?). In answering these questions, students need to develop a variety of skills that relate to direction, distance, and position in space. Students develop the ability to navigate first by noticing landmarks, then by building knowledge of a route (a connected series of landmarks), and finally by putting many routes and locations into a kind of mental map (Clements 1999b).

Teachers should extend young students' knowledge of relative position in space through conversations, demonstrations, and stories. When students act out the story of the three billy goats and illustrate *over* and *under*, *near* and *far*, and *between*, they are learning about location, space, and shape. Gradually students should distinguish navigation ideas such as *left* and *right* along with the concepts of distance and measurement. As they build three-dimensional models and read maps of their own environments, students can discuss which blocks are used to represent various objects like a desk or a file cabinet. They can mark paths on the model, such as from a table to the wastebasket, with masking tape to emphasize the shape of the path. Teachers should help students relate their models to other representations by drawing a map of the same room that includes the path. In similar activities, older students should develop map skills that include making route maps and using simple coordinates to locate their school on a city map (Liben and Downs 1989).

Computers can help students abstract, generalize, and symbolize their experiences with navigating. For example, students might "walk out" objects such as a rectangular-shaped rug and then use a computer program to make a rectangle on the computer screen. When students measure the rug with footprints and create a computer-generated rectangle with the same relative dimensions, they are exploring scaling and similarity. Some computer programs allow students to navigate through mazes or maps. Teachers should encourage students to move beyond trial and error as a strategy for moving through desired paths to visualizing, describing, and justifying the moves they need to make. Using these programs, students can learn orientation, direction, and measurement concepts.

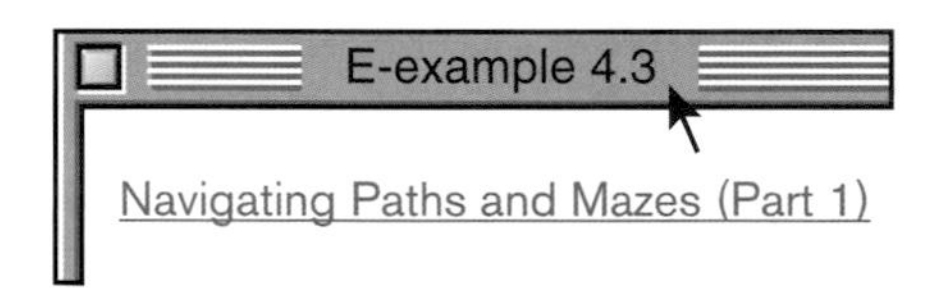

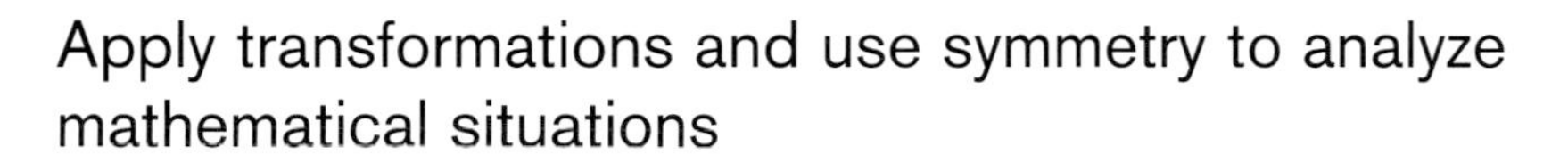

Apply transformations and use symmetry to analyze mathematical situations

Students can naturally use their own physical experiences with shapes to learn about transformations such as slides (translations), turns (rotations), and flips (reflections). They use these movements intuitively when they solve puzzles, turning the pieces, flipping them over, and experimenting with new arrangements. Students using interactive computer programs, with shapes often have to choose a motion to solve a puzzle. These actions are explorations with transformations and are an important part of spatial learning. They help students become conscious of the motions and encourage them to predict the results of changing a shape's position or orientation but not its size or shape.

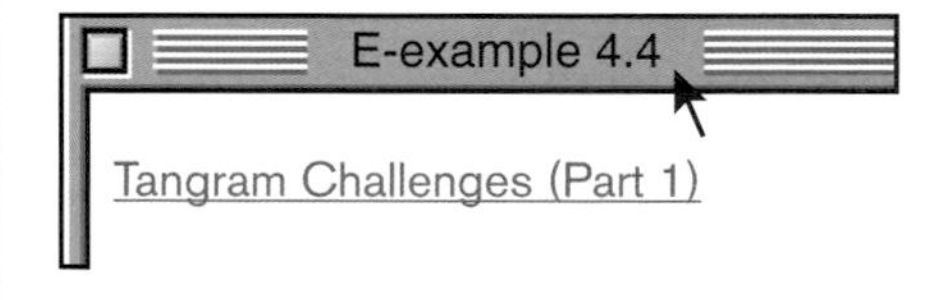

Teachers should choose geometric tasks that are accessible to all students and sufficiently open-ended to engage students with a range of interests. For example, a second-grade teacher might instruct the class to find all the different ways to put five squares together so that one edge of each square coincides with an edge of at least one other square (see fig. 4.15). The task should include keeping a record of the pentominoes that are identified and developing a strategy for recognizing when they are transformations of another pentomino. Teachers can

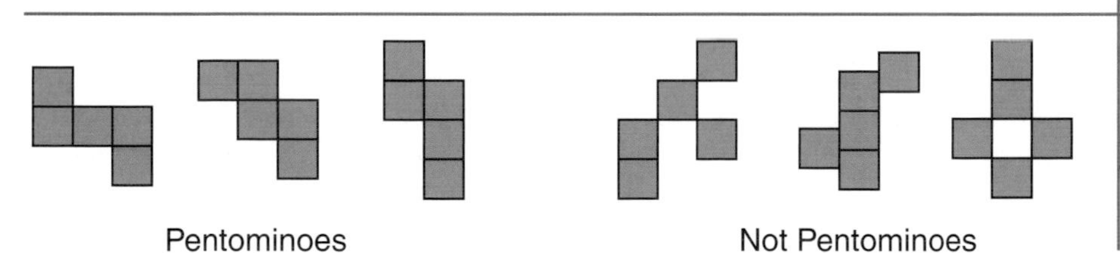

Fig. **4.15.**
Examples of pentominoes and nonpentominoes

Fig. **4.16.**
Symmetries can be found in designs.

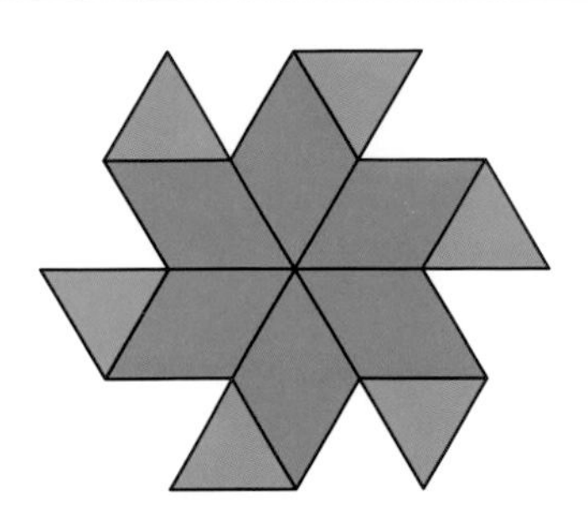

Rotational symmetry

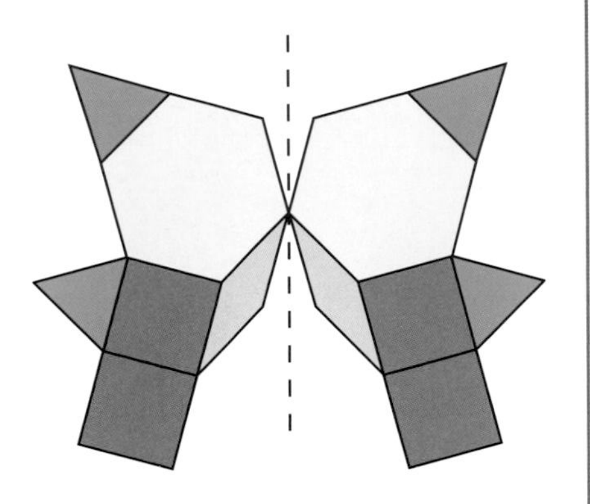

Line symmetry

encourage students to develop strategies for being systematic by asking, “How will you know if each pentomino is different from all the others? Are you certain you have identified all the possibilities?” They can challenge students to predict which of their pentominoes, if cut out of grid paper, would fold into an open box and then verify (or reject) their predictions by cutting them out and trying to fold them into boxes.

Teachers should guide students to recognize, describe, and informally prove the symmetric characteristics of designs through the materials they supply and the questions they ask. Students can use pattern blocks to create designs with line and rotational symmetry (see fig. 4.16) or use paper cutouts, paper folding, and mirrors to investigate lines of symmetry.

Use visualization, spatial reasoning, and geometric modeling to solve problems

Spatial visualization can be developed by building and manipulating first concrete and then mental representations of shapes, relationships, and transformations. Teachers should plan instruction so that students can explore the relationships of different attributes or change one characteristic of a shape while preserving others. In the activity in figure 4.17, students are holding a long loop of yarn so that each student’s hand serves as a vertex of the triangle. In this arrangement, students experiment with changing a shape by increasing the number of sides while the perimeter is unchanged. Conversations about what they notice and how to change from one shape to another allow students to hear different points of view and at the same time give teachers insight into their students’ understanding. Work with concrete shapes, illustrated in this activity, lays a valuable foundation for spatial sense. To further develop students’ abilities, teachers might ask them to see in their “mind’s eye” the shapes that would result when a shape is flipped or when a square is cut diagonally from corner to corner. Thus, many shape and transformation activities build spatial reasoning if students are asked to imagine,

Fig. **4.17.**
Making a string triangle

predict, experiment, and check the results of the work themselves (see fig. 4.18).

Classroom activities that enhance visualization, such as asking students to recall the configuration of the dots on dominoes and determine the number of dots without counting, can promote spatial memory. A teacher could place objects (such as scissors, a pen, a leaf, a paper clip, and a block) on the overhead projector, show the objects briefly, and ask students to name the objects they glimpsed. Or the teacher might have students close their eyes; she could then take one object away and ask which one was removed.

In another "quick image" activity, students can be briefly shown a simple configuration such as the one in figure 4.19 projected on a screen and then asked to reproduce it. The configuration is shown again for a couple of seconds, and they are encouraged to modify their drawings. The process may be repeated several times so that they have opportunities to evaluate and self-correct their work (Yackel and Wheatley 1990). Asking, "What did you see? How did you decide what to draw?" is likely to elicit different explanations, such as "three triangles," "a sailboat sinking," "a square with two lines through it," "a y in a box," and "a sandwich that has been cut into three pieces." Students who can see the configuration in several ways may have more mathematical knowledge and power than those who are limited to one perspective.

Spatial visualization and reasoning can be fostered in navigation activities when teachers ask students to visualize the path they just walked from the library and describe it by specifying landmarks along the route or when students talk about how solid geometric shapes look from different perspectives. Teachers should ask students to identify structures from various viewpoints and to match views of the same structure portrayed from different perspectives. Using a variety of magazine photographs, older students might discuss the location of the photographers when they took each one.

Teachers should help students forge links among geometry, measurement, and number by choosing activities that encourage them to use knowledge from previous lessons to solve new problems. The story of second graders estimating cranberries to fill a jar, described in the "Connections" section of this chapter, illustrates a lesson in which students use their understanding of number, measurement, geometry, and data to complete the tasks. When teachers point out geometric shapes in nature or in architecture, students' awareness of geometry in the environment is increased. When teachers invite students to discover why most fire hydrants have pentagonal caps rather than square or hexagonal ones or why balls can roll in straight lines but cones roll to one side, they are encouraging them to apply their geometric understandings. When students are asked to visualize numbers geometrically by modeling various arrangements of the same number with square tiles, they also are making connections to area. Making and drawing such rectangular arrays of squares help primary-grades students learn to organize space and shape, which is important to their later understanding of grids and coordinate systems (Battista et al. 1998).

Fig. **4.18.**

Paper cutting can aid spatial visualization and reasoning.

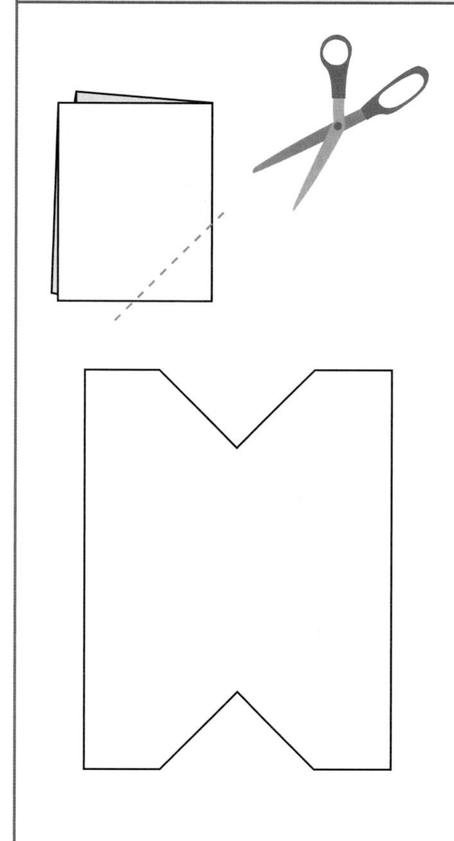

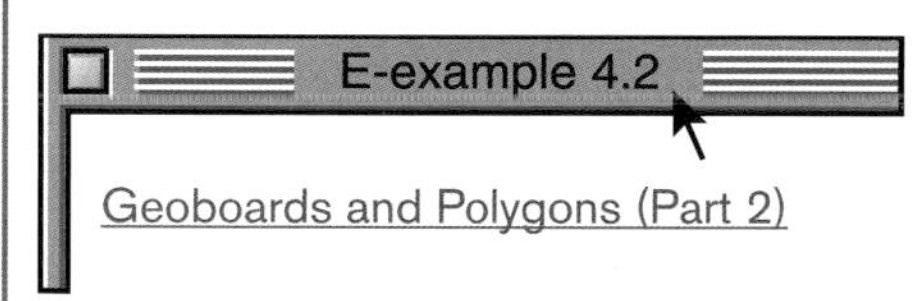

Fig. **4.19.**

In a quick-image activity, students try to reproduce this image, which has briefly been projected on a screen.

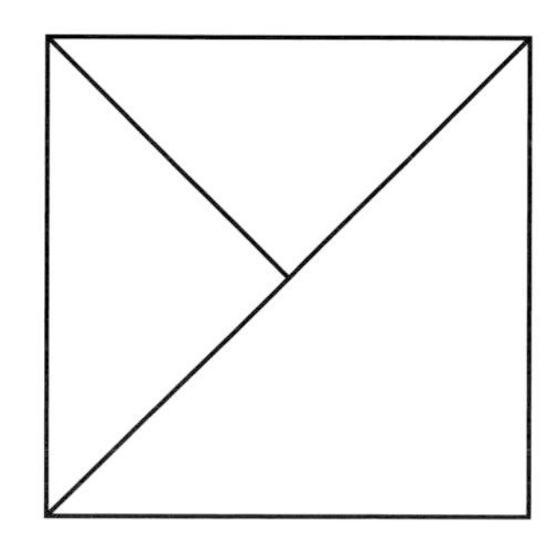

Measurement Standard for Grades Pre-K–2

Instructional programs from prekindergarten through grade 12 should enable all students to—

Expectations

In prekindergarten through grade 2 all students should–

Standard	Expectations
Understand measurable attributes of objects and the units, systems, and processes of measurement	• recognize the attributes of length, volume, weight, area, and time; • compare and order objects according to these attributes; • understand how to measure using nonstandard and standard units; • select an appropriate unit and tool for the attribute being measured.
Apply appropriate techniques, tools, and formulas to determine measurements	• measure with multiple copies of units of the same size, such as paper clips laid end to end; • use repetition of a single unit to measure something larger than the unit, for instance, measuring the length of a room with a single meterstick; • use tools to measure; • develop common referents for measures to make comparisons and estimates.

Measurement

Measurement is one of the most widely used applications of mathematics. It bridges two main areas of school mathematics—geometry and number. Measurement activities can simultaneously teach important everyday skills, strengthen students' knowledge of other important topics in mathematics, and develop measurement concepts and processes that will be formalized and expanded in later years.

Teaching that builds on students' intuitive understandings and informal measurement experiences helps them understand the attributes to be measured as well as what it means to measure. A foundation in measurement concepts that enables students to use measurement systems, tools, and techniques should be established through direct experiences with comparing objects, counting units, and making connections between spatial concepts and number.

Understand measurable attributes of objects and the units, systems, and processes of measurement

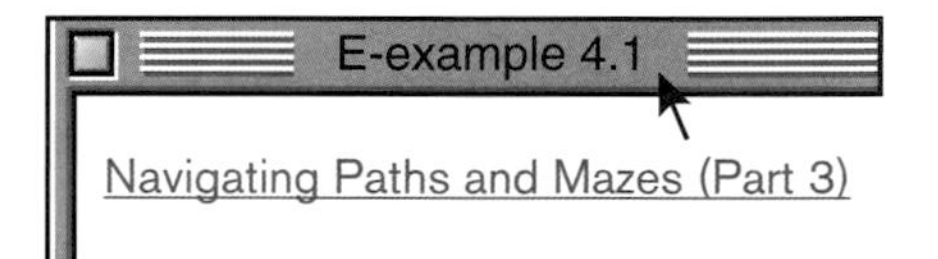

Children should begin to develop an understanding of attributes by looking at, touching, or directly comparing objects. They can determine who has more by looking at the size of piles of objects or identify which of two objects is heavier by picking them up. They can compare shoes, placing them side by side, to check which is longer. Adults should help young children recognize attributes through their conversations. "That is a *deep* hole." "Let's put the toys in the *large* box." "That is a *long* piece of rope." In school, students continue to learn about attributes as they describe objects, compare them, and order them by different attributes. Seeing order relationships, such as that the soccer ball is bigger than the baseball but smaller than the beach ball, is important in developing measurement concepts.

Teachers should guide students' experiences by making the resources for measuring available, planning opportunities to measure, and encouraging students to explain the results of their actions. Discourse builds students' conceptual and procedural knowledge of measurement and gives teachers valuable information for reporting progress and planning next steps. The same conversations and questions that help students build vocabulary help teachers learn about students' understandings and misconceptions. For example, when students measure the length of a desk with rods, the teacher might ask what would happen if they used rods that were half as long. Would they need more rods or fewer rods? If students are investigating the height of a table, the teacher might ask what measuring tools would be appropriate and why.

Measurement is one of the most widely used applications of mathematics.

Although a conceptual foundation for measuring many different attributes should be developed during the early years, linear measurements are the main emphasis. Measurement experiences should include direct comparisons as well as the use of nonstandard and standard units. For example, teachers might ask young students to find objects in the room that are about as long as their foot or to measure the length of a table with connecting cubes. Later they can supply standard measurement tools, such as rulers, to measure classroom plants and use those measurements to chart the plants' growth.

Students need opportunities to expand their beginning understandings of attributes.

Students need opportunities to expand their beginning understandings of attributes other than those related to linear measures and area. Preschool children learn about volume as they pour sand or water from one container to another. In the classroom, they should continue to explore the capacity of various containers by direct comparisons or by counting the number of scoops or cups required to fill each one. They also should experiment with filling larger containers with the contents of smaller ones and conjecture whether a quantity may be too much for a proposed container.

Young students should also have experiences with weighing objects. Balances help them understand comparative weights and reinforce the concept of equality; for example, they can predict that two cubes will weigh the same as twenty links and then test their prediction. Or they can measure equal weights of clay for an art project or compare the weights of different-sized blocks. Scales permit students to assign numerical values to the weights of objects (as rulers allow them to assign numerical values for linear measures) and allow them to begin using standard measures in meaningful ways. By the end of second grade, students should relate standard measures such as kilograms or pounds to the attribute "how heavy."

Another emphasis at this level should be on developing concepts of time and the ways it is measured. When students use calendars or sequence events in stories, they are using measures of time in a real context. Opportunities arise throughout the school day for teachers to focus on time and its measurement through short conversations with their students. A teacher might say, for example, "Look at the clock. It's one o'clock—time for gym! It is just like the picture of the clock on our schedule." As teachers call attention to the clock, many young students will learn to tell time. However, this is less important than their understanding patterns of minutes, hours, days, weeks, and months.

The measurement process is identical, in principle, for measuring any attribute: choose a unit, compare that unit to the object, and report the number of units. The number of units can be determined by *iterating* the unit (repeatedly laying the unit against the object) and counting the iterations or by using a measurement tool. For example, students can tile a space and count the number of tiles to find the area. For linear measurement, they can record their height by using a meterstick.

Teachers should provide many hands-on opportunities for students to choose tools—some with nonstandard and others with standard units—for measuring different attributes. Students should learn that rods and rulers with centimeters and inches may be used to measure length. They should recognize that different units give different levels of precision for their measurements. Although for many measurement tasks students will use nonstandard units, it is appropriate for them to experiment with and use standard measures such as centimeters and meters and inches and feet by the end of grade 2.

The measurement process is identical, in principle, for measuring any attribute.

Apply appropriate techniques, tools, and formulas to determine measurements

If students initially explore measurement with a variety of units, nonstandard as well as standard, they will develop an understanding of the nature of units. For example, if some students measure the width of a door using pencils and others use large paper clips, the number of paper clips will be different from the number of pencils. If some students use small paper clips, then the width of the door will measure yet a different number of units. Similarly, when students cover an area, some using dominoes and others using square tiles, they will recognize that "domino measurements" have different values from "tile measurements." Such experiences and discussions can create an awareness of the need for standard units and tools and of the fact that different measuring tools will yield different numerical measurements of the same object.

Estimation activities are an early application of number sense.

Measurement concepts and skills can develop together as students position multiple copies of the same units without leaving spaces between them or as they measure by iterating one unit without overlapping or leaving gaps. Both types of experiences are necessary. Similarly, using rulers, students learn concepts and procedures, including accurate alignment (e.g., ignoring the leading edge at the beginning of many rulers), starting at zero, and focusing on the lengths of the units rather than only on the numbers on the ruler. By emphasizing the question "What are you counting?" teachers help students focus on the meaning of the measurements they are making.

Teachers cannot assume that students understand measurement fully even when they are able to tell how long an object is when it is aligned with a ruler. Using tools accurately and questioning when measurements may not be accurate require concepts and skills that develop over extended periods through many varied experiences. Consider the following episode drawn from a classroom experience:

> A teacher had given her class a list of things to measure; because she was interested in finding out how the students will approach the task, she had left the choice of measuring tools up to them. Mari was using a ruler when the teacher stopped by the desk to observe her measuring her book. "It's twelve inches," Mari said as she wrote the measurement on the recording sheet. Next she measured her pencil, which was noticeably shorter than the book. The teacher observed that Mari's hand slipped as she was aligning her ruler with the pencil. Mari made no comment but recorded this measurement as twelve inches also.
>
> "I notice that you wrote that each of these is twelve inches," said the teacher. "I'm confused. The book looks much longer than the pencil to me. What do you think?"
>
> Mari pushed both items close together and studied them. "You're right," she said. "The book is longer, but they are both twelve inches."
>
> In her anecdotal records, the teacher noted what happened in order to address the issue in future lessons and conversations with Mari and the class.

Estimation activities are an early application of number sense; they focus students' attention on the attributes being measured, the process of measuring, the sizes of units, and the value of referents. Thus estimating measurements contributes to students' development of spatial sense, number concepts, and skills. Because precise measurements are not always needed to answer questions, students should realize that it is often appropriate to report a measurement as an estimate.

Data Analysis and Probability Standard for Grades Pre-K–2

Instructional programs from prekindergarten through grade 12 should enable all students to—

Expectations

In prekindergarten through grade 2 all students should–

Standard	Expectations
Formulate questions that can be addressed with data and collect, organize, and display relevant data to answer them	• pose questions and gather data about themselves and their surroundings; • sort and classify objects according to their attributes and organize data about the objects; • represent data using concrete objects, pictures, and graphs.
Select and use appropriate statistical methods to analyze data	• describe parts of the data and the set of data as a whole to determine what the data show.
Develop and evaluate inferences and predictions that are based on data	• discuss events related to students' experiences as likely or unlikely.
Understand and apply basic concepts of probability	

Data Analysis and Probability

Informal comparing, classifying, and counting activities can provide the mathematical beginnings for developing young learners' understanding of data, analysis of data, and statistics. The types of activities needed and appropriate for kindergartners vary greatly from those for second graders; however, throughout the pre-K–2 years, students should pose questions to investigate, organize the responses, and create representations of their data. Through data investigations, teachers should encourage students to think clearly and to check new ideas against what they already know in order to develop concepts for making informed decisions.

As students' questions become more sophisticated and their data sets larger, their use of traditional representations should increase. By the end of the second grade, students should be able to organize and display their data through both graphical displays and numerical summaries. They should be using counts, tallies, tables, bar graphs, and line plots. The titles and labels for their displays should clearly identify what the data represent. As students work with numerical data, they should begin to sort out the meaning of the different numbers—those that represent values ("I have four people in my family") and those that represent how often a value occurs in a data set (frequency) ("Nine children have four people in their families"). They should discuss when conclusions about data from one population might or might not apply to data from another population. Considerations like these are the precursors to understanding the notion of inferences from samples.

Ideas about probability at this level should be informal and focus on judgments that children make because of their experiences. Activities that underlie experimental probability, such as tossing number cubes or dice, should occur at this level, but the primary purpose for these activities is focused on other strands, such as number.

The main purpose of collecting data is to answer questions when the answers are not immediately obvious.

Formulate questions that can be addressed with data and collect, organize, and display relevant data to answer them

The main purpose of collecting data is to answer questions when the answers are not immediately obvious. Students' natural inclination to ask questions must be nurtured. At the same time, teachers should help them develop ways to gather information to answer these questions so that they learn when and how to make decisions on the basis of data. As children enter school and their interests extend from their immediate surroundings to include other environments, they must learn how to keep track of multiple responses to their questions and those posed by others. Students also should begin to refine their questions to get the information they need.

Organizing data into categories should begin with informal sorting experiences, such as helping to put away groceries. These experiences and the conversations that accompany them focus children's attention on the attributes of objects and help develop an understanding of "things that go together," while building a vocabulary for describing attributes and for classifying according to criteria. Young students should continue activities that focus on attributes of objects and data so that by the second grade, they can sort and classify simultaneously, using more than one attribute.

Students should learn through multiple experiences that how data are gathered and organized depends on the questions they are trying to answer. For example, when students are asked to put a counter into a bowl to indicate whether they vote for a class trip to the zoo or to the museum, the responses are organized as the data are gathered (see fig. 4.20). To address a particular question such as "What is your favorite beverage served in the school cafeteria?" real objects such as containers for chocolate milk, plain milk, or juice can be collected, organized, and displayed. At other times, pictures of objects, counters, name cards, or tallies can be contributed by students, organized, and then displayed to indicate preferences.

Fig. **4.20.**

Students can contribute counters to bowls to vote.

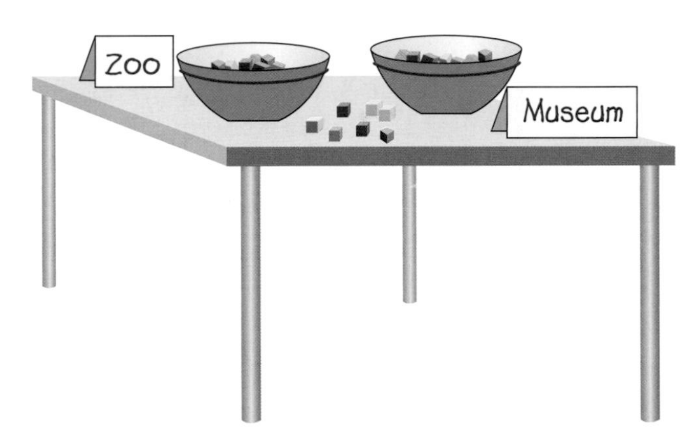

Methods used by students in different grades to investigate the number of pockets in their clothing provide an example of students' growth in data investigations during the period through grade 2. Younger students might count pockets (Burns 1996). They could survey their classmates and gather data by listing names, asking how many pockets, and noting that number beside each name. Together the class could create one large graph to show the data about all the students by coloring a bar on the graph to represent the number of pockets for each student (see fig. 4.21). In the second grade, however, students might decide to count the number of classmates who have various numbers of pockets (see fig. 4.22). Their methods of gathering the information, organizing it, and displaying the data are likely to be different because they are grouping the data—three students have two pockets, five students have four pockets, and so on. They will have to think carefully about the meaning of all the numbers—some represent the value of a piece of data and some represent how many times that value occurs.

Students do not automatically refine their questions, consider alternative ways of collecting information, or choose the most appropriate way to organize and display data; these skills are acquired through experience, class discussions, and teachers' guidance. Take, for example, the following episode drawn from a classroom experience:

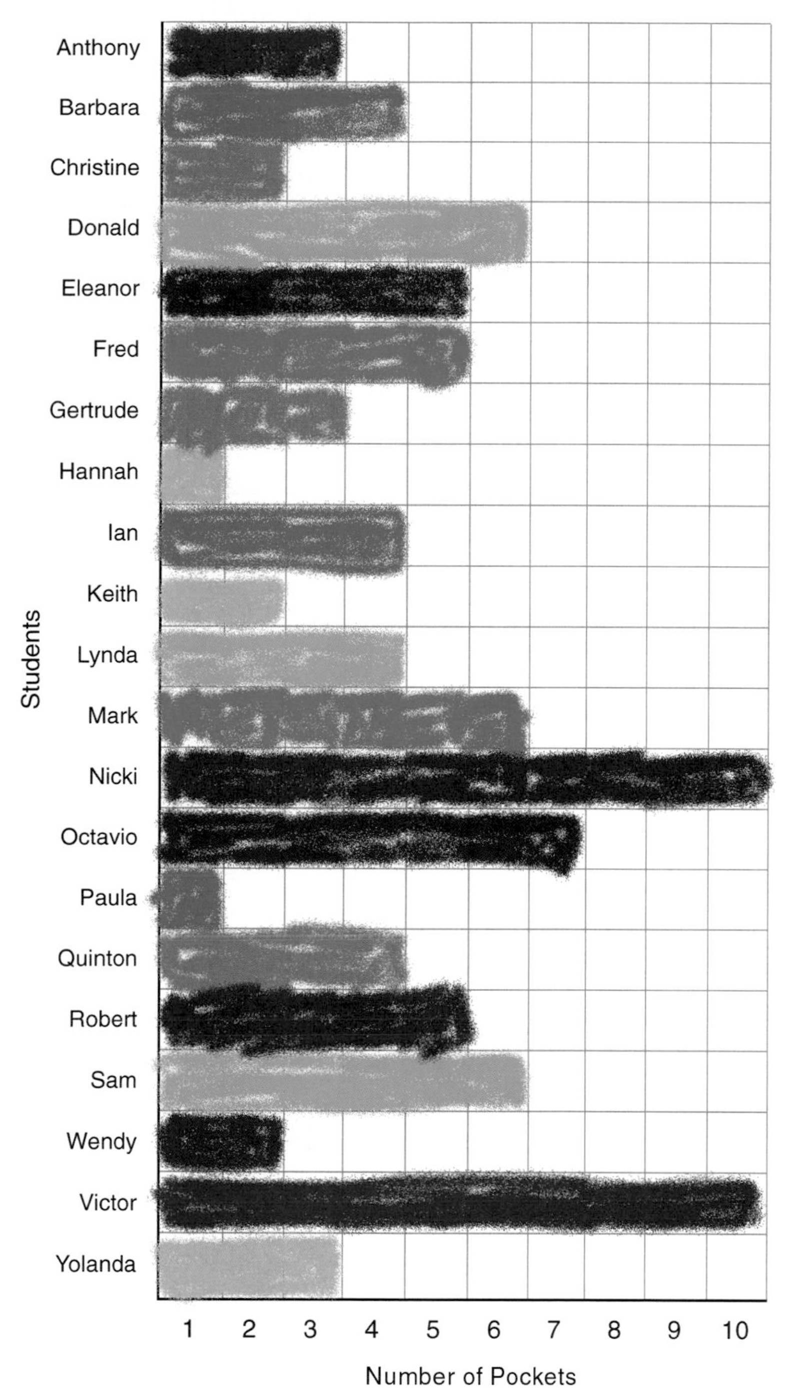

Fig. **4.21.**

A bar graph illustrating the number of pockets in kindergarten students' clothes

Number of Students										
						×				
						×				
				×		×				
				×		×				
		×	×	×	×	×		×		
		×	×	×	×	×		×	×	
	×	×	×	×	×	×		×	×	×
Number of Pockets	1	2	3	4	5	6	7	8	9	10

Fig. **4.22.**

A line-plot graph of the number of students in a second-grade class who have from one to ten pockets

The students had become interested in the question of whether more families had cars with two doors or four doors. As they planned, the students had to decide if trucks should be included. What about vans with four doors or station wagons with five doors? After the class had settled on common categories, different groups of students kept track of the data in different ways. One group put cubes in different cups that represented the different categories. Another group recorded the data using tallies. A third group of students made a list of families with cars with two doors and those with cars with four doors without attempting to organize the information or agree on the results of their data collection. The teacher used the students' work for a class discussion about which groups were able to answer the question they had posed.

Students' representations should be discussed, shared with classmates, and valued because they reflect the students' understandings. These representations afford teachers opportunities to assess students' understandings and to initiate class discussions about important issues related to representing data. Misconceptions that arise because of students' representations of data offer situations for new learning and instruction. A teacher asked first-grade students to fold a piece of paper in half and cut out a heart (adapted from University of North Carolina Mathematics and Science Education Network [1997, p. 19]). When the students sorted their hearts into three columns according to size (see fig. 4.23a), some of them stated that the large hearts represented the most popular choice because that column was the tallest. A teacher could use a class discussion of the difference between the sizes and the numbers of hearts as an early experience with scale and as an opportunity for the students to plan how to revise the graph to convey the data more accurately. By pasting their hearts on equal-sized pieces of paper, the students could create a new graph, shown in figure 4.23b.

Fig. **4.23.**

A misleading data display and its subsequent revision (source of (a): University of North Carolina Mathematics and Science Education 1997, p. 19)

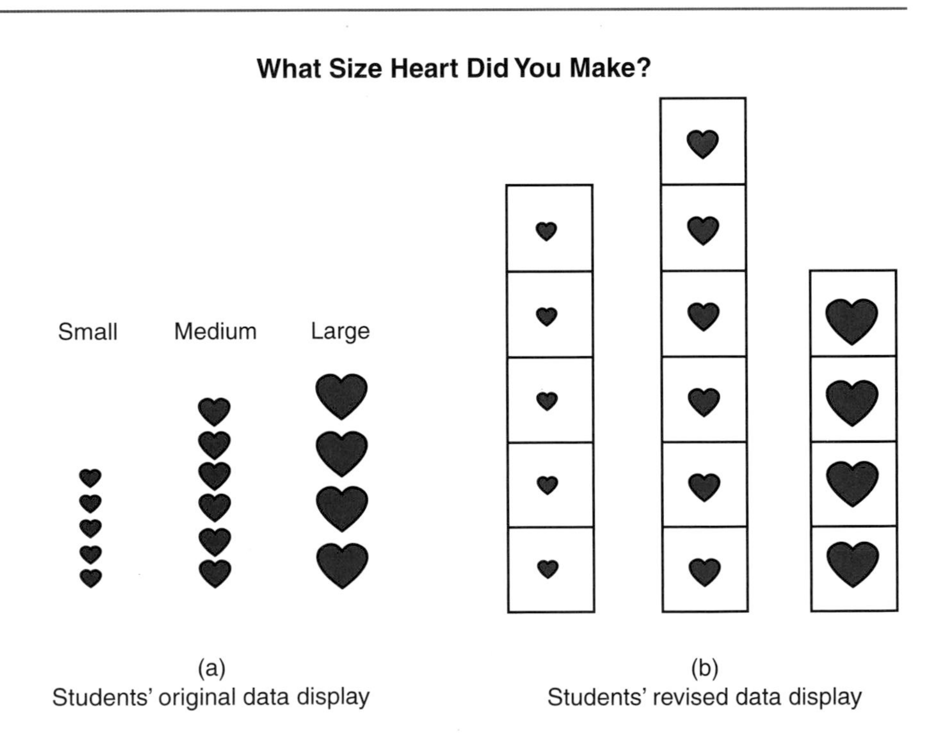

Data investigations can encourage students to wrestle with counting issues.

Select and use appropriate statistical methods to analyze data

Through their data investigations, young students should develop the idea that data, charts, and graphs give information. When data are displayed in an organized manner, class discussions should focus on what the graph or other representation conveys and whether the data help answer the specific questions that were posed. Teachers should encourage students to compare parts of the data ("The same number of children have dogs as have cats") and make statements about the data as a whole ("Most students in the class have lost only two teeth").

By the end of grade 2, students should begin to question inappropriate statements about data, as illustrated in this classroom conversation: Two students, interested in how many of their classmates watched a particular television show, surveyed only their friends and reported their results to the class. "You didn't ask me and I watched it!" one girl complained. Another student said, "Wait a minute, you didn't ask me and I *didn't* watch it. I bet most kids didn't watch it."

Data investigations can encourage students to wrestle with counting issues that are fundamental to all data collection: Whom do I count? How can I be certain I have counted each piece of data once and only once?

The concept of sample is difficult for young students. Most of their data gathering is for full populations, such as their own class. With guidance, students can begin to recognize when conclusions about one population cannot be applied to another, as demonstrated in the following hypothetical example: A teacher reads a book about whistling to a first-grade class. The students decide to survey the class and discover that eight students can whistle and nineteen cannot. When the teacher asks the class to title the chart they have created, the students agree that

Teachers should address the beginnings of probability through informal activities.

an appropriate title would be "Most Children Cannot Whistle." The teacher then asks, "What do you think would happen if we asked the fourth graders?" The students repeat the survey and discover that almost every fourth grader can whistle, so they decide to retitle the graph "Number of Students in Our Class Who Can Whistle."

Develop and evaluate inferences and predictions that are based on data

Inference and prediction are more-advanced aspects of this Standard. The development of these concepts requires work with sampling that begins in the next grade band. As appropriate beginnings for these concepts, however, teachers should encourage informal discussions about whether or not students in other classes would reach a similar conclusion.

Understand and apply basic concepts of probability

At this level, probability experiences should be informal and often take the form of answering questions about the likelihood of events, using such vocabulary as *more likely* or *less likely*. Young students enjoy thinking about impossible events and often encounter them in the books they are learning to read. Questions about more and less likely events should come from the students' experiences, and the answers will often depend on the community and its location. During the winter, the question "Is it likely to snow tomorrow?" has quite different answers in Toronto and San Diego.

Teachers should address the beginnings of probability through informal activities with spinners or number cubes that reinforce conceptions in other Standards, primarily number. For example, as students repeatedly toss two dice or number cubes and add the results of each toss, they may begin to keep track of the results. They will realize that a sum of 1 is impossible, that a sum of 2 or 12 is rare, and that the sums 6, 7, and 8 are fairly common. Through discussion, they may realize that their observations have something to do with the number of ways to get a particular sum from two dice, but the exact calculation of the probabilities should occur in higher grades.

Problem Solving Standard for Grades Pre-K–2

Instructional programs from prekindergarten through grade 12 should enable all students to—

Build new mathematical knowledge through problem solving

Solve problems that arise in mathematics and in other contexts

Apply and adapt a variety of appropriate strategies to solve problems

Monitor and reflect on the process of mathematical problem solving

Problem solving is a hallmark of mathematical activity and a major means of developing mathematical knowledge. It is finding a way to reach a goal that is not immediately attainable. Problem solving is natural to young children because the world is new to them, and they exhibit curiosity, intelligence, and flexibility as they face new situations. The challenge at this level is to build on children's innate problem-solving inclinations and to preserve and encourage a disposition that values problem solving. Teachers should encourage students to use the new mathematics they are learning to develop a broad range of problem-solving strategies, to pose (formulate) challenging problems, and to learn to monitor and reflect on their own ideas in solving problems.

What should problem solving look like in prekindergarten through grade 2?

Problem solving in the early years should involve a variety of contexts, from problems related to daily routines to mathematical situations arising from stories. Students in the same classroom are likely to have very different mathematical understandings and skills; the same situation that is a problem for one student may elicit an automatic response from another. For instance, when first-grade students were working in small groups to create models of animals with geometric solids, some had difficulty seeing the parts of animals as geometric shapes. Other students readily saw that they could use seven rectangular prisms to make a giraffe (see figure 4.24). Similarly, the question "How many books would there be on the shelf if Marita put six books on it and Al put three more on it?" may not be a problem for the student who knows the basic number combination 6 and 3 and its connection with the question. For the student who has not yet learned the number combination and may not yet know how to represent the task symbolically, this problem presents an opportunity to learn the skills needed to solve similar problems.

Solving problems gives students opportunities to use and extend their knowledge of concepts in each of the Content Standards. For example, many problems relate to classification, shape, or space: Which blocks will fit on this shelf? Will this puzzle piece fit in the space that remains? How are these figures alike and how are they different? In answering these questions, students are using spatial-visualization skills and their knowledge of transformations. Other problems support students' development of number sense and understanding of operations: How many more days until school vacation? There are 43 cards in this group; how many packets of 10 can we make? If there are 26 students in our class and 21 are here today, how many are absent? When young students solve problems that involve comparing and completing collections by using counting strategies, they develop a better understanding of addition and subtraction and the relationship between these operations.

Posing problems, that is, generating new questions in a problem context, is a mathematical disposition that teachers should nurture and develop. Through asking questions and identifying what information is essential, students can organize their thoughts, as the following episode drawn from classroom observations demonstrates:

> Lei wanted to know all the ways to cover the yellow hexagon using pattern blocks. At first she worked with the blocks using fairly undirected trial and error. Gradually she became more methodical and placed the various arrangements in rows. The teacher showed her a pattern-block program on the class computer and how to "glue" the pattern-block designs together on the screen. Lei organized the arrangements by the numbers of blocks used and began predicting which attempts would be transformations of other arrangements even before she completed the hexagons (see fig. 4.25). The next challenge Lei set for herself was to see if she could create a hexagonal figure using only the orange squares. She had experimented with square blocks and could not make a hexagon. "But," she explained to her teacher, "it might be different on the computer," indicating that she felt the computer was a powerful problem-solving tool.

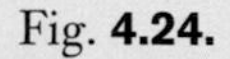

Fig. **4.24.**

A block giraffe made from seven rectangular prisms (Adapted from Russell, Clements, and Sarama [1998, p. 115])

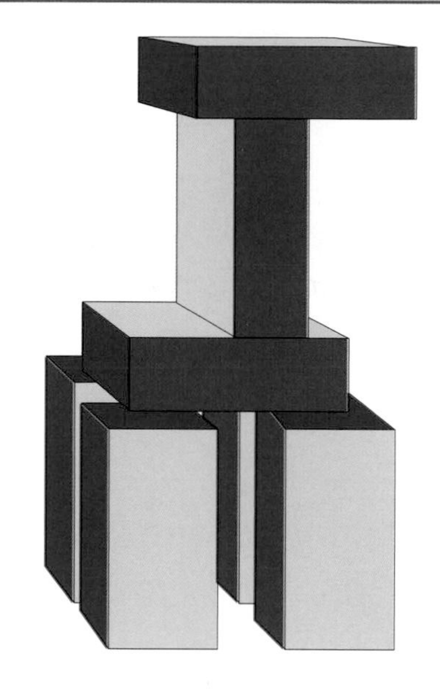

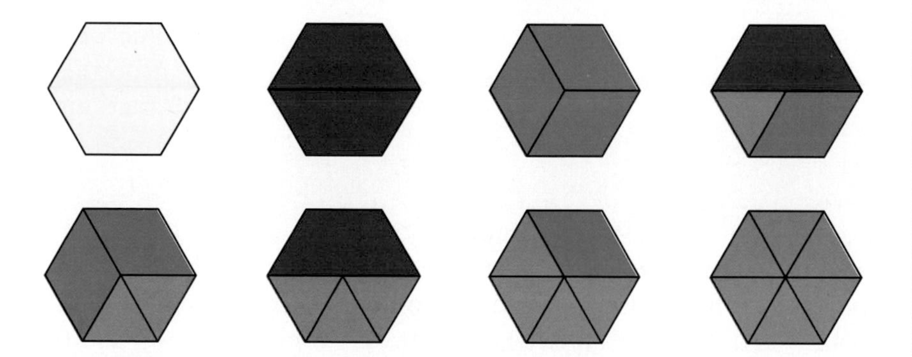

Fig. **4.25.**

Organizing arrangements that make hexagons

> Kyle was certain that he could find more arrangements for hexagons than Lei had found. Other students joined the discussions between Lei and Kyle. When this activity created a great demand for "turns" with the pattern blocks and the computer, the teacher took advantage of the class's interest by having students discuss how they would know when an arrangement of blocks was a duplicate and how they might keep a written record of their work.

Kyle's participation illustrates that students are persistent when problems are interesting and challenging. Their interest also stimulates curiosity in other students.

Sharing gives students opportunities to hear new ideas and compare them with their own.

Students working together often begin to solve problems one way and, before reaching a solution, change their strategies. In addition, as they create and modify their strategies, students often recognize the need to learn more mathematics. The following episode, drawn from classroom observation, illustrates how teachers can make a problem mathematically rich.

> Several first-grade classes in the same school were planting a garden in the school courtyard. The students wanted each class to have the same amount of space for planting; thus, how to divide the area into three equal parts was greatly debated. A walkway, two shade trees, and several benches complicated the discussions. The students began to list all their concerns and the questions they needed to answer before dividing the area for the garden: How big can the garden be? Do the three sections have to be the same shape? How can we be sure each class has the same amount of space?
>
> The teachers drew a large map for each class and indicated the approximate location and amount of space taken by trees, walkway, and benches. In one class the students decided they wanted rectangular gardens and needed to measure the courtyard to figure out how large the rectangles could be. After many measurements and much debate, they cut out three rectangles that were four feet by nine feet to show how big each garden could be. When they were not certain how to use this information on their map, the teacher showed them a scale on a road map and how map scales are used. She suggested appropriate dimensions for the three rectangles, which they cut and glued to their map.
>
> The second class began with a discussion of what "the same area" means. They used large-grid paper squares and taped them to their map to allocate the maximum space for gardening, counting carefully to be certain each class had the same number of squares even though the shapes of the regions were different. This group also needed to learn about scale to actually make a plan for marking off the gardens outside.
>
> Before voting on how to mark off the gardens, the two groups presented their plans to all the classes.

Deciding how to share land for a garden is an example of a classroom-based problem that facilitates students' development of problem-solving strategies. The task was complex. The students struggled with how to share the area equally, how to measure, and how to communicate their ideas. However, the project was rich with proposed strategies, counterproposals, and opportunities for the teachers to introduce new mathematics.

Children's literature is helpful in setting a context for both student-generated and teacher-posed problems. For example, after reading *1 Hunter* (Hutchins 1982) to her class, a second-grade teacher asked students to figure out how many animals, including the hunter, were in the story. Figure 4.26 illustrates several approaches used by the students.

Sharing gives students opportunities to hear new ideas and compare them with their own and to justify their thinking. As students struggle with problems, seeing a variety of successful solutions improves their

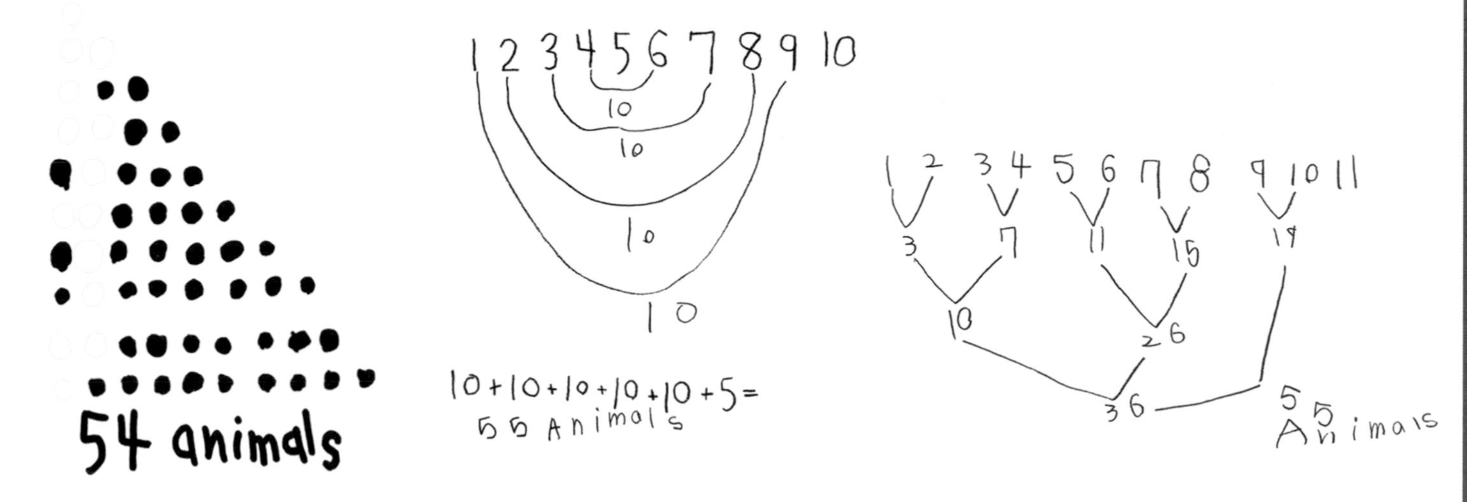

Fig. **4.26.**

Determining the number of animals in *1 Hunter*

chance of learning useful strategies and allows them to determine if some strategies are more flexible and efficient. When the teacher invited the students to explain their solutions to the *1 Hunter* problem, several of them discovered their counting or computational errors and made corrections during the presentations. Explaining their pictorial and written solutions helped them articulate their thinking and make it precise.

What should be the teacher's role in developing problem solving in prekindergarten through grade 2?

The decisions that teachers make about problem-solving opportunities influence the depth and breadth of students' mathematics learning. Teachers must be clear about the mathematics they want their students to accomplish as they structure situations that are both problematic and attainable for a wide range of students. They make important decisions about when to probe, when to give feedback that affirms what is correct and identifies what is incorrect, when to withhold comments and plan similar tasks, and when to use class discussions to advance the students' mathematical thinking. By allowing time for thinking, believing that young students can solve problems, listening carefully to their explanations, and structuring an environment that values the work that students do, teachers promote problem solving and help students make their strategies explicit.

Instead of teaching problem solving separately, teachers should embed problems in the mathematics-content curriculum. When teachers integrate problem solving into the context of mathematical situations, students recognize the usefulness of strategies. Teachers should choose specific problems because they are likely to prompt particular strategies and allow for the development of certain mathematical ideas. For example, the problem "I have pennies, dimes, and nickels in my pocket. If I take three coins out of my pocket, how much money could I have taken?" can help children learn to think and record their work.

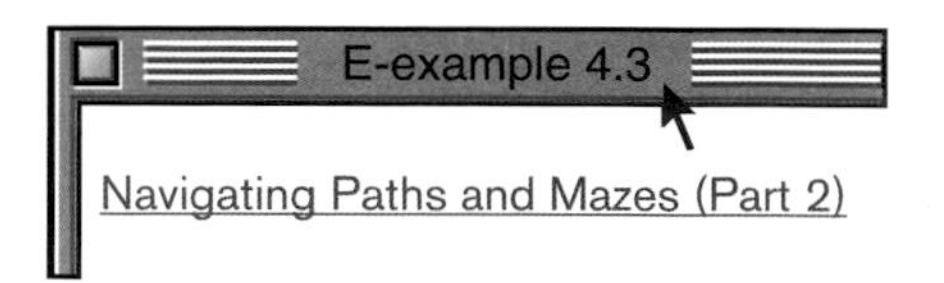

Assessing students' abilities to solve problems is more difficult than evaluating computational skills. However, it is imperative that teachers

E-example 4.4

Tangram Challenges (Part 2)

gather evidence in a variety of ways, such as through students' work and conversations, and use that information to plan how to help individual students in a whole-class context. Knowing students' interests allows teachers to formulate problems that extend the mathematical thinking of some students and that also reinforce the concepts learned by other students who have not yet reached the same understandings. Classrooms in which students have ready access to materials such as counters, calculators, and computers and in which they are encouraged to use a wide variety of strategies support thinking that results in multiple levels of understanding.

Two examples illustrate how conversations with students give teachers useful information about students' thinking. Both examples have been drawn from observations of students.

Katie, a kindergarten student, said that her sister in third grade had taught her to multiply. "Give me a problem," she said. The teacher asked, "How much is three times four?" There was a long pause before Katie replied, "Twelve!" When the teacher asked how she knew, Katie responded, "I counted ducks in my head—three groups with four ducks." Katie, while demonstrating an additive understanding of multiplication by counting the ducks in each group, was also exhibiting an interest in, and readiness for, mathematics that is traditionally a focus in the higher grades. Luis, a second grader, demonstrated fluency with composing and decomposing numbers when he announced that he could figure out multiplication. His teacher asked, "Can you tell me four times seven?" Luis was quiet for a few moments, and then he gave the answer twenty-eight. When the teacher asked how he got twenty-eight, Luis replied, "Seven plus three is ten, and four more is fourteen; six more is twenty and one more is twenty-one; seven more is twenty-eight." Luis's approach also built on additive thinking but with a far more sophisticated use of number relationships. He added $7 + 7 + 7 + 7$ mentally by breaking the sevens into parts to complete tens along the way.

Students are intrigued with calculators and computers and can be challenged by the mathematics that technology makes available to them, as shown in the following episode, adapted from Riedesel (1980, pp. 74–75):

> Erik, a very capable kindergarten student, observed his teacher using a calculator and asked how it worked. The teacher showed him how to compute simple additions. Erik took the calculator to the math corner and a few minutes later loudly proclaimed, "Five plus four equals nine. Hey, this thing got it right!" A few minutes later, he walked over to the teacher and they had the following conversation:
>
> *Erik*: What does this button mean?
>
> *Teacher*: That's called the "square root." It's a pretty difficult idea in math.
>
> *Erik*: OK. (*He wanders away, but not for long.*) But this is a disaster! I pressed 2, then the square-root key, and I got a whole lot of numbers.
>
> *Teacher*: Try using 1. (*Erik tries this.*)
>
> *Erik*: That just gives 1 back.

Teacher: Try 4. (*Erik notes that the result is 2 and asks why. The teacher tells him to get the square tiles and put out one.*) Is that a square? (*Erik nods.*) Try to add more tiles beside this one until it is a square again. (*Erik adds one tile.*) Is that a square?

Erik: No, it's a rectangle. (*The teacher asks how he could make it into a square, and Erik adds two more tiles.*)

Teacher: How many tiles are there in all? (*Erik responds that there are four.*) Good. Press 4 on the calculator. How long is the bottom of the square?

Erik: Two.

Teacher: And here at the left side?

Erik: Two there, too.

Teacher: Press the square-root key.

Erik: Hey, it comes out 2!

The teacher challenged Erik to add more tiles until he made another square. Erik built a 3×3 array, counted the total tiles, entered this number into the calculator, and pressed the square-root key. He found that the result was the number of rows and also the number of tiles in each row. Erik kept building squares until at a 9×9 array he said his eyes hurt. The teacher asked him what he had found out.

Erik: Well, if you make a square, then all you have to do is count the tiles and press that number and the square-root key and the calculator tells you how many tiles there are on each side.

Teacher: Good work! What else does that number mean?

Erik: It means that there are that many rows and that many tiles in each row. (*The teacher congratulates Erik on figuring this out.*) Yeah, I guess if you want to learn something really bad, you can. Tomorrow, I'm going to go up to one hundred!

Teachers should ask students to reflect on, explain, and justify their answers.

Teachers should ask students to reflect on, explain, and justify their answers so that problem solving both leads to and confirms students' understanding of mathematical concepts. For example, following an estimation activity in a first-grade class, students learned that there were eighty-three marbles in a jar. There were twenty-five students in the class, so the teacher asked how many marbles each child could get. Graham said, "Three." When the teacher asked how he knew, Graham replied, "Eighty-three is just a little more than seventy-five, so we only get three. There are four quarters in a dollar. There are three quarters in seventy-five cents. So we can only get three."

Teachers must make certain that problem solving is not reserved for older students or those who have "got the basics." Young students can engage in substantive problem solving and in doing so develop basic skills, higher-order-thinking skills, and problem-solving strategies (Cobb et al. 1991; Trafton and Hartman 1997).

Reasoning and Proof STANDARD for Grades Pre-K–2

Instructional programs from prekindergarten through grade 12 should enable all students to—

- Recognize reasoning and proof as fundamental aspects of mathematics
- Make and investigate mathematical conjectures
- Develop and evaluate mathematical arguments and proofs
- Select and use various types of reasoning and methods of proof

Young students are just forming their store of mathematical knowledge, but even the youngest can reason from their own experiences (Bransford, Brown, and Cocking 1999). Although young children are working from a small knowledge base, their logical reasoning begins before school and is continually modified by their experiences. Teachers should maintain an environment that respects, nurtures, and encourages students so that they do not give up their belief that the world, including mathematics, is supposed to make sense.

Although they have yet to develop all the tools used in mathematical reasoning, young students have their own ways of finding mathematical results and convincing themselves that they are true. Two important elements of reasoning for students in the early grades are pattern-recognition and classification skills. They use a combination of ways of justifying their answers—perception, empirical evidence, and short chains of deductive reasoning grounded in previously accepted facts. They make conjectures and reach conclusions that are logical and defensible from their perspective. Even when they are struggling, their responses reveal the sense they are making of mathematical situations.

Young students naturally generalize from examples (Carpenter and Levi 1999), so teachers should guide them to use examples and counterexamples to test whether their generalizations are appropriate. By the end of second grade, students should be using this method for testing their conjectures and those of others.

What should reasoning and proof look like in prekindergarten through grade 2?

The ability to reason systematically and carefully develops when students are encouraged to make conjectures, are given time to search for evidence to prove or disprove them, and are expected to explain and justify their ideas. In the beginning, perception may be the predominant method of determining truth: nine markers spread far apart may be seen as “more” than eleven markers placed close together. Later, as students develop their mathematical tools, they should use empirical approaches such as matching the collections, which leads to the use of more-abstract methods such as counting to compare the collections. Maturity, experiences, and increased mathematical knowledge together promote the development of reasoning throughout the early years.

Creating and describing patterns offer important opportunities for students to make conjectures and give reasons for their validity, as the following episode drawn from classroom experience demonstrates.

> The student who created the pattern shown in figure 4.27 proudly announced to her teacher that she had made four patterns in one. “Look,” she said, “there’s triangle, triangle, circle, circle, square, square. That’s one pattern. Then there’s small, large, small, large, small, large. That’s the second pattern. Then there’s thin, thick, thin, thick, thin, thick. That’s the third pattern. The fourth pattern is blue, blue, red, red, yellow, yellow.”
>
> Her friend studied the row of blocks and then said, “I think there are just two patterns. See, the shapes and colors are an AABBCC pattern. The sizes are an ABABAB pattern. Thick and thin is an ABABAB pattern, too. So you really only have two different patterns.” The first student considered her friend’s argument and replied, “I guess you’re right—but so am I!”

Fig. **4.27.**

Four patterns in one

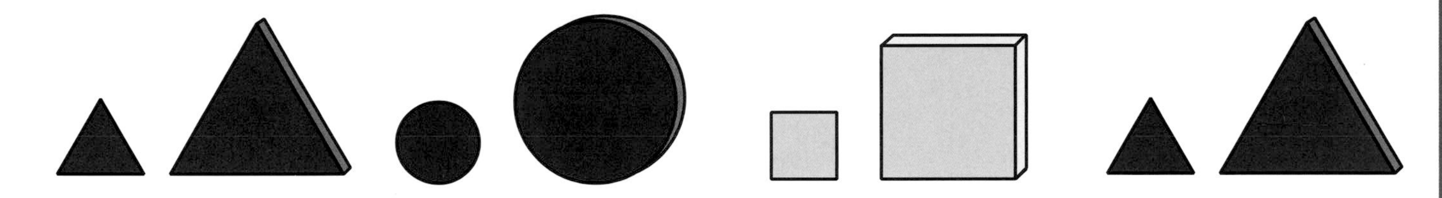

Being able to explain one’s thinking by stating reasons is an important skill for formal reasoning that begins at this level.

Finding patterns on a hundred board allows students to link visual patterns with number patterns and to make and investigate conjectures. Teachers extend students’ thinking by probing beyond their initial observations. Students frequently describe the changes in numbers or the visual patterns as they move down columns or across rows. For example, asked to color every third number beginning with 3 (see figure 4.28), different students are likely to see different patterns: “Some rows have three and some have four,” or “The pattern goes sideways to the left.” Some students, seeing the diagonals in the pattern, will no longer count by threes in order to complete the pattern. Teachers need to ask these students to explain to their classmates how they know what to color without counting. Teachers also extend students’ mathematical reasoning by posing new questions and asking for arguments to support their answers. “You found patterns when counting by twos, threes, fours, fives, and tens on the hundred board. Do you think there will be patterns if you count by sixes, sevens, eights, or nines? What about counting by elevens or fifteens or by any numbers?” With calculators, students could extend their explorations of these and other numerical patterns beyond 100.

Students’ reasoning about classification varies during the early years. For instance, when kindergarten students sort shapes, one student may pick up a big triangular shape and say, “This one is big,” and then put it with other large shapes. A friend may pick up another big triangular shape, trace its edges, and say, “Three sides—a triangle!” and then put

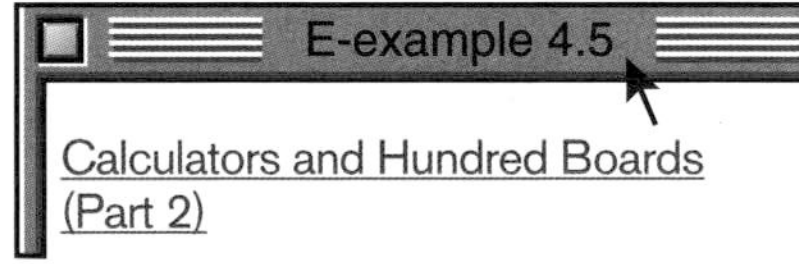

Fig. **4.28.**

Patterns on a hundred board

1	2	3	4	5	6	7	8	9	10
11	12	13	14	15	16	17	18	19	20
21	22	23	24	25	26	27	28	29	30
31	32	33	34	35	36	37	38	39	40
41	42	43	44	45	46	47	48	49	50
51	52	53	54	55	56	57	58	59	60
61	62	63	64	65	66	67	68	69	70
71	72	73	74	75	76	77	78	79	80
81	82	83	84	85	86	87	88	89	90
91	92	93	94	95	96	97	98	99	100

it with other triangles. Both of these students are focusing on only one property, or attribute. By second grade, however, students are aware that shapes have multiple properties and should suggest ways of classifying that will include multiple properties.

By the end of second grade, students also should use properties to reason about numbers. For example, a teacher might ask, "Which number does not belong and why: 3, 12, 16, 30?" Confronted with this question, a student might argue that 3 does not belong because it is the only single-digit number or is the only odd number. Another student might say that 16 does not belong because "you do not say it when counting by threes." A third student might have yet another idea and state that 30 is the only number "you say when counting by tens."

Students must explain their chains of reasoning in order to see them clearly and use them more effectively; at the same time, teachers should model mathematical language that the students may not yet have connected with their ideas. Consider the following episode, adapted from Andrews (1999, pp. 322–23):

> One student reported to the teacher that he had discovered "that a triangle equals a square." When the teacher asked him to explain, the student went to the block corner and took two half-unit (square) blocks, two half-unit triangular (triangle) blocks, and one unit (rectangle) block (shown in fig. 4.29). He said, "If these two [square half-units] are the same as this one unit and these two [triangular half-units] are the same as this one [unit], then this square has to be the same as this triangle!"

Even though the student's wording—that shapes were "equal"—was not correct, he was demonstrating powerful reasoning as he used the blocks to justify his idea. In situations such as this, teachers could point to the faces of the two smaller blocks and respond, "You discovered that

Fig. **4.29.**

A student's explanation of the equal areas of square and triangular block faces

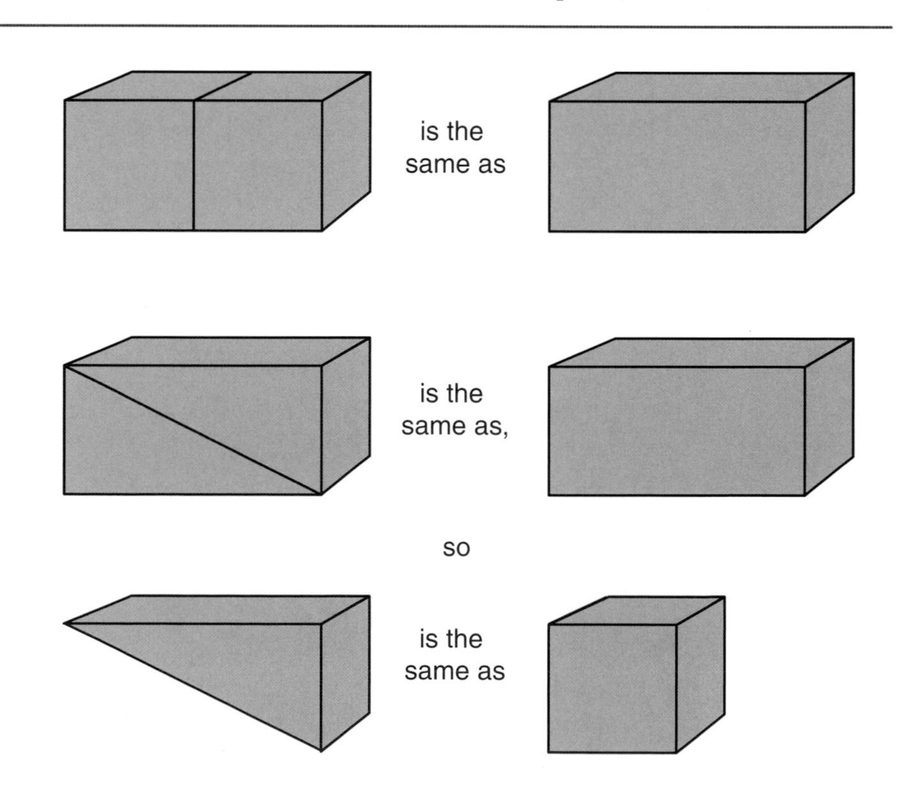

the area of this square equals the area of this triangle because each of them is half the area of the same larger rectangle."

What should be the teacher's role in developing reasoning and proof in prekindergarten through grade 2?

Teachers should create learning environments that help students recognize that all mathematics can and should be understood and that they are expected to understand it. Classrooms at this level should be stocked with physical materials so that students have many opportunities to manipulate objects, identify how they are alike or different, and state generalizations about them. In this environment, students can discover and demonstrate general mathematical truths using specific examples. Depending on the context in which events such as the one illustrated by figure 4.29 take place, teachers might focus on different aspects of students' reasoning and continue conversations with different students in different ways. Rather than restate the student's discovery in more-precise language, a teacher might pose several questions to determine whether the student was thinking about equal areas of the faces of the blocks, or about equal volumes. Often students' responses to inquiries that focus their thinking help them phrase conclusions in more-precise terms and help the teacher decide which line of mathematical content to pursue.

Teachers should help students recognize that all mathematics can and should be understood.

Teachers should prompt students to make and investigate mathematical conjectures by asking questions that encourage them to build on what they already know. In the example of investigating patterns on a hundred board, for instance, teachers could challenge students to consider other ideas and make arguments to support their statements: "If we extended the hundred board by adding more rows until we had a thousand board, how would the skip-counting patterns look?" or "If we made charts with rows of six squares or rows of fifteen squares to count to a hundred, would there be patterns if we skip-counted by twos or fives or by any numbers?"

Through discussion, teachers help students understand the role of nonexamples as well as examples in informal proof, as demonstrated in a study of young students (Carpenter and Levi 1999, p. 8). The students seemed to understand that number sentences like $0 + 5869 = 5869$ were always true. The teacher asked them to state a rule. Ann said, "Anything with a zero can be the right answer." Mike offered a counterexample: "No. Because if it was 100 + 100 that's 200." Ann understood that this invalidated her rule, so she rephrased it, "I said, umm, if you have a zero in it, it can't be like 100, because you want just plain zero like $0 + 7 = 7$."

The students in the study could form rules on the basis of examples. Many of them demonstrated the understanding that a single example was not enough and that counterexamples could be used to disprove a conjecture. However, most students experienced difficulty in giving justifications other than examples.

From the very beginning, students should have experiences that help them develop clear and precise thought processes. This development of reasoning is closely related to students' language development and is dependent on their abilities to explain their reasoning rather than just give the answer. As students learn language, they acquire basic logic words, including *not*, *and*, *or*, *all*, *some*, *if…then*, and *because*. Teachers should help students gain familiarity with the language of logic by using such words frequently. For example, a teacher could say, "You may choose an apple or a banana for your snack" or "If you hurry and put on your jacket, then you will have time to swing." Later, students should use the words modeled for them to describe mathematical situations: "If six green pattern blocks cover a yellow hexagon, then three blues also will cover it, because two greens cover one blue."

Teachers can understand students' thinking when they listen carefully to students' explanations.

Sometimes students reach conclusions that may seem odd to adults, not because their reasoning is faulty, but because they have different underlying beliefs. Teachers can understand students' thinking when they listen carefully to students' explanations. For example, on hearing that he would be "Star of the Week" in half a week, Ben protested, "You can't have half a week." When asked why, Ben said, "Seven can't go into equal parts." Ben had the idea that to divide 7 by 2, there could be two groups of 3, with a remainder of 1, but at that point Ben believed that the number 1 could not be divided.

Teachers should encourage students to make conjectures and to justify their thinking empirically or with reasonable arguments. Most important, teachers need to foster ways of justifying that are within the reach of students, that do not rely on authority, and that gradually incorporate mathematical properties and relationships as the basis for the argument. When students make a discovery or determine a fact, rather than tell them whether it holds for all numbers or if it is correct, the teacher should help the students make that determination themselves. Teachers should ask such questions as "How do you know it is true?" and should also model ways that students can verify or disprove their conjectures. In this way, students gradually develop the abilities to determine whether an assertion is true, a generalization valid, or an answer correct and to do it on their own instead of depending on the authority of the teacher or the book.

Communication Standard for Grades Pre-K–2

Instructional programs from prekindergarten through grade 12 should enable all students to—

Organize and consolidate their mathematical thinking through communication

Communicate their mathematical thinking coherently and clearly to peers, teachers, and others

Analyze and evaluate the mathematical thinking and strategies of others

Use the language of mathematics to express mathematical ideas precisely

Language, whether used to express ideas or to receive them, is a very powerful tool and should be used to foster the learning of mathematics. Communicating about mathematical ideas is a way for students to articulate, clarify, organize, and consolidate their thinking. Students, like adults, exchange thoughts and ideas in many ways—orally; with gestures; and with pictures, objects, and symbols. By listening carefully to others, students can become aware of alternative perspectives and strategies. By writing and talking with others, they learn to use more-precise mathematical language and, gradually, conventional symbols to express their mathematical ideas. Communication makes mathematical thinking observable and therefore facilitates further development of that thought. It encourages students to reflect on their own knowledge and their own ways of solving problems. Throughout the early years, students should have daily opportunities to talk and write about mathematics. They should become increasingly effective in communicating what they understand through their own notation and language as well as in conventional ways.

What should communication look like during prekindergarten through grade 2?

Children begin to communicate mathematically very early in their lives. They want *more* milk, a *different* toy, or *three* books. The communication abilities of most children have developed tremendously before they enter kindergarten. This growth is determined to a large extent by the children's maturity, how language is modeled for them, and their opportunities and experiences. Verbal interaction with families and caregivers is a primary means for promoting the development of early mathematical vocabulary.

Language is as important to learning mathematics as it is to learning to read. As students enter school, their opportunities to communicate are expanded by new learning resources, enriched uses of language, and experiences with classmates and teachers. Students' developing communication skills can be used to organize and consolidate their mathematical thinking. Teachers should help students learn how to talk about mathematics, to explain their answers, and to describe their strategies. Teachers can encourage students to reflect on class conversations and to "talk about talking about mathematics" (Cobb, Wood, and Yackel 1994).

An important step in communicating mathematical thinking to others is organizing and clarifying one's ideas. When students struggle to communicate ideas clearly, they develop a better understanding of their own thinking. Working in pairs or small groups enables students to hear different ways of thinking and refine the ways in which they explain their own ideas. Having students share the results of their small-group findings gives teachers opportunities to ask questions for clarification and to model mathematical language. Students in prekindergarten through grade 2 should be encouraged to listen attentively to each other, to question others' strategies and results, and to ask for clarification so that their mathematical learning advances.

Adequate time and interesting mathematical problems and materials, including calculators and computer applications, encourage conversation and learning among young students, as demonstrated in the following episode, drawn from a classroom experience:

> Rosalinda, usually a quiet child, was very excited to learn how to skip-count to 100 on the calculator. However, she was puzzled when counting to 100 by threes. "It always goes over 100!" she exclaimed. The teacher encouraged Rosalinda and her partner to investigate the phenomenon. Over several days, the students talked together about why the calculator did not display 100 when they counted by threes. They used the hundred board and counters along with their calculator and concluded that equal groups of twos could be made with 100 counters but not equal groups of threes. The investigation resulted in a chart that Rosalinda and her partner made to explain to the class what they had figured out and how the calculator had supported their conclusions.

Experiences such as this help students see themselves as problem posers and also see how tools such as calculators can be used to support their mathematical investigations.

Manipulating objects and drawing pictures are natural ways that students communicate in prekindergarten through grade 2, but they also learn to explain their answers in writing, to use diagrams and charts,

E-example 4.5

Calculators and Hundred Boards (Part 1)

and to express ideas with mathematical symbols. Their language should become more precise as they use words such as *angles* and *faces* instead of *corners* and *sides*. Opportunities to express their ideas encourage students to organize and consolidate their mathematical thinking.

Young students' abilities to talk and listen are usually more advanced than their abilities to read and write, especially in the early years of this grade band. Therefore, teachers must be diligent in providing experiences that allow varied forms of communication as a natural component of mathematics class, as the following episode, adapted from Andrews (1996, p. 293), demonstrates:

> A kindergarten teacher read a story about a family's journey across the country. She asked the students to make maps to show the route taken by the family. As they worked in groups, some students incorporated letters or other symbols into the work. One group drew pictures of each landmark. Another group asked the teacher to help them label parts of their map and numbered each step along the way. As each group shared its work with the class, the teacher asked what changes they would make next time that might improve their work. The maps were hung in the hall, giving the teacher opportunities to question the students about the mathematical ideas of space and navigation they had used in creating the maps.

What should be the teacher's role in developing communication in prekindergarten through grade 2?

Teachers can create and structure mathematically rich environments for students in a number of ways. They should present problems that challenge students mathematically, but they should also let students know they believe that the students can solve them. They should expect students to explain their thinking and should give students many opportunities to talk with, and listen to, their peers. Teachers should recognize that learning to analyze and reflect on what is said by others is essential in developing an understanding of both content and process. When it is difficult for young learners to follow the reasoning of a classmate, teachers can help by guiding students to rephrase their reasoning in words that are easier for themselves and others to understand. Teachers should model appropriate conventional vocabulary and help students build such vocabulary on the basis of shared knowledge and processes.

Teachers must be diligent in providing experiences that allow varied forms of communication.

Teachers should support students' mathematics learning through the languages that they bring to school; they should also help them develop standard English vocabulary and mathematical terms that will enable them to communicate better with others. Students should be encouraged and respected when they use their native language as well as English in their mathematical communications. If possible, the mathematical terms should be displayed in both English and the native languages. Students who are not yet proficient in English can be paired with other students who speak the same language and with bilingual students or community volunteers, who can support the communication of ideas to the teacher and the rest of the class.

Teachers also need to be aware that the patterns of communication between students and adults in the school may not necessarily match the patterns of communication in students' homes. For example, patterns of questioning can be very different. In some cultures, adults generally do not ask questions when the answer is known; they ask questions primarily to seek information that they do not have. In school, however, teachers frequently ask questions to which the answer is known. Students who are not accustomed to such questions can be bewildered, since it is obvious that the teacher already knows the answers. Similarly, in some cultures, people routinely interrupt one another in conversations, whereas in others, interruptions are considered extremely rude. Students from the first group may unduly dominate class discussions. In other cultures, children are expected not to ask questions but to learn by observation. A student from such a group may be uncomfortable asking questions in class (Bransford, Brown, and Cocking 1999). Teachers need to be aware of the cultural patterns in their students' communities in order to provide equitable opportunities for them to communicate about mathematical thinking.

Teachers should model appropriate conventional vocabulary.

Building a community of learners, where students exchange mathematical ideas not only with the teacher but also with one another, should be a goal in every classroom. Consider the following example, which has been drawn from a classroom experience:

> A teacher asked, "How many books do I need to return to the library if I have three nonfiction and four fiction books?" A student who seldom shared his answers with the class voluntarily responded, "Seven." In the past, if the teacher asked him how he figured out a problem, the student just shrugged his shoulders. This time, however, the teacher decided to involve another student and asked her, "How do you think Maury figured it out?" The second student held up three fingers as she answered, "I think he did it this way. I know there are three, so I just put up four more fingers and then count them all." This prompted Maury to respond, "I did it a different way. I just knew that three and three make six and then I counted one more."

The teacher thus set the stage for two students to explain their methods and for all their classmates to hear and discuss the two ways of thinking about the same problem.

Just as teachers accept multiple forms of communication from their students, so they should also communicate with the students in a variety of ways to ensure maximum success for all. For example, because not all children at this level are able to follow written instructions, teachers could decide to read instructions or to draw pictures to represent the sequence and contents of a task.

It is the responsibility of the teacher to recognize appropriate times to make connections between invented symbols and standard notation. When students present their own representations of their mathematical knowledge, the presentations are often unique and creative. For instance, figure 4.30 is a kindergartner's notation to remember that a jar held ten and a half scoops. Teachers must seek to understand what students are trying to communicate and use that information to advance individual students' learning and that of the class as a whole. The use of mathematical symbols should follow, not precede, other ways of communicating mathematical ideas. In this way, teachers help young students relate their everyday language to mathematical language and symbols in a meaningful way.

Fig. **4.30.**

A child's notation for 10 1/2

Connections Standard for Grades Pre-K–2

Instructional programs from prekindergarten through grade 12 should enable all students to—

Recognize and use connections among mathematical ideas

Understand how mathematical ideas interconnect and build on one another to produce a coherent whole

Recognize and apply mathematics in contexts outside of mathematics

The most important connection for early mathematics development is between the intuitive, informal mathematics that students have learned through their own experiences and the mathematics they are learning in school. All other connections—between one mathematical concept and another, between different mathematics topics, between mathematics and other fields of knowledge, and between mathematics and everyday life—are supported by the link between the students' informal experiences and more-formal mathematics. Students' abilities to experience mathematics as a meaningful endeavor that makes sense rests on these connections.

When young students use the relationships in and among mathematical content and processes, they advance their knowledge of mathematics and extend their ability to apply concepts and skills more effectively. Understanding connections eliminates the barriers that separate the mathematics learned in school from the mathematics learned elsewhere. It helps students realize the beauty of mathematics and its function as a means of more clearly observing, representing, and interpreting the world around them.

Teachers can facilitate these connections in several ways: They should spotlight the many situations in which young students encounter mathematics in and out of school. They should make explicit the connections between and among the mathematical ideas students are developing, such as subtraction with addition, measurement with number and geometry, or representations with algebra and problem solving. They should plan lessons so that skills and concepts are taught not as isolated topics but rather as valued, connected, and useful parts of students' experiences.

What should connections look like in prekindergarten through grade 2?

Young children often connect new mathematical ideas with old ones by using concrete objects. When a preschool child holds up three fingers and asks an adult, "Am I this many years old?" he is trying to connect the word *three* with the number that represents his age through a set of concrete objects, his fingers. Teachers should encourage students to use their own strategies to make connections among mathematical ideas, the vocabulary associated with the ideas, and the ways the ideas are represented. For example, students frequently use objects and counting strategies as they develop their understanding of addition and subtraction and connect the two operations.

Students can better understand relationships among the many aspects of mathematics as they engage in purposeful activities. Often activities that include making estimates provide links among concepts in multiple Standards (Roche 1996). Consider the following example:

> In a second-grade class in which students were investigating filling jars with scoops of cranberries, the teacher had students first estimate the number of scoops needed. They organized the estimates into a graph and talked about the range of numbers. After pouring some scoops into one jar as a referent, the students refined their estimates and talked about how and why the range was narrowed. Throughout the lesson, which included several more activities, students worked in groups and repeatedly came back to whole-class discussions that involved mathematics expectations about number and operations, data analysis, and connections.

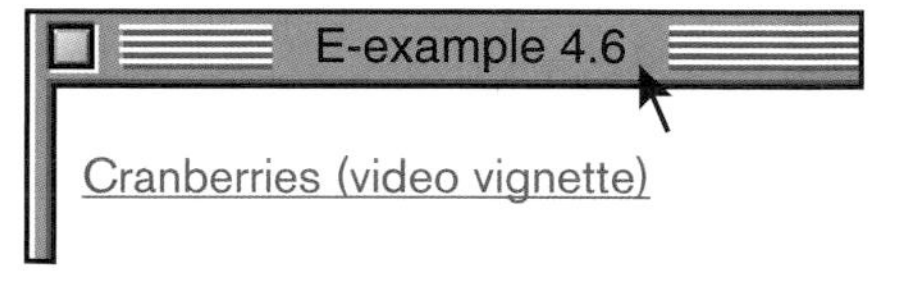

In another example, the process of making a string of cubes by using only two colors of connecting cubes helps students understand addition. It also involves pattern and relates ideas from number and geometry. As students try to find different ways of building a string of four cubes, if order does not matter and there are only two different colors, they may discover that only five different solutions are possible. Further investigation can reveal that there are six ways to build a string of five cubes (see fig. 4.31). Using this knowledge, some students may generalize and predict that a string of six cubes can be built in seven different ways.

Fig. **4.31.**

Building a string of five cubes by using only two colors

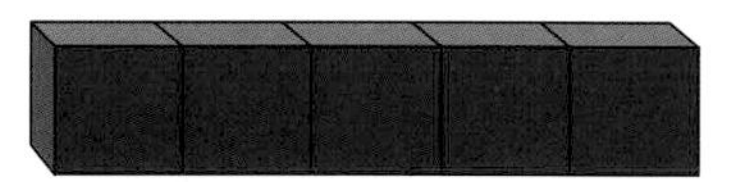
Five red

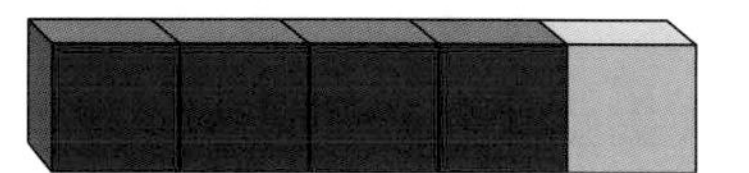
Four red and one yellow

Three red and two yellow

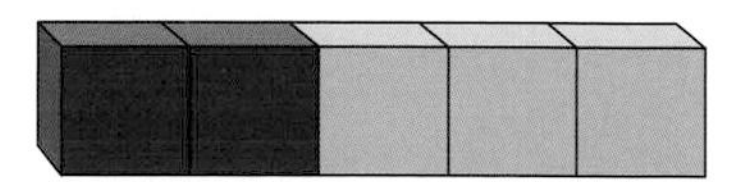
Two red and three yellow

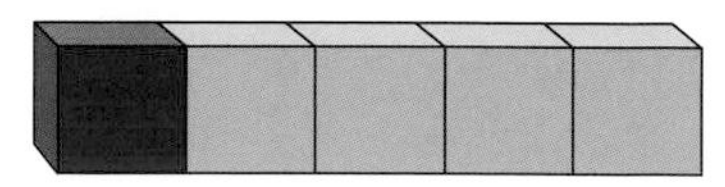
One red and four yellow

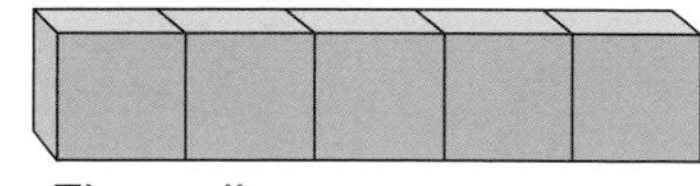
Five yellow

Mathematics is embedded in many activities that young students do throughout the day. For example, in physical education class, children may count the number of times they can jump rope successfully. They may measure and compare the amount of time the members of a team take to sprint from one end of the gym to the other. Concepts involving geometric shapes are reinforced as children form a circle to play a game, take their places around the outer edge of a parachute, or assume their positions to play a game on the field. Students explore symmetry in art class as they make hearts by folding and cutting paper. They classify leaves collected on a nature walk in science lessons. In music, they sing songs rich in patterns.

In one first-grade classroom, children used the mathematics of patterns to investigate and quantify syllables in names, as related in this episode drawn from a classroom experience:

> The teacher clapped out a student's name (one clap for each syllable) and asked the students if they could figure out whose name she was clapping. They realized that her clapping matched the names of several students. When the class began to try to determine which students had the same number of claps in their names, the teacher drew the chart shown in figure 4.32 on the board. She added students' names as the class identified the number of beats in a name.
>
> One student stated that he could not find a name with seven beats. Another student disagreed and illustrated seven beats by including

Fig. **4.32.**

A chart showing the number of beats in students' names

3 beats	4 beats	5 beats	6 beats
John Gosha	Sarah Andrews	Bobby Erickson	Sylvia McPherson
Timmy Simms	Carlos Sanchez	Anisha Johnson	Alyssa Huerrero
Ana Jones	Becky Tott	Theodore Hopkins	Stephanie Abramson

the beats in her middle name in her total. Other students then began to experiment with middle names and nicknames to match the number of beats in other names. One student looked at the chart and questioned whether John Gosha and Timmy Simms were actually "beat twins." Even though the number of beats was the same, the names sounded different. This encouraged the students to record their name-beat patterns in a more specific way using the number of syllables in each word to help them write different equations with the same sums (see fig. 4.33).

Fig. **4.33.**

Recording name-beat patterns

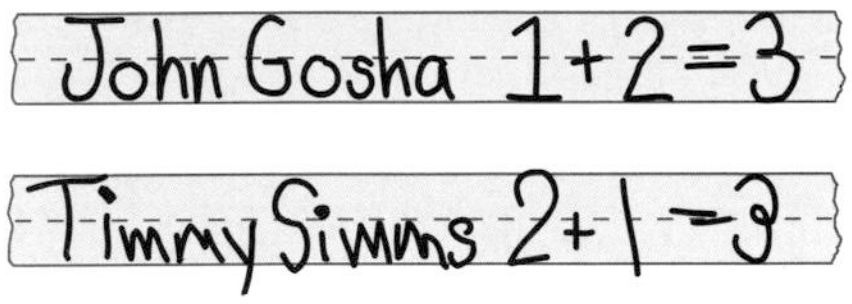

John Gosha 1+2=3

Timmy Simms 2+1=3

As the students determined the best ways to describe name beats through number, they also were reinforcing their arithmetic skills. They were also actually creating a function that assigned a number to each student's name. Such interplay, in which mathematics illuminates a situation and the situation illuminates mathematics, is an important aspect of mathematical connections.

Seeing the usefulness of mathematics contributes to students' success in situations requiring mathematical solutions. When students measure the field for the hundred-yard relay, they truly know the purpose of learning to measure. Determining when they have saved enough allowance money to purchase a prized toy helps students realize the usefulness and importance of the knowledge of counting, of addition, and of the value of coins. Pointing to the hexagonal pattern in a honeycomb illustrates the use of mathematical ideas in describing nature. Surveying friends and family members about a favorite vacation site gives meaning and purpose to data collection. Observing the patterns on the fences in town demonstrates how mathematics is used in construction. These associations add purpose and pleasure to the learning of mathematics.

What should be the teacher's role in developing connections in prekindergarten through grade 2?

In classrooms where connecting mathematical ideas is a focus, lessons are fluid and take many different formats. Teachers should ensure that links are made between routine school activities and mathematics by asking questions that emphasize the mathematical aspects of situations. They should plan tasks in new contexts that revisit topics previously taught, enabling students to forge new links between previously learned mathematical concepts and procedures and new applications, always with an eye on their mathematics goals. When teachers help students make explicit connections—mathematics to other mathematics and mathematics to other content areas—they are helping students learn to think mathematically.

Often connections are best made when students are challenged to apply mathematics learning in extended projects and investigations. As they formulate questions and design inquiries, students decide on methods of gathering and recording information and plan representations to communicate the data and help them make reasonable conjectures and interpretations.

Teachers can find in the way a child interprets mathematical situations clues to that child's understanding. They should listen to students in order to assess the connections students bring to their situation, and they should use this information to plan activities that will further students' mathematical knowledge and skills and establish new and different connections. A teacher can, for example, notice the different understandings represented by different students' comments or questions about pattern blocks: one student may ask the teacher to name a pattern-block shape; a second student might build a pattern-block design and direct the teacher's attention to its symmetry; a third student may explain to the teacher that two pattern-block trapezoids together make a hexagon.

Teachers should capitalize on unexpected learning opportunities.

It is the responsibility of the teacher to help students see and experience the interrelation of mathematical topics, the relationships between mathematics and other subjects, and the way that mathematics is embedded in the students' world. Teachers should capitalize on unexpected learning opportunities, such as the lesson in which syllables in students' names were recorded as equations. They should ask questions that direct students' thinking and present tasks that help students see how ideas are related.

Representation Standard for Grades Pre-K–2

Instructional programs from prekindergarten through grade 12 should enable all students to—

Create and use representations to organize, record, and communicate mathematical ideas

Select, apply, and translate among mathematical representations to solve problems

Use representations to model and interpret physical, social, and mathematical phenomena

Young students use many varied representations to build new understandings and express mathematical ideas. Representing ideas and connecting the representations to mathematics lies at the heart of understanding mathematics. Teachers should analyze students' representations and carefully listen to their discussions to gain insights into the development of mathematical thinking and to enable them to provide support as students connect their languages to the conventional language of mathematics. The goals of the Communication Standard are closely linked with those of this Standard, with each set contributing to and supporting the other.

Students in prekindergarten through grade 2 represent their thoughts about, and understanding of, mathematical ideas through oral and written language, physical gestures, drawings, and invented and conventional symbols (Edwards, Gandini, and Forman 1993). These representations are methods for communicating as well as powerful tools for thinking. The process of linking different representations, including technological ones, deepens students' understanding of mathematics because of the connections they make between ideas and the ways the ideas can be expressed. Teachers can gain insight into students' thinking and their grasp of mathematical concepts by examining, questioning, and interpreting their representations. Although a striking aspect of children's mathematical development in the pre-K–2 years is their growth in using standard mathematical symbols, teachers at this level should encourage students to use multiple representations, and they should assess the level of mathematical understanding conveyed by those representations.

What should representation look like in prekindergarten through grade 2?

Young students represent their mathematical ideas and procedures in many ways. They use physical objects such as their own fingers, natural language, drawings, diagrams, physical gestures, and symbols. Through interactions with these representations, other students, and the teacher, students develop their own mental images of mathematical ideas. Although the representations that children use may not be those traditionally used by adults, students' representations provide a record of their efforts to understand mathematics and also make their understanding available to others.

Representations make mathematical ideas more concrete and available for reflection. Students can represent ideas with objects that can be moved and rearranged. Such concrete representations lay the foundation for the later use of symbols. Students' representations are often insightful and many times resemble more-conventional representations. For example, a second grader working with place-value mats and base-ten blocks can represent 103. The student might point to the blocks and tap at the empty column, explaining, "One hundred, (tap), three." The tap helps the student connect the zero with the empty tens column.

The following account of a lesson, drawn from a classroom experience, illustrates that what children do and say as they find answers and represent their thinking gives teachers information about their levels of understanding.

> When a first-grade teacher read *Rooster's Off to See the World* (Carle 1971), the students' representations of the number of animals going off to see the world varied (see fig. 4.34). Two cats, then three frogs, four turtles, and five fish joined the rooster for a total of fifteen animals. To find how many went on the trip, some students drew the animals and numbered them. Two students modeled the animals with counters, counted, and wrote "15" on their papers. Other students used more traditional notations, although their representations revealed different ways of thinking. One student declared the answer to be zero, because all the animals had gone home when it got dark.

Fig. **4.34.**

Three students' representations for the numbers of animals that went off to see the world

5 fishes
+4 turtels
9
+3 frogs
12
+2 cats
14
+1 raster
15
-5
10
-4
6
-3
3
-2
1
-1
0

I added the animals the I sutracted cause they went away.

Student 1

Student 2

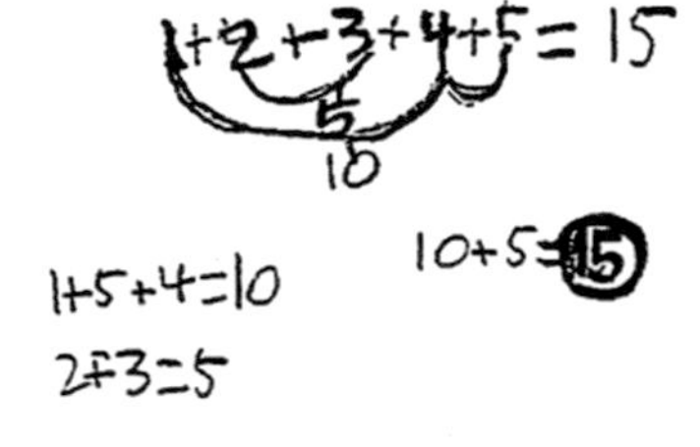

rooster plus turtles plus fish= 10 animals
cats plus frogs = 5animals fishes
10 animals plus 5animals = 15 animals

Student 3

Fig. **4.35.**

A student's representation of the number of animals that went off to see the world

The teacher was puzzled, however, by the student whose answer was 21 (see fig. 4.35). The teacher asked the student to explain. The student responded that she had noticed that there were fireflies in the story on the page where the animals decided to turn around and go home. She couldn't count how many, but she thought there were six because that was the pattern, so she drew the animals and added. The teacher asked about the list at the top right of the student's paper, and she responded that she had made the tallies to show how many, and there were 21.

Seeing similarities in the ways to represent different situations is an important step toward abstraction.

Representations help students recognize the common mathematical nature of different situations. Students might represent the following three scenarios by writing $5 - 3 = 2$. The first problem is to determine the number of objects left after three objects have been taken from a collection of five. The second problem is, How much taller is a tower of five cubes than a tower of three cubes? The third problem asks the number of balls that must be put into a box if it is to have five balls and there are already three in the box. Students could also represent the situations as $3 + \square = 5$. Seeing similarities in the ways to represent different situations is an important step toward abstraction.

Students use representations to organize their thinking. Representations can carry some of the burden of remembering by letting students record intermediate steps in a process. For example, a student trying to find the number of wheels in four bicycles and three tricycles drew the picture shown in figure 4.36. In the first row, the student represented the number of wheels on the bicycles and in the second, the number of

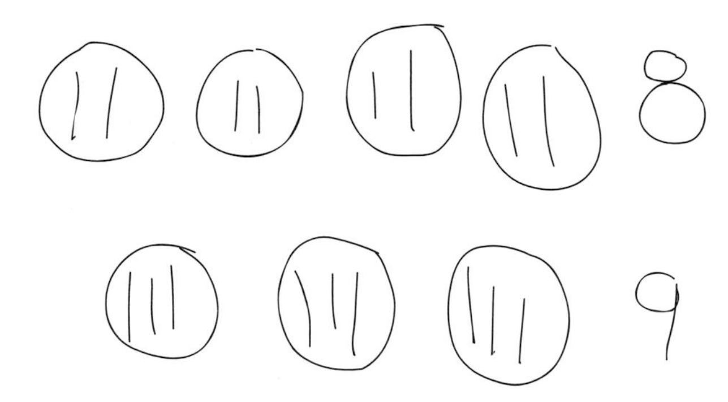

Fig. **4.36.**

A student's representation of the number of wheels on four bicycles and three tricycles (Adapted from Flores [1997, p. 86.])

wheels on the tricycles. Thus the student was able to compute the sums separately and add them together. The representation served as a placeholder for thoughts that were not yet internalized.

Understanding and using mathematical concepts and procedures is enhanced when students can translate between different representations of the same idea. In doing so, students appreciate that some representations highlight features of the problem in a better way or make it easier to understand certain properties. For example, a student who represents three groups of four squares as an array and uses skip-counting (4, 8, 12), repeated addition (4 + 4 + 4), and oral language to describe the representation as three rows of four squares is laying the groundwork for understanding multiplication and its properties as well as the area of a rectangle.

What should be the teacher's role in developing representation in prekindergarten through grade 2?

A major responsibility of teachers is to create a learning environment in which students' use of multiple representations is encouraged, supported, and accepted by their peers and adults. Teachers should guide students to develop and use multiple representations effectively. Students will thus develop their own perceptions, create their own evidence, structure their own analytical processes, and become confident and competent in their use of mathematics.

A major responsibility of teachers is to create a learning environment in which students' use of multiple representations is encouraged.

Teachers at this level need to listen to what students say, thoughtfully observe their mathematical activities, analyze their recordings, and reflect on the implications of the observations and analyses. Using representations helps students remember what they did and explain their reasoning. Representations furnish a record of students' thinking that shows both the answer and the process, and they assist teachers in formulating questions that can help students reflect on their processes and products and advance their understanding of concepts and procedures. The information gathered from these multiple sources makes possible a clearer assessment of what students understand and what mathematical ideas are still developing.

Students should be encouraged to share their different representations to help them consider other perspectives and ways of explaining their thinking. Teachers can model conventional ways of representing mathematical situations, but it is important for students to use representations that are meaningful to them. Transitions to conventional notations should be connected to the methods and thinking of the students. For example, when students use blocks or mental computation to solve a problem like "Find the sum of 17 + 25," they frequently add the tens first. Teachers can write the intermediate steps for students as 10 + 20 = 30, 7 + 5 = 12, and 30 + 12 = 42. Students should see their method recorded both horizontally and vertically and should develop their own ways of keeping track of their work that are clear to them (see fig. 4.37).

Fig. **4.37.**
Recording a method to find 17 + 25 in two different ways

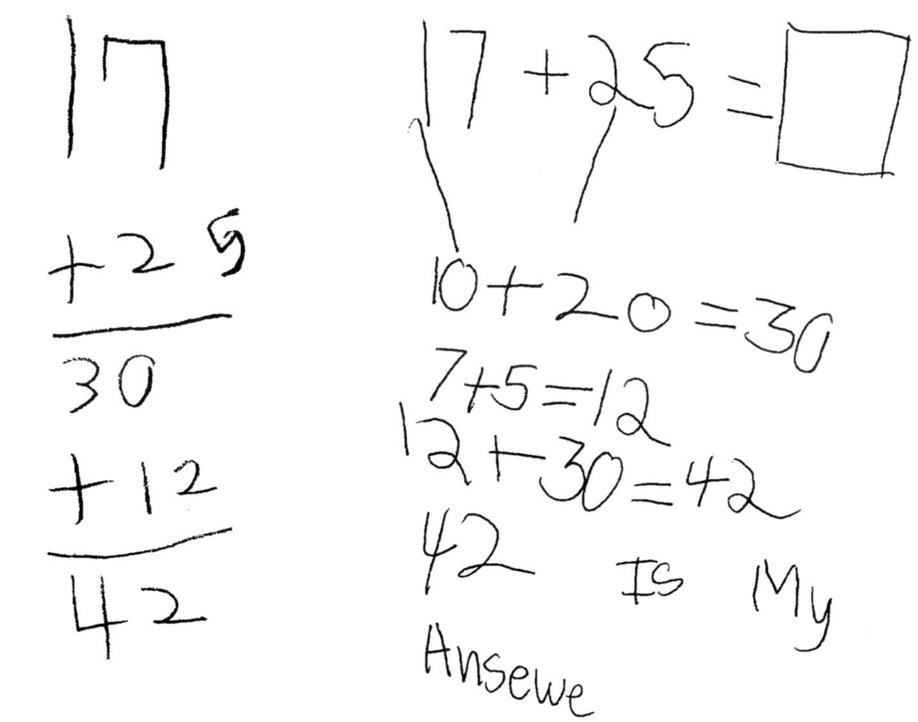

It is important for students to use representations that are meaningful to them.

Through class discussions of students' ways of thinking and recording, teachers can lay foundations for students' understanding of conventional ways of representing the process of adding numbers. Equally important, students' work and conversations about their representations can reveal the extent to which they understand their use of symbols.

Written work often does not reveal a student's entire thinking, as the following hypothetical story about Armando demonstrates:

> Armando does not show marks to cross out any digits or write a small "1" to represent "borrowing" with paper-and-pencil subtraction, but he consistently writes correct answers (see fig. 4.38). When the teacher probes, Armando explains that he has learned a different way to subtract at home. The teacher asks him to explain his method. "From 8 to 14—that is 6, and we need to add 1 to the 2 because we used 14 instead of 4." He writes 6 in the units place and continues, "From 3 to 7 is 4," and he writes 4 in the tens column. The teacher rephrases the second part of the method, emphasizing place value: "So, you add 10 to the 20 and then subtract 70 – 30." Realizing that the method is based on a property the class has recently discussed, that the same number can be added to both terms

of a difference and the result does not change, she invites the class to talk about the process Armando is using. After some discussion, one student explains, "You are adding 10 to 74 because you really did 14 – 8, and you also added 10 to 28 because you did 70 – 30. So the answer is the same."

Teachers should help students understand that representations are tools to model and interpret phenomena of a mathematical nature that are found in different contexts. Teachers should help students represent aspects of situations in mathematical terms, possibly by using more than one representation. Technology may help students who are challenged by oral or written communication find greater success. The processing schema required in some computer programs can aid students in showing what they know. For example, when a student changes the representation of a number with base-ten blocks on the screen, the computer shows how the corresponding symbols change.

It is important that teachers realize and teach students that any representations, not only those created by students, are subject to multiple interpretations. Drawings, charts, graphs, and diagrams, for instance, can be read in different ways. Therefore, teachers should not assume that students understand a diagram or equation the same way adults do. Communicating the intended meaning and using alternative representations can enhance understanding by students and teachers alike.

Fig. **4.38.**

Finding 74 – 28 without "borrowing"

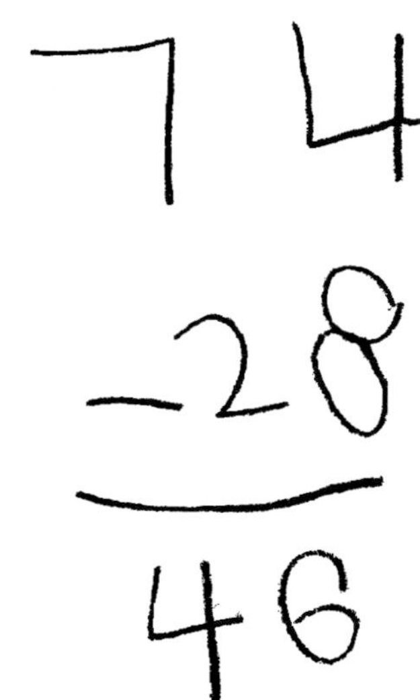

Nearly three-quarters of U.S. fourth graders report liking mathematics.

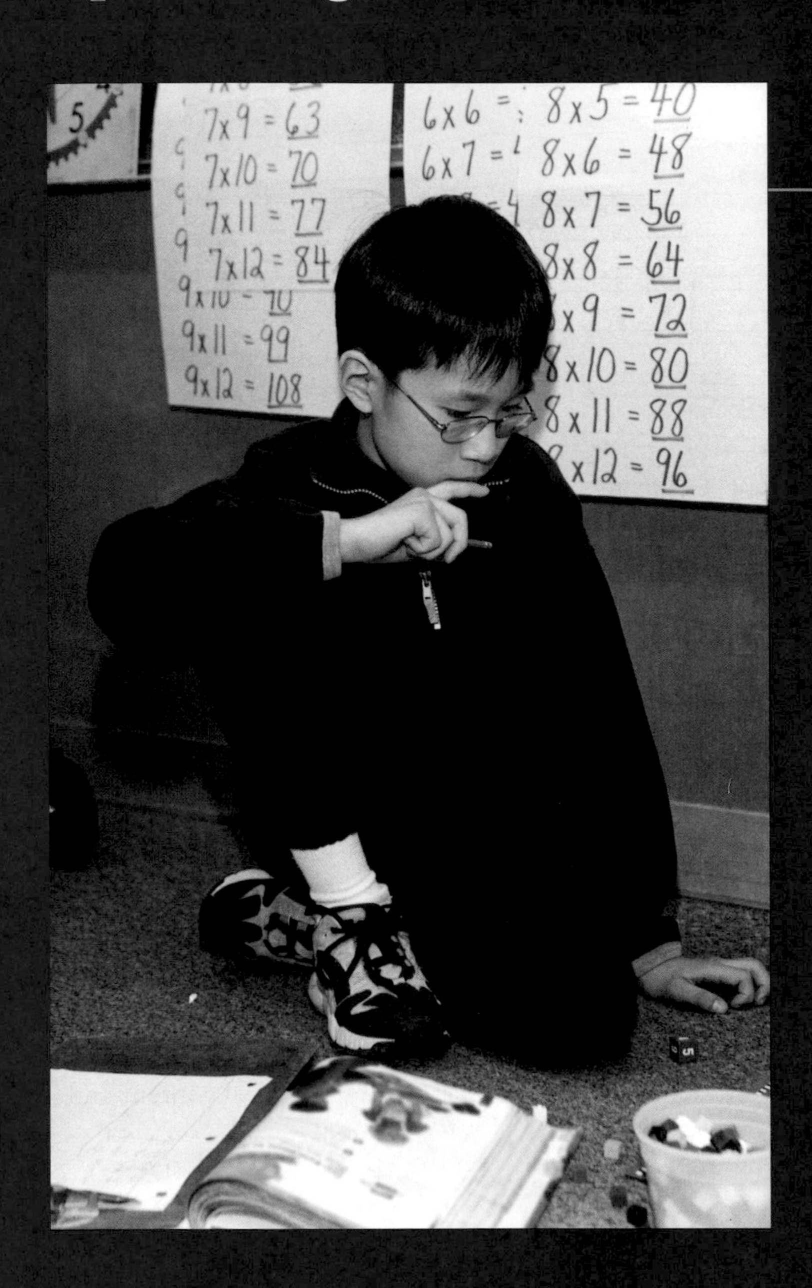

Instruction at this level must be active and intellectually stimulating.

Standards for Grades 3–5

Most students enter grade 3 with enthusiasm for, and interest in, learning mathematics. In fact, nearly three-quarters of U.S. fourth graders report liking mathematics (Silver, Strutchens, and Zawojewski 1997). They find it practical and believe that what they are learning is important. If the mathematics studied in grades 3–5 is interesting and understandable, the increasingly sophisticated mathematical ideas at this level can maintain students' engagement and enthusiasm. But if their learning becomes a process of simply mimicking and memorizing, they can soon begin to lose interest. Instruction at this level must be active and intellectually stimulating and must help students make sense of mathematics.

This chapter presents a challenging set of mathematical content and processes that students in grades 3–5 can and should learn. It also emphasizes teaching that fosters and builds on students' mathematical understanding and thinking. The Content and Process Standards described here form the basis for a significant and interconnected mathematics curriculum. Interwoven through these Standards are three central mathematical themes—multiplicative reasoning, equivalence, and computational fluency. They are briefly discussed here and elaborated on throughout the chapter.

Students entering grade 3 should have a good grasp of, and much experience with, additive reasoning. Their understanding of whole numbers is often based on an additive model—a sequence of numbers used to count in different ways—and their computing strategies usually involve counting on or counting back. In grades 3–5, multiplicative reasoning should become a focus. Multiplicative reasoning is more than just doing multiplication or division. It is about understanding situations in which multiplication or divi-

Three central mathematical themes are discussed—multiplicative reasoning, equivalence, and computational fluency.

Mathematics learning is both about making sense of mathematical ideas and about acquiring skills and insights to solve problems.

sion is an appropriate operation. It involves a way of viewing situations and thinking about them (Thompson forthcoming). For example, to estimate the height of an adult, students might use their own heights as a benchmark and then think of the situation from an additive perspective (the adult is about 50 centimeters taller than the student) or a multiplicative perspective (the adult is one quarter again as tall as the student).

In grades 3–5, multiplicative reasoning emerges and should be discussed and developed through the study of many different mathematical topics. Students' understanding of the base-ten number system is deepened as they come to understand its multiplicative structure. That is, 484 is 4×100 plus 8×10 plus 4×1 as well as a collection of 484 individual objects. Multiplicative reasoning is further developed as students use a geometric model of multiplication, such as a rectangular array, and adapt this model for computing the area of shapes and the volume of solids. They also begin to reason algebraically with multiplication, looking for general patterns. For example, they explore problems such as, What is the effect of doubling one factor and halving the other in a multiplication problem? The focus on multiplicative reasoning in grades 3–5 provides foundational knowledge that can be built on as students move to an emphasis on proportional reasoning in the middle grades.

Equivalence should be another central idea in grades 3–5. Students' ability to recognize, create, and use equivalent representations of numbers and geometric objects should expand. For example, 3/4 can be thought of as a half and a fourth, as 6/8, or as 0.75; a parallelogram can be transformed into a rectangle with equal area by cutting and pasting; 8×25 can be thought of as $8 \times 5 \times 5$ or as 4×50; and three feet is the same as thirty-six inches, or one yard. Students should extend their use of equivalent forms of numbers as they develop new strategies for computing and should recognize that different representations of numbers are helpful for different purposes. Likewise, they should explore when and how shapes can be decomposed and reassembled and what features of the shapes remain unchanged. Equivalence also takes center stage as students study fractions and as they relate fractions, decimals and percents. Examining equivalences provides a way to explore algebraic ideas, including properties such as commutativity and associativity.

A major goal in grades 3–5 is the development of computational fluency with whole numbers. *Fluency* refers to having efficient, accurate, and generalizable methods (algorithms) for computing that are based on well-understood properties and number relationships. Some of these methods are performed mentally, and others are carried out using paper and pencil to facilitate the recording of thinking. Students should come to view algorithms as tools for solving problems rather than as the goal of mathematics study. As students develop computational algorithms, teachers should evaluate their work, help them recognize efficient algorithms, and provide sufficient and appropriate practice so that they become fluent and flexible in computing. Students in these grades should also develop computational-estimation strategies for situations that call for an estimate and as a tool for judging the reasonableness of solutions.

This set of Standards reinforces the dual goals that mathematics learning is both about making sense of mathematical ideas and about acquiring skills and insights to solve problems. The calculator is an important tool in reaching these goals in grades 3–5 (Groves 1994). However,

calculators do not replace fluency with basic number combinations, conceptual understanding, or the ability to formulate and use efficient and accurate methods for computing. Rather, the calculator should support these goals by enhancing and stimulating learning. As a student works on problems involving many or complex computations, the calculator is an efficient computational tool for applying the strategies determined by the student. The calculator serves as a tool for enabling students to focus on the problem-solving process. Calculators can also provide a means for highlighting mathematical patterns and relationships. For example, using the calculator to skip-count by tenths or hundredths highlights relationships among decimal numbers. For example, 4 is one-tenth more that 3.9, or 2.49 is one-hundredth less than 2.5. Students at this age should begin to develop good decision-making habits about when it is useful and appropriate to use other computational methods, rather than reach for a calculator. Teachers should create opportunities for these decisions as well as make judgments about when and how calculators can be used to support learning.

Teachers in grades 3–5 make decisions every day that influence their students' opportunities to learn and the quality of that learning. The classroom environment they create, the attention to various topics of mathematics, and the tools they and their students use to explore mathematical ideas are all important in helping students in grades 3–5 gain increased mathematical maturity. In these grades teachers should help students learn to work together as part of building a mathematical community of learners. In such a community, students' ideas are valued and serve as a source of learning, mistakes are seen not as dead ends but rather as potential avenues for learning, and ideas are valued because they are mathematically sound rather than because they are argued strongly or proposed by a particular individual (Hiebert et al. 1997). A classroom environment that would support the learning of mathematics with meaning should have several characteristics: students feel comfortable making and correcting mistakes; rewards are given for sustained effort and progress, not the number of problems completed; and students think through and explain their solutions instead of seeking or trying to recollect the "right" answer or method (Cobb et al. 1988). Creating a

Students' ideas should be valued and serve as a source of learning.

Teachers in grades 3–5 often must seek ways to advance their own understanding.

classroom environment that fosters mathematics as sense making requires the careful attention of the teacher. The teacher establishes the model for classroom discussion, making explicit what counts as a convincing mathematical argument. The teacher also lays the groundwork for students to be respectful listeners, valuing and learning from one another's ideas even when they disagree with them.

Because of the increasing mathematical sophistication of the curriculum in grades 3–5, the development of teachers' expertise is particularly important. Teachers need to understand both the mathematical content for teaching and students' mathematical thinking. However, teachers at this level are usually called on to teach a variety of disciplines in addition to mathematics. Many elementary teacher preparation programs require minimal attention to mathematics content knowledge. Given their primary role in shaping the mathematics learning of their students, teachers in grades 3–5 often must seek ways to advance their own understanding.

Many different professional development models emphasize the enhancement of teachers' mathematical knowledge. Likewise, schools and districts have developed strategies for strengthening the mathematical expertise in their instructional programs. For example, some elementary schools identify a mathematics teacher-leader (someone who has particular interest and expertise in mathematics) and then support that teacher's continuing development and create a role for him or her to organize professional development events for colleagues. Such activities can include grade-level mathematics study groups, seminars and workshops, and coaching and modeling in the classroom. Other schools use mathematics specialists in the upper elementary grades. These are elementary school teachers with particular interest and expertise in mathematics who assume primary responsibility for teaching mathematics to a group of students—for example, all the fourth graders in a school. This strategy allows some teachers to focus on a particular content area rather than to attempt being an expert in all areas.

Ensuring that the mathematics outlined in this chapter is learned by all students in grades 3–5 requires a commitment of effort by teachers to continue to be mathematical learners. It also implies that districts, schools, and teacher preparation programs will develop strategies to identify current and prospective elementary school teachers for specialized mathematics preparation and assignment. Each of the models outlined here—mathematics teacher-leaders and mathematics specialists—should be explored as ways to develop and enhance students' mathematics education experience. For successful implementation of these Standards, it is essential that the mathematical expertise of teachers be developed, whatever model is used.

Number and Operations

Standard for Grades 3–5

Instructional programs from prekindergarten through grade 12 should enable all students to—

Expectations

In grades 3–5 all students should–

Standard	Expectations
Understand numbers, ways of representing numbers, relationships among numbers, and number systems	• understand the place-value structure of the base-ten number system and be able to represent and compare whole numbers and decimals; • recognize equivalent representations for the same number and generate them by decomposing and composing numbers; • develop understanding of fractions as parts of unit wholes, as parts of a collection, as locations on number lines, and as divisions of whole numbers; • use models, benchmarks, and equivalent forms to judge the size of fractions; • recognize and generate equivalent forms of commonly used fractions, decimals, and percents; • explore numbers less than 0 by extending the number line and through familiar applications; • describe classes of numbers according to characteristics such as the nature of their factors.
Understand meanings of operations and how they relate to one another	• understand various meanings of multiplication and division; • understand the effects of multiplying and dividing whole numbers; • identify and use relationships between operations, such as division as the inverse of multiplication, to solve problems; • understand and use properties of operations, such as the distributivity of multiplication over addition.
Compute fluently and make reasonable estimates	• develop fluency with basic number combinations for multiplication and division and use these combinations to mentally compute related problems, such as 30×50; • develop fluency in adding, subtracting, multiplying, and dividing whole numbers; • develop and use strategies to estimate the results of whole-number computations and to judge the reasonableness of such results; • develop and use strategies to estimate computations involving fractions and decimals in situations relevant to students' experience; • use visual models, benchmarks, and equivalent forms to add and subtract commonly used fractions and decimals; • select appropriate methods and tools for computing with whole numbers from among mental computation, estimation, calculators, and paper and pencil according to the context and nature of the computation and use the selected method or tool.

Number and Operations

In grades 3–5, students' development of number sense should continue, with a focus on multiplication and division. Their understanding of the meanings of these operations should grow deeper as they encounter a range of representations and problem situations, learn about the properties of these operations, and develop fluency in whole-number computation. An understanding of the base-ten number system should be extended through continued work with larger numbers as well as with decimals. Through the study of various meanings and models of fractions—how fractions are related to each other and to the unit whole and how they are represented—students can gain facility in comparing fractions, often by using benchmarks such as 1/2 or 1. They also should consider numbers less than zero through familiar models such as a thermometer or a number line.

When students leave grade 5, they should be able to solve problems involving whole-number computation and should recognize that each operation will help them solve many different types of problems. They should be able to solve many problems mentally, to estimate a reasonable result for a problem, to efficiently recall or derive the basic number combinations for each operation, and to compute fluently with multidigit whole numbers. They should understand the equivalence of fractions, decimals, and percents and the information each type of representation conveys. With these understandings and skills, they should be able to develop strategies for computing with familiar fractions and decimals.

Students who understand the structure of numbers and the relationships among numbers can work with them flexibly.

Understand numbers, ways of representing numbers, relationships among numbers, and number systems

In grades 3–5, students' study and use of numbers should be extended to include larger numbers, fractions, and decimals. They need to develop strategies for judging the relative sizes of numbers. They should understand more deeply the multiplicative nature of the number system, including the structure of 786 as 7×100 plus 8×10 plus 6×1. They should also learn about the position of this number in the base-ten number system and its relationship to benchmarks such as 500, 750, 800, and 1000. They should explore the effects of operating on numbers with particular numbers, such as adding or subtracting 10 or 100 and multiplying or dividing by a power of 10. In order to develop these understandings, students should explore whole numbers using a variety of models and contexts. For example, a third-grade class might explore the size of 1000 by skip-counting to 1000, building a model of 1000 using ten hundred charts, gathering 1000 items such as paper clips and developing efficient ways to count them, or using strips that are 10 or 100 centimeters long to show the length of 1000 centimeters.

Students who understand the structure of numbers and the relationships among numbers can work with them flexibly (Fuson 1992). They recognize and can generate equivalent representations for the same number. For example, 36 can be thought of as $30 + 6$, $20 + 16$, 9×4, $40 - 4$, three dozen, or the square of 6. Each form is useful for a particular situation. Thinking of 36 as $30 + 6$ may be useful when multiplying by 36, whereas thinking of it as 6 sixes or 9 fours is helpful when considering equal shares. Students need to have many experiences decomposing and composing numbers in order to solve problems flexibly.

During grades 3–5, students should build their understanding of fractions as parts of a whole and as division. They will need to see and explore a variety of models of fractions, focusing primarily on familiar fractions such as halves, thirds, fourths, fifths, sixths, eighths, and tenths. By using an area model in which part of a region is shaded, students can see how fractions are related to a unit whole, compare fractional parts of a whole, and find equivalent fractions. They should develop strategies for ordering and comparing fractions, often using benchmarks such as 1/2 and 1. For example, fifth graders can compare fractions such as 2/5 and 5/8 by comparing each with 1/2—one is a little less than 1/2 and the other is a little more. By using parallel number lines, each showing a unit fraction and its multiples (see fig. 5.1), students can see fractions as numbers, note their relationship to 1, and see relationships among fractions, including equivalence. They should also begin to understand that between any two fractions, there is always another fraction.

Fig. **5.1.**

Parallel number lines with unit fractions and their multiples

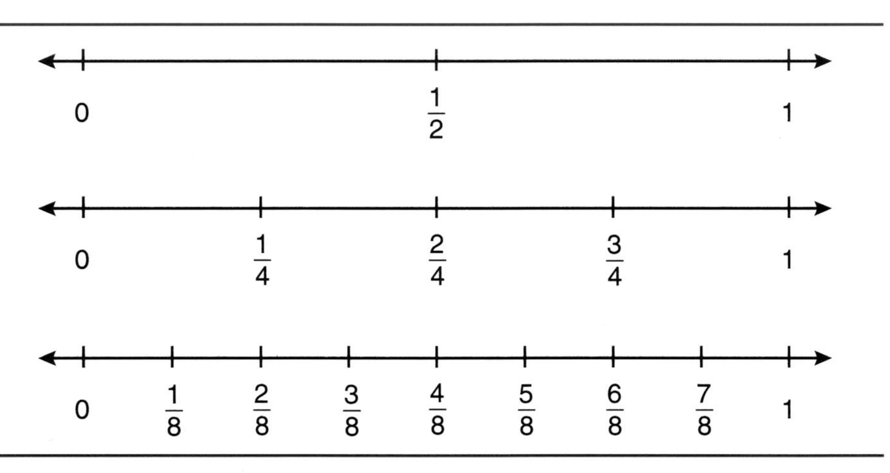

Students in these grades should use models and other strategies to represent and study decimal numbers. For example, they should count by tenths (one-tenth, two-tenths, three-tenths, ...) verbally or use a calculator to link and relate whole numbers with decimal numbers. As students continue to count orally from nine-tenths to ten-tenths to eleven-tenths and see the display change from 0.9 to 1.0 to 1.1, they see that ten-tenths is the same as one and also how it relates to 0.9 and 1.1. They should also investigate the relationship between fractions and decimals, focusing on equivalence. Through a variety of activities, they should understand that a fraction such as 1/2 is equivalent to 5/10 and that it has a decimal representation (0.5). As they encounter a new meaning of a fraction—as a quotient of two whole numbers ($1/2 = 1 \div 2 = 0.5$)—they can also see another way to arrive at this equivalence. By using the calculator to carry out the indicated division of familiar fractions like 1/4, 1/3, 2/5, 1/2, and 3/4, they can learn common fraction-decimal equivalents. They can also learn that some fractions can be expressed as terminating decimals but others cannot.

E-example 5.1

Communication through Games

Students should understand the meaning of a percent as part of a whole and use common percents such as 10 percent, 33 1/3 percent, or

50 percent as benchmarks in interpreting situations they encounter. For example, if a label indicates that 36 percent of a product is water, students can think of this as about a third of the product. By studying fractions, decimals, and percents simultaneously, students can learn to move among equivalent forms, choosing and using an appropriate and convenient form to solve problems and express quantities.

Negative integers should be introduced at this level through the use of familiar models such as temperature or owing money. The number line is also an appropriate and helpful model, and students should recognize that points to the left of 0 on a horizontal number line can be represented by numbers less than 0.

Throughout their study of numbers, students in grades 3–5 should identify classes of numbers and examine their properties. For example, integers that are divisible by 2 are called *even numbers* and numbers that are produced by multiplying a number by itself are called *square numbers*. Students should recognize that different types of numbers have particular characteristics; for example, square numbers have an odd number of factors and prime numbers have only two factors.

Understand meanings of operations and how they relate to one another

In grades 3–5, students should focus on the meanings of, and relationship between, multiplication and division. It is important that students understand what each number in a multiplication or division expression represents. For example, in multiplication, unlike addition, the factors in the problem can refer to different units. If students are solving the problem 29×4 to find out how many legs there are on 29 cats, 29 is the number of cats (or number of groups), 4 is the number of legs on each cat (or number of items in each group), and 116 is the total number of legs on all the cats. Modeling multiplication problems with pictures, diagrams, or concrete materials helps students learn what the factors and their product represent in various contexts.

Students should focus on the meanings of, and relationship between, multiplication and division.

Students should consider and discuss different types of problems that can be solved using multiplication and division. For example, if there are 112 people traveling by bus and each bus can hold 28 people, how many buses are needed? In this case, $112 \div 28$ indicates the number of groups (buses), where the total number of people (112) and the size of each group (28 people in each bus) are known. In a different problem, students might know the number of groups and need to find how many items are in each group. If 112 people divide themselves evenly among four buses, how many people are on each bus? In this case, $112 \div 4$ indicates the number of people on each bus, where the total number of people and the number of groups (buses) are known. Students need to recognize both types of problems as division situations, should be able to model and solve each type of problem, and should know the units of the result: Is it 28 buses or 28 people per bus? Students in these grades will also encounter situations where the result of division includes a remainder. They should learn the meaning of a remainder by modeling division problems and exploring the size of remainders given a particular divisor. For example, when dividing groups of counters into sets of 4, what remainders could there be for groups of different sizes?

Computational fluency refers to having efficient and accurate methods for computing.

Students can extend their understanding of multiplication and division as they consider the inverse relationship between the two operations. Another way their knowledge can grow is through new multiplicative situations such as rates (3 candy bars for 59 cents each), comparisons (the book weighs 4 times as much as the tablet), and combinations (the number of outfits possible from 3 shirts and 2 pairs of shorts). Examining the effect of multiplying or dividing numbers can also lead to a deeper understanding of these operations. For example, dividing 28 by 14 and comparing the result to dividing 28 by 7 can lead to the conjecture that the smaller the divisor, the larger the quotient. With models or calculators, students can explore dividing by numbers between 0 and 1, such as 1/2, and find that the quotient is larger than the original number. Explorations such as these help dispel common, but incorrect, generalizations such as "division always makes things smaller."

Further meaning for multiplication should develop as students build and describe area models, showing how a product is related to its factors. The area model is important because it helps students develop an understanding of multiplication properties (Graeber and Campbell 1993). Using area models, properties of operations such as the commutativity of multiplication become more apparent. Other relationships can be seen by decomposing and composing area models. For example, a model for 20×6 can be split in half and the halves rearranged to form a 10×12 rectangle, showing the equivalence of 10×12 and 20×6. The distributive property is particularly powerful as the basis of many efficient multiplication algorithms. For example, figure 5.2 shows the strategies three students might use to compute 7×28—all involving the distributive property.

Fig. **5.2.**

Three strategies for computing 7×28 using the distributive property

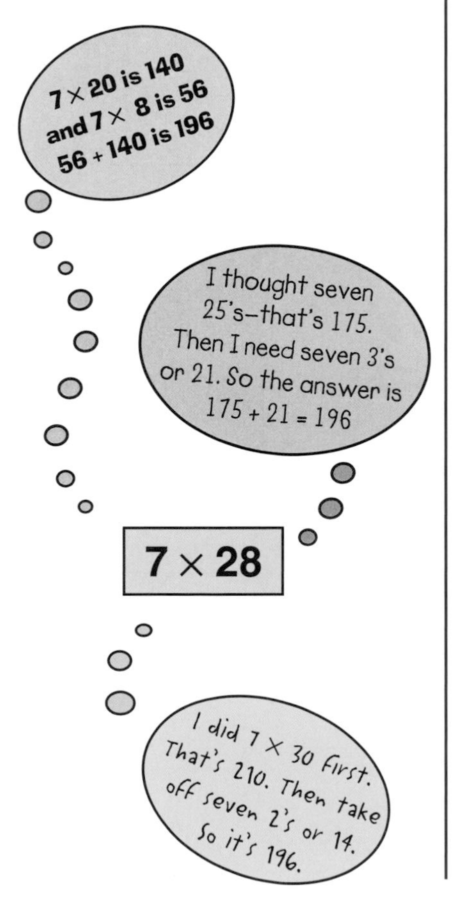

Compute fluently and make reasonable estimates

By the end of this grade band, students should be computing fluently with whole numbers. *Computational fluency* refers to having efficient and accurate methods for computing. Students exhibit computational fluency when they demonstrate flexibility in the computational methods they choose, understand and can explain these methods, and produce accurate answers efficiently. The computational methods that a student uses should be based on mathematical ideas that the student understands well, including the structure of the base-ten number system, properties of multiplication and division, and number relationships.

A significant amount of instructional time should be devoted to rational numbers in grades 3–5. The focus should be on developing students' conceptual understanding of fractions and decimals—what they are, how they are represented, and how they are related to whole numbers—rather than on developing computational fluency with rational numbers. Fluency in rational-number computation will be a major focus of grades 6–8.

Fluency with whole-number computation depends, in large part, on fluency with basic number combinations—the single-digit addition and multiplication pairs and their counterparts for subtraction and division. Fluency with the basic number combinations develops from well-understood meanings for the four operations and from a focus on thinking strategies (Thornton 1990; Isaacs and Carroll 1999). By working on many multiplication problems with a variety of models for multiplication, students should initially learn and become fluent with some of the "easier" combinations. For example, many students will readily learn basic number combinations such as 3×2 or 4×5 or the squares of numbers, such as 4×4 or 5×5. Through skip-counting, using area models, and relating unknown combinations to known ones, students will learn and become fluent with unfamiliar combinations. For example, 3×4 is the same as 4×3; 6×5 is 5 more than 5×5; 6×8 is double 3×8. Because division is the inverse of multiplication, students can use the multiplication combinations to learn division combinations. For example, $24 \div 6$ can be thought of as $6 \times ? = 24$. If by the end of the fourth grade, students are not able to use multiplication and division strategies efficiently, then they must either develop strategies so that they are fluent with these combinations or memorize the remaining "harder" combinations. Students should also learn to apply the single-digit basic number combinations to related problems, for example, using 5×6 to compute 50×6 or 5000×600.

Research suggests that by solving problems that require calculation, students develop methods for computing and also learn more about operations and properties (McClain, Cobb, and Bowers 1998; Schifter 1999). As students develop methods to solve multidigit computation problems, they should be encouraged to record and share their methods. As they do so, they can learn from one another, analyze the efficiency and generalizability of various approaches, and try one another's methods. In the past, common school practice has been to present a single algorithm for each operation. However, more than one efficient and accurate computational algorithm exists for each arithmetic operation. In addition, if given the opportunity, students naturally invent methods to compute that make sense to them (Fuson forthcoming; Madell 1985). The following episode, drawn from unpublished classroom observation notes, illustrates how one teacher helped students analyze and compare their computational procedures for division:

As students develop methods to solve multidigit computation problems, they should be encouraged to record and share their methods.

> Students in Ms. Spark's fifth-grade class were sharing their solutions to a homework problem, $728 \div 34$. Ms. Sparks asked several students to put their work on the board to be discussed. She deliberately chose students who had approached the problem in several different ways. As the students put their work on the board, Ms. Sparks circulated among the other students, checking their homework.
>
> Henry had written his solution:
>
> 34×10=340
> 34×20=680
>
> 680 + 34 = 114   728 − 714 = 14
>
> Henry explained to the class, "Twenty 34s plus one more is 21. I knew I was pretty close. I didn't think I could add any more 34s, so I subtracted 714 from 728 and got 14. Then I had 21 remainder 14."

Michaela showed her solution:

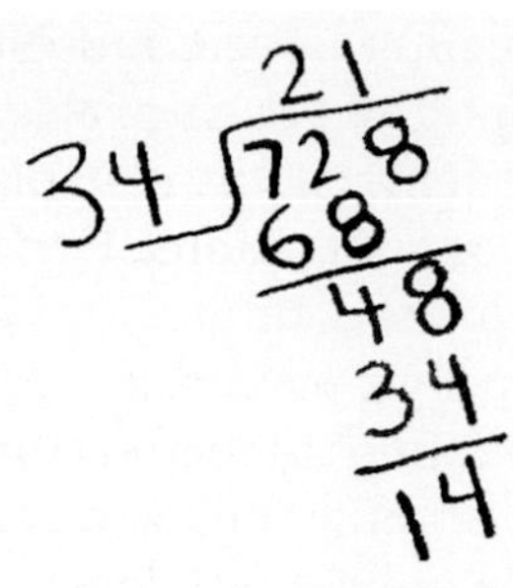

Michaela says, "34 goes into 72 two times and that's 68. You gotta minus that, bring down the 8, then 34 goes into 48 one time."

Ricky: I don't know how to do that.

Michaela: You divide, then you multiply, you subtract, then you bring down.

Ricky: I still don't get it.

Ms. Sparks: Does anyone see any parts of Michaela's and Henry's work that are similar?

Christy: They both did 728 divided by 34.

Ms. Sparks: Right, they both did the same problem. Do you see any parts of the ways they solved the problem that look similar?

Fanshen: (*Hesitantly*) Well, there's a 680 in Henry's and a 68 in Michaela's.

Ms. Sparks: So, what is that 68, Michaela?

Michaela: Um, it's the 2×34.

Ms. Sparks: Oh, is that 2×34? (*Ms. Sparks waits. Lots of silence.*) So, I don't get what you're saying about 2 times 34. What does this 2 up here in the 21 represent?

Samir: It's 20.

Henry: But 20 times 34 is 680, not 68.

Ms. Sparks: So what if I wrote a 0 here to show that this is 680? Does that help you see any more similarities?

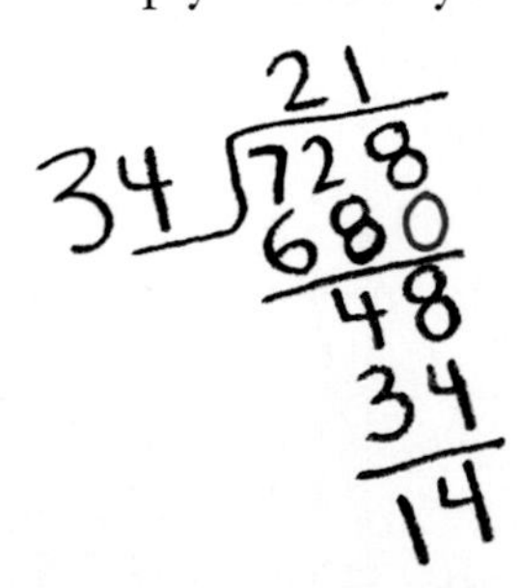

Maya: They both did twenty 34s first.

Rita: I get it. Then Michaela did, like, how many more are left, and it was 48, and then she could do one more 34.

Ms. Sparks saw relationships between the two methods described by students, but she doubted that any of her students would initially see these relationships. Through her questioning, she helped students focus on the ways in which both Michaela's and Henry's methods used multiplication to find the total number of 34s in 728 and helped students clarify what quantities were represented by the notation in Michaela's solution. As the class continues their study of division, Ms. Sparks should encourage this type of explanation and discussion in order to help the students understand, explain, and justify their computational strategies.

As students move from third to fifth grade, they should consolidate and practice a small number of computational algorithms for addition, subtraction, multiplication, and division that they understand well and can use routinely. Many students enter grade 3 with methods for adding and subtracting numbers. In grades 3–5 they should extend these methods to adding and subtracting larger numbers and learn to record their work systematically and clearly. Having access to more than one method for each operation allows students to choose an approach that best fits the numbers in a particular problem. For example, 298×42 can be thought of as $(300 \times 42) - (2 \times 42)$, whereas 41×16 can be computed by multiplying 41×8 to get 328 and then doubling 328 to get 656. Although the expectation is that students develop fluency in computing with whole numbers, frequently they should use calculators to solve complex computations involving large numbers or as part of an extended problem.

Many students are likely to develop and use methods that are not the same as the conventional algorithms (those widely taught in the United States). For example, many students and adults use multiplication to solve division problems or add starting with the largest place rather than with the smallest. The conventional algorithms for multiplication and division should be investigated in grades 3–5 as one efficient way to calculate. Regardless of the particular algorithm used, students should be able to explain their method and should understand that many methods exist. They should also recognize the need to develop efficient and accurate methods.

Estimation serves as an important companion to computation.

As students acquire conceptual grounding related to rational numbers, they should begin to solve problems using strategies they develop or adapt from their whole-number work. At these grades, the emphasis should not be on developing general procedures to solve all decimal and fraction problems. Rather, students should generate solutions that are based on number sense and properties of the operations and that use a variety of models or representations. For example, in a fourth-grade class, students might work on this problem:

> Jamal invited seven of his friends to lunch on Saturday. He thinks that each of the eight people (his seven guests and himself) will eat one and a half sandwiches. How many sandwiches should he make?

Students might draw a picture and count up the number of sandwiches, or they might use reasoning based on their knowledge of number and operations—for example, "That would be eight whole sandwiches and eight half sandwiches; since two halves make a whole sandwich, the eight halves will make four more sandwiches, so Jamal needs to make twelve sandwiches."

Estimation serves as an important companion to computation. It provides a tool for judging the reasonableness of calculator, mental, and paper-and-pencil computations. However, being able to compute exact answers does not automatically lead to an ability to estimate or judge the reasonableness of answers, as Reys and Yang (1998) found in their

The teacher plays an important role in helping students develop and select an appropriate computational tool.

work with sixth and eighth graders. Students in grades 3–5 will need to be encouraged to routinely reflect on the size of an anticipated solution. Will 7 × 18 be smaller or larger than 100? If 3/8 of a cup of sugar is needed for a recipe and the recipe is doubled, will more than or less than one cup of sugar be needed? Instructional attention and frequent modeling by the teacher can help students develop a range of computational estimation strategies including flexible rounding, the use of benchmarks, and front-end strategies. Students should be encouraged to frequently explain their thinking as they estimate. As with exact computation, sharing estimation strategies allows students access to others' thinking and provides many opportunities for rich class discussions.

The teacher plays an important role in helping students develop and select an appropriate computational tool (calculator, paper-and-pencil algorithm, or mental strategy). If a teacher models the choices she makes and thinks aloud about them, students can learn to make good choices. For example, determining the cost of four notebooks priced at $0.75 is an easy mental problem (two notebooks cost $1.50, so four notebooks cost $3.00). Adding the cost of all the school supplies purchased by the class is a problem in which using a calculator makes sense because of the amount of data. Dividing the cost of the class pizza party ($45) by the number of students (25) is an appropriate time to make an estimate (a little less than $2 each) or to use a paper-and-pencil algorithm or a calculator if a more precise answer is needed.

Algebra Standard for Grades 3–5

Instructional programs from prekindergarten through grade 12 should enable all students to—

Expectations

In grades 3–5 all students should–

Understand patterns, relations, and functions	• describe, extend, and make generalizations about geometric and numeric patterns; • represent and analyze patterns and functions, using words, tables, and graphs.
Represent and analyze mathematical situations and structures using algebraic symbols	• identify such properties as commutativity, associativity, and distributivity and use them to compute with whole numbers; • represent the idea of a variable as an unknown quantity using a letter or a symbol; • express mathematical relationships using equations.
Use mathematical models to represent and understand quantitative relationships	• model problem situations with objects and use representations such as graphs, tables, and equations to draw conclusions.
Analyze change in various contexts	• investigate how a change in one variable relates to a change in a second variable; • identify and describe situations with constant or varying rates of change and compare them.

Algebra

Although *algebra* is a word that has not commonly been heard in grades 3–5 classrooms, the mathematical investigations and conversations of students in these grades frequently include elements of algebraic reasoning. These experiences and conversations provide rich contexts for advancing mathematical understanding and are also an important precursor to the more formalized study of algebra in the middle and secondary grades. In grades 3–5, algebraic ideas should emerge and be investigated as students—

- identify or build numerical and geometric patterns;
- describe patterns verbally and represent them with tables or symbols;
- look for and apply relationships between varying quantities to make predictions;
- make and explain generalizations that seem to always work in particular situations;
- use graphs to describe patterns and make predictions;
- explore number properties;
- use invented notation, standard symbols, and variables to express a pattern, generalization, or situation.

Understand patterns, relations, and functions

In grades 3–5, students should investigate numerical and geometric patterns and express them mathematically in words or symbols. They should analyze the structure of the pattern and how it grows or changes, organize this information systematically, and use their analysis to develop generalizations about the mathematical relationships in the pattern. For example, a teacher might ask students to describe patterns they see in the "growing squares" display (see fig. 5.3) and express the patterns in mathematical sentences. Students should be encouraged to explain these patterns verbally and to make predictions about what will happen if the sequence is continued.

Fig. **5.3.**

Expressing "growing squares" in mathematical sentences (Adapted from Burton et al. 1992, p. 6)

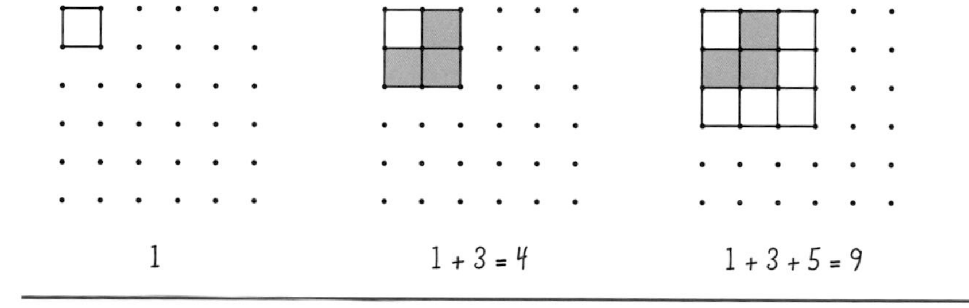

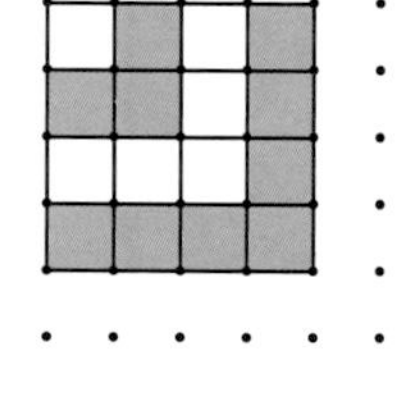

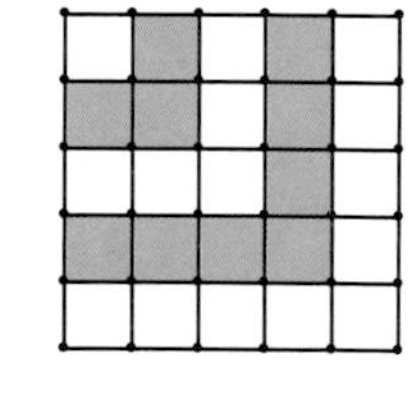

In this example, one student might notice that the area changes in a predictable way—it increases by the next odd number with each new square. Another student might notice that the previous square always fits into the "corner" of the next-larger square. This observation might lead to a description of the area of a square as equal to the area of the previous square plus "its two sides and one more." A student might represent his thinking as in figure 5.4.

Fig. **5.4.**

A possible student observation about the area of the 5 × 5 square in the "growing squares" pattern

Area of a 5×5 square = area of a 4×4 square + 4 + 4 + 1

Fig. **5.5.**

Finding surface areas of towers of cubes

What is the surface area of each tower of cubes (include the bottom)? As the towers get taller, how does the surface area change?

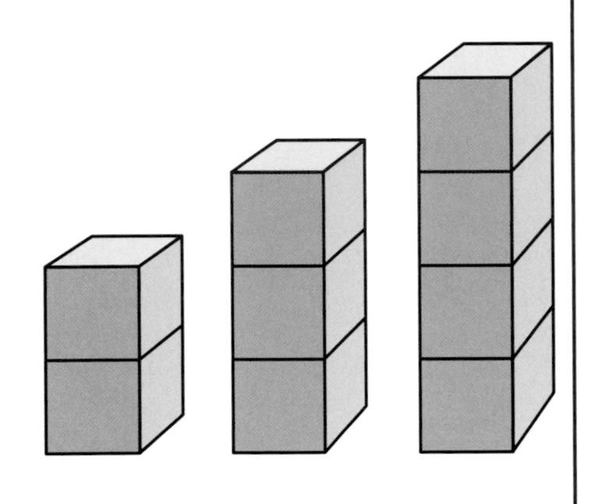

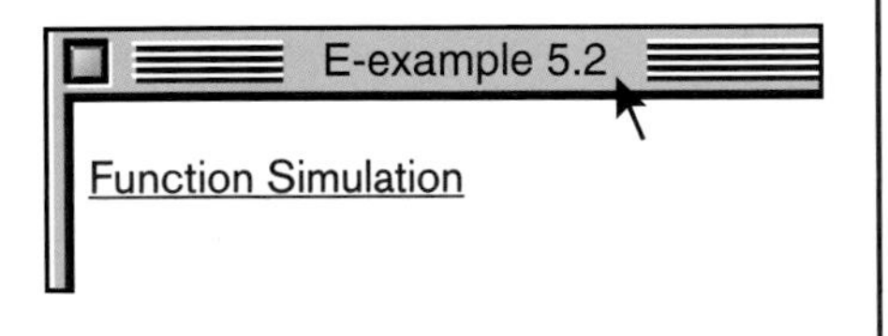

Fig. **5.6.**

A table used in the "tower of cubes" problem

Number of cubes (N)	Surface area in square units (S)
1	6
2	10
3	14
4	18

Examples like this one give the teacher important opportunities to engage students in thinking about how to articulate and express a generalization—"How can we talk about how this pattern works for a square of any size?" Students in grade 3 should be able to predict the next element in a sequence by examining a specific set of examples. By the end of fifth grade, students should be able to make generalizations by reasoning about the structure of the pattern. For example, a fifth-grade student might explain that "if you add the first n odd numbers, the sum is the same as $n \times n$."

As they study ways to measure geometric objects, students will have opportunities to make generalizations based on patterns. For example, consider the problem in figure 5.5. Fourth graders might make a table (see fig. 5.6) and note the iterative nature of the pattern. That is, there is a consistent relationship between the surface area of one tower and the next-bigger tower: "You add four to the previous number." Fifth graders could be challenged to justify a general rule with reference to the geometric model, for example, "The surface area is always four times the number of cubes plus two more because there are always four square units around each cube and one extra on each end of the tower." Once a relationship is established, students should be able to use it to answer questions like, "What is the surface area of a tower with fifty cubes?" or "How many cubes would there be in a tower with a surface area of 242 square units?"

In this example, some students may use a table to organize and order their data, and others may use connecting cubes to model the growth of an arithmetic sequence. Some students may use words, but others may use numbers and symbols to express their ideas about the functional relationship. Students should have many experiences organizing data and examining different representations. Computer simulations are an interactive way to explore functional relationships and the various ways they are represented. In a simulation of two runners along a track, students can control the speed and starting point of the runners and can view the results by watching the race and examining a table and graph of the time-versus-distance relationship. Students need to feel comfortable using various techniques for organizing and expressing ideas about relationships and functions.

Represent and analyze mathematical situations and structures using algebraic symbols

In grades 3–5, students can investigate properties such as commutativity, associativity, and distributivity of multiplication over addition. Is 3×5 the same as 5×3? Is 15×27 equal to 27×15? Will reversing the factors always result in the same product? What if one of the factors is a decimal number (e.g., 1.5×6)? An area model can help students see that two factors in either order have equal products, as represented by congruent rectangles with different orientations (see fig. 5.7).

An area model can also be used to investigate the distributive property. For example, the representation in figure 5.8 shows how 8×14 can be decomposed into 8×10 and 8×4.

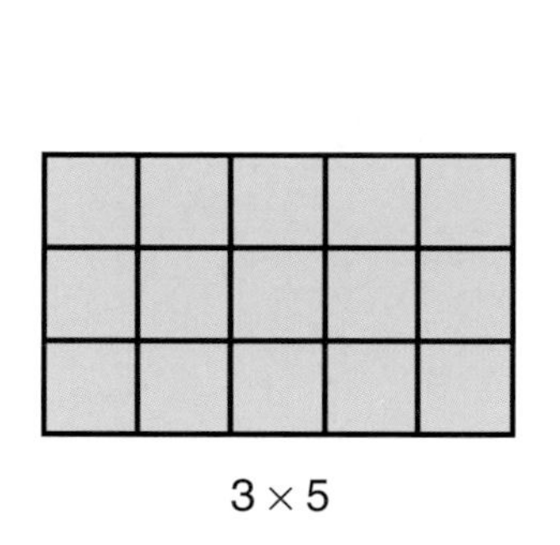

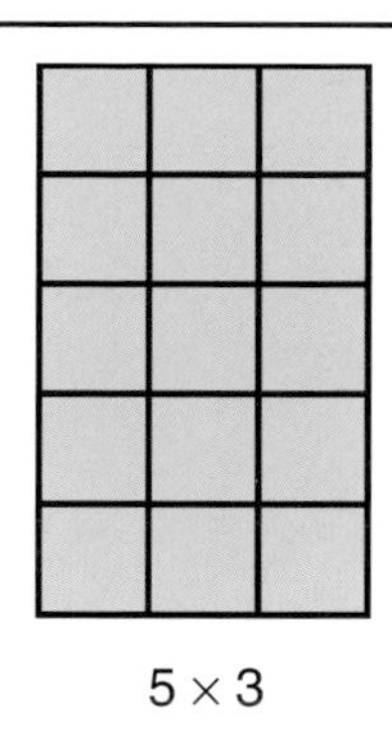

Fig. **5.7.**

Area models illustrating the commutative property of multiplication

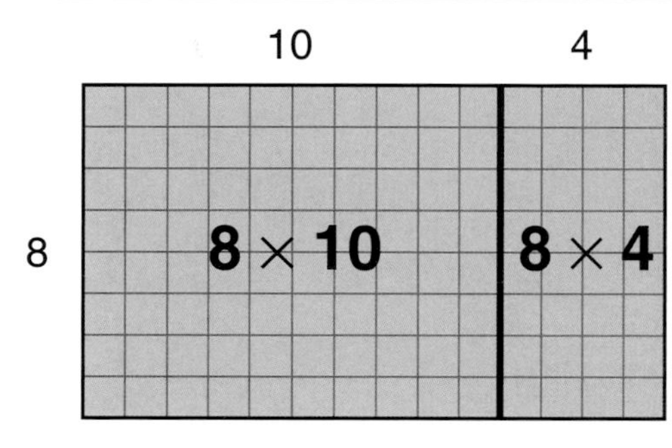

Fig. **5.8.**

Area model showing the distributive property of multiplication

As students learn about the meaning of multiplication and develop strategies to solve multiplication problems, they will begin to use properties such as distributivity naturally (Schifter 1999). However, discussion about the properties themselves, as well as how they serve as tools for solving a range of problems, is important if students are to add strength to their intuitive notions and advance their understanding of multiplicative structures. For example, students might explore questions such as these: Why can't 24×32 be solved by adding the results of 20×30 and 4×2? If a number is tripled, then tripled again, what is the relationship of the result to the original number? Analyzing the properties of operations gives students opportunities to extend their thinking and to build a foundation for applying these understandings to other situations.

At this grade band the idea and usefulness of a variable (represented by a box, letter, or symbol) should also be emerging and developing more fully. As students explore patterns and note relationships, they should be encouraged to represent their thinking. In the example showing the sequence of squares that grow (fig. 5.3), students are beginning to use the idea of a variable as they think about how to describe a rule for finding the area of any square from the pattern they have observed. As students become more experienced in investigating, articulating, and justifying generalizations, they can begin to use variable notation and equations to represent their thinking. Teachers will need to model how to represent thinking in the form of equations. In this way, they can help students connect the ways they are describing their findings to mathematical notation. For example, a student's description of the surface area of a cube tower of any size ("You get the surface area by multiplying the number of cubes by 4 and adding 2") can be recorded by the teacher as $S = 4 \times n + 2$. Students should also understand the use of a variable as a placeholder in an expression or equation. For example, they should explore the role of n in the equation $80 \times 15 = 40 \times n$ and be able to find the value of n that makes the equation true.

Use mathematical models to represent and understand quantitative relationships

Historically, much of the mathematics used today was developed to model real-world situations, with the goal of making predictions about those situations. As patterns are identified, they can be expressed numerically, graphically, or symbolically and used to predict how the pattern will continue. Students in grades 3–5 develop the idea that a mathematical model has both descriptive and predictive power.

Students in these grades can model a variety of situations, including geometric patterns, real-world situations, and scientific experiments. Sometimes they will use their model to predict the next element in a pattern, as students did when they described the area of a square in terms of the previous smaller square (see fig. 5.3). At other times, students will be able to make a general statement about how one variable is related to another variable: If a sandwich costs $3, you can figure out how many dollars any number of sandwiches costs by multiplying that number by 3 (two sandwiches cost $6, three sandwiches cost $9, and so forth). In this case, students have developed a model of a proportional relationship: the value of one variable (total cost, C) is always three times the value of the other (number of sandwiches, S), or $C = 3 \cdot S$.

In modeling situations that involve real-world data, students need to know that their predictions will not always match observed outcomes for a variety of reasons. For example, data often contain measurement error, experiments are influenced by many factors that cause fluctuations, and some models may hold only for a certain range of values. However, predictions based on good models should be reasonably close to what actually happens.

Students in grades 3–5 develop the idea that a mathematical model has both descriptive and predictive power.

Students in grades 3–5 should begin to understand that different models for the same situation can give the same results. For example, as a group of students investigates the relationship between the number of cubes in a tower and its surface area, several models emerge. One student thinks about each side of the tower as having the same number of units of surface area as the number of cubes (n). There are four sides and an extra unit on each end of the tower, so the surface area is four times the number of cubes plus two ($4 \cdot n + 2$). Another student thinks about how much surface area is contributed by *each* cube in the tower: each end cube contributes five units of surface area and each "middle" cube contributes four units of surface area. Algebraically, the surface area would be $2 \cdot 5 + (n - 2) \cdot 4$. For a tower of twelve cubes, the first student thinks, "4 times 12, that's 48, plus 2 is 50." The second student thinks, "The two end cubes each have 5, so that's 10. There are 10 more cubes. They each have 4, so that's 40. 40 plus 10 is 50." Students in this grade band may not be able to show how these solutions are algebraically equivalent, but they can recognize that these different models lead to the same solution.

Analyze change in various contexts

Change is an important mathematical idea that can be studied using the tools of algebra. For example, as part of a science project, students might plant seeds and record the growth of a plant. Using the data represented in the table and graph (fig. 5.9), students can describe how the rate of growth varies over time. For example, a student might express the rate of growth in this way: "My plant didn't grow for the first four days, then it grew slowly for the next two days, then it started to grow faster, then it slowed down again." In this situation, students are focusing not simply on the height of the plant each day, but on what has happened between the recorded heights. This work is a precursor to later, more focused attention on what the slope of a line represents, that is, what the steepness of the line shows about the rate of change. Students should have opportunities to study situations that display different patterns of change—change that occurs at a constant rate, such as someone walking at a constant speed, and rates of change that increase or decrease, as in the growing-plant example.

Fig. **5.9.**

A table and graph showing growth of a plant

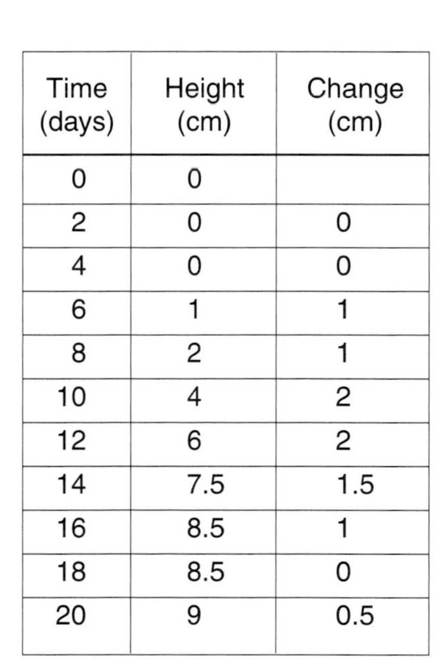

Time (days)	Height (cm)	Change (cm)
0	0	
2	0	0
4	0	0
6	1	1
8	2	1
10	4	2
12	6	2
14	7.5	1.5
16	8.5	1
18	8.5	0
20	9	0.5

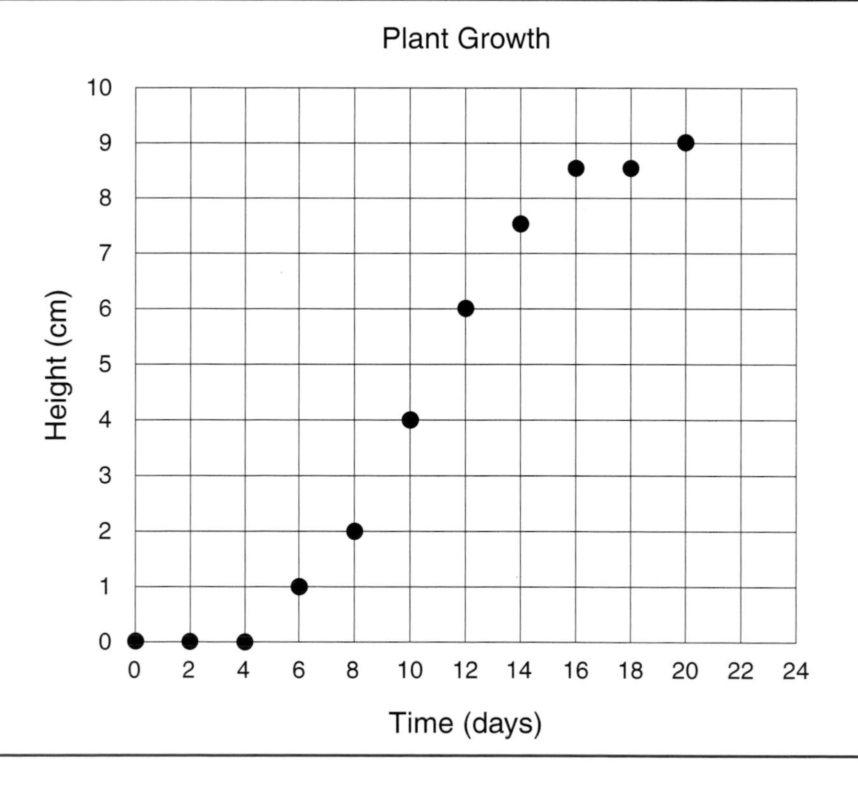

Geometry STANDARD for Grades 3–5

Instructional programs from prekindergarten through grade 12 should enable all students to—	**Expectations** In grades 3–5 all students should–
Analyze characteristics and properties of two- and three-dimensional geometric shapes and develop mathematical arguments about geometric relationships	• identify, compare, and analyze attributes of two- and three-dimensional shapes and develop vocabulary to describe the attributes; • classify two- and three-dimensional shapes according to their properties and develop definitions of classes of shapes such as triangles and pyramids; • investigate, describe, and reason about the results of subdividing, combining, and transforming shapes; • explore congruence and similarity; • make and test conjectures about geometric properties and relationships and develop logical arguments to justify conclusions.
Specify locations and describe spatial relationships using coordinate geometry and other representational systems	• describe location and movement using common language and geometric vocabulary; • make and use coordinate systems to specify locations and to describe paths; • find the distance between points along horizontal and vertical lines of a coordinate system.
Apply transformations and use symmetry to analyze mathematical situations	• predict and describe the results of sliding, flipping, and turning two-dimensional shapes; • describe a motion or a series of motions that will show that two shapes are congruent; • identify and describe line and rotational symmetry in two- and three-dimensional shapes and designs.
Use visualization, spatial reasoning, and geometric modeling to solve problems	• build and draw geometric objects; • create and describe mental images of objects, patterns, and paths; • identify and build a three-dimensional object from two-dimensional representations of that object; • identify and build a two-dimensional representation of a three-dimensional object; • use geometric models to solve problems in other areas of mathematics, such as number and measurement; • recognize geometric ideas and relationships and apply them to other disciplines and to problems that arise in the classroom or in everyday life.

Geometry

The reasoning skills that students develop in grades 3–5 allow them to investigate geometric problems of increasing complexity and to study geometric properties. As they move from grade 3 to grade 5, they should develop clarity and precision in describing the properties of geometric objects and then classifying them by these properties into categories such as rectangle, triangle, pyramid, or prism. They can develop knowledge about how geometric shapes are related to one another and begin to articulate geometric arguments about the properties of these shapes. They should also explore motion, location, and orientation by, for example, creating paths on a coordinate grid or defining a series of flips and turns to demonstrate that two shapes are congruent. As students investigate geometric properties and relationships, their work can be closely connected with other mathematical topics, especially measurement and number.

As students sort, build, draw, model, trace, measure, and construct, their capacity to visualize geometric relationships will develop.

The study of geometry in grades 3–5 requires thinking *and* doing. As students sort, build, draw, model, trace, measure, and construct, their capacity to visualize geometric relationships will develop. At the same time they are learning to reason and to make, test, and justify conjectures about these relationships. This exploration requires access to a variety of tools, such as graph paper, rulers, pattern blocks, geoboards, and geometric solids, and is greatly enhanced by electronic tools that support exploration, such as dynamic geometry software.

Analyze characteristics and properties of two- and three-dimensional geometric shapes and develop mathematical arguments about geometric relationships

In the early grades, students will have classified and sorted geometric objects such as triangles or cylinders by noting general characteristics. In grades 3–5, they should develop more-precise ways to describe shapes, focusing on identifying and describing the shape's properties and learning specialized vocabulary associated with these shapes and properties. To consolidate their ideas, students should draw and construct shapes, compare and discuss their attributes, classify them, and develop and consider definitions on the basis of a shape's properties, such as that a rectangle has four straight sides and four square corners. For example, many students in these grades will easily name the first two shapes in figure 5.10 as rectangles but will need to spend more time discussing why the third one is also a rectangle—indeed, a special kind of rectangle.

In grades 3–5, teachers should emphasize the development of mathematical arguments. As students' ideas about shapes evolve, they should

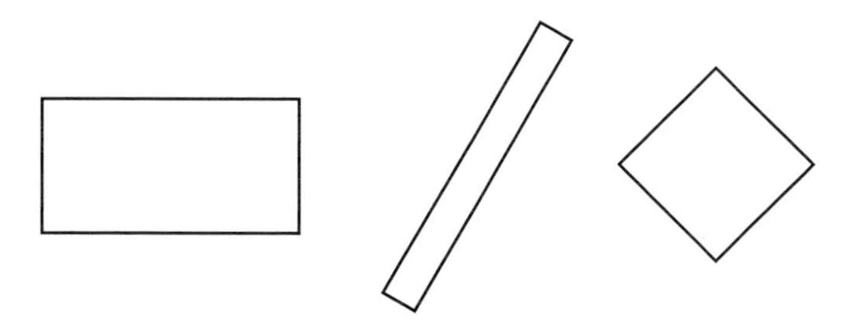

Fig. **5.10.**
Examples of rectangles

Fig. **5.11.**

The relationship between the areas of a rectangle and a nonrectangular parallelogram with equal bases and heights

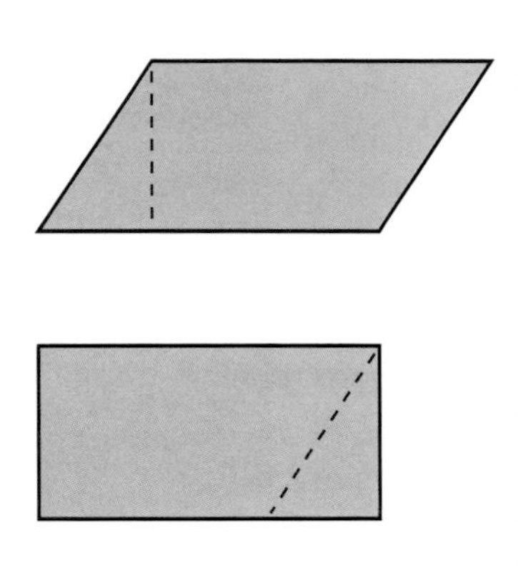

formulate conjectures about geometric properties and relationships. Using drawings, concrete materials, and geometry software to develop and test their ideas, they can articulate clear mathematical arguments about why geometric relationships are true. For example: "You can't possibly make a triangle with two right angles because if you start with one side of the triangle across the bottom, the other two sides go straight up. They're parallel, so they can't possibly ever meet, so you can't get it to be a triangle."

When students subdivide, combine, and transform shapes, they are investigating relationships among shapes. For example, a fourth-grade class might investigate the relationship between a rectangle and a nonrectangular parallelogram with equal bases and heights (see fig. 5.11) by asking, "Does one of these shapes have a larger area than the other?" One student might cut the region formed by the parallelogram as shown in figure 5.11 and then rearrange the pieces so that the parallelogram visually matches the rectangle. This work can lead to developing a general conjecture about the relationship between the areas of rectangles and parallelograms with the same base and height. The notion that shapes that look different can have equal areas is a powerful one that leads eventually to the development of general methods (formulas) for finding the area of a particular shape, such as a parallelogram. In this investigation, students are building their ideas about the properties of classes of shapes, formulating conjectures about geometric relationships, exploring how geometry and measurement are related, and investigating the shapes with equal area.

An understanding of congruence and similarity will develop as students explore shapes that in some way look alike. They should come to understand congruent shapes as those that exactly match and similar shapes as those that are related by "magnifying" or "shrinking." For example, consider the following problem involving the creation of shapes with a particular set of properties:

> Make a triangle with one right angle and two sides of equal length. Can you make more than one triangle with this set of properties? If so, what is the relationship of the triangles to one another?

Fig. **5.12.**

Right triangles with two sides of equal length

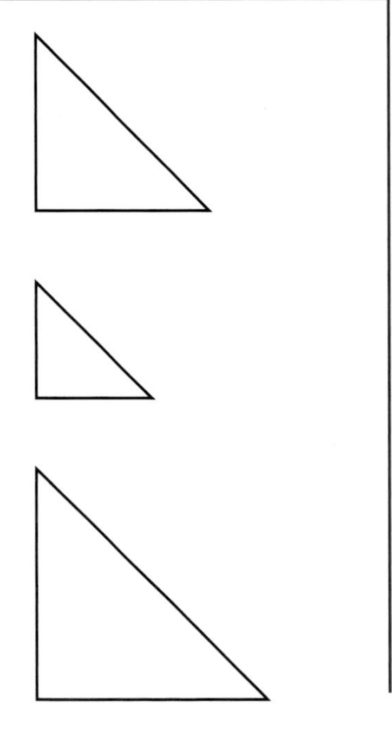

As students make triangles with the stipulated properties (see fig. 5.12), they will see that although these triangles share a common set of characteristics (one right angle and a pair of sides of equal length), they are not all the same size. However, they are all related in that they look alike; that is, one is just a smaller or larger version of the other. The triangles are similar. Although students will not develop a full understanding of similarity until the middle grades, when they focus on proportionality, in grades 3–5 they can begin to think about similarity in terms of figures that are related by the transformations of magnifying or shrinking.

When discussing shapes, students in grades 3–5 should be expanding their mathematical vocabulary by hearing terms used repeatedly in context. As they describe shapes, they should hear, understand, and use mathematical terms such as *parallel*, *perpendicular*, *face*, *edge*, *vertex*, *angle*, *trapezoid*, *prism*, and so forth, to communicate geometric ideas with greater precision. For example, as students develop a more sophisticated understanding of how geometric shapes can be the same or different, the everyday meaning of *same* is no longer sufficient, and they begin to need words such as *congruent* and *similar* to explain their thinking.

Specify locations and describe spatial relationships using coordinate geometry and other representational systems

In grades 3–5, the ideas about location, direction, and distance that were introduced in prekindergarten through grade 2 can be developed further. For instance, students can give directions for moving from one location to another in their classroom, school, or neighborhood; use maps and grids; and learn to locate points, create paths, and measure distances within a coordinate system. Students can first navigate on grids by using landmarks. For example, the map in figure 5.13 can be used to explore questions like these: What is the shortest possible route from the school to the park along the streets (horizontal and vertical lines of the grid)? How do you know? Can there be several different "shortest paths," each of which is equal in length? If so, how many different "shortest paths" are there? What if you need to start at the school, go to the park to pick up your little sister, stop at the store, and visit the library—in what order should you visit these locations to minimize the distance traveled? In this activity, students are using grids and developing fundamental ideas and strategies for navigating them, an important component of discrete mathematics.

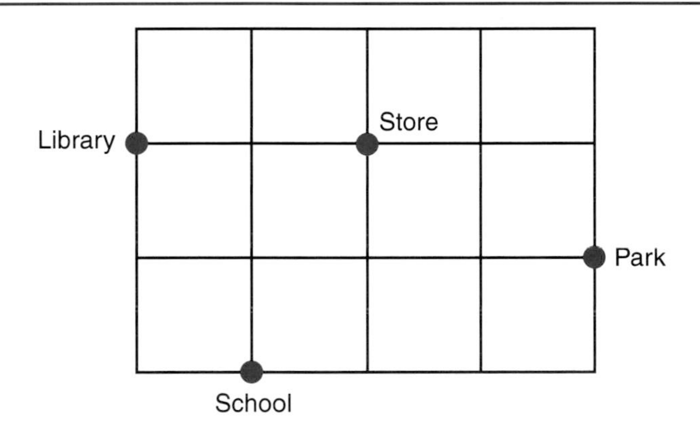

Fig. **5.13.**

A map for exploring questions about navigation

Students at this level also should learn how to use two numbers to name points on a coordinate grid and should realize that a pair of numbers corresponds to a particular point on the grid. Using coordinates, they can specify paths between locations and examine the symmetry, congruence, and similarity of shapes drawn on the grid. They can also explore methods for measuring the distance between locations on the grid. As students' ideas about the number system expand to include negative numbers, they can work in all four quadrants of the Cartesian plane.

Apply transformations and use symmetry to analyze mathematical situations

Students in grades 3–5 should consider three important kinds of transformations: reflections, translations, and rotations (flips, slides, and turns). Younger students generally "prove" (convince themselves) that two shapes are congruent by physically fitting one on top of the other, but students in grades 3–5 can develop greater precision as they describe the motions needed to show congruence ("turn it 90°" or "flip it vertically, then rotate it 180°"). They should also be able to visualize what will happen when a shape is rotated or reflected and predict the result.

Students in grades 3–5 can explore shapes with more than one line of symmetry. For example:

> In how many ways can you place a mirror on a square so that what you see in the mirror looks exactly like the original square? Is this true for all squares?
>
> Can you make a quadrilateral with exactly two lines of symmetry? One line of symmetry? No lines of symmetry? If so, in each case, what kind of quadrilateral is it?

Although younger students often create figures with rotational symmetry with, for example, pattern blocks, they have difficulty describing the regularity they see. In grades 3–5, they should be using language about turns and angles to describe designs such as the one in figure 5.14: "If you turn it 180 degrees about the center, it's exactly the same" or "It would take six equal small turns to get back to where you started, but you can't tell where you started unless you mark it because it looks the same after each small turn."

Fig. **5.14.**

Pattern with rotational symmetry

Use visualization, spatial reasoning, and geometric modeling to solve problems

Students in grades 3–5 should examine the properties of two- and three-dimensional shapes and the relationships among shapes. They should be encouraged to reason about these properties by using spatial relationships. For instance, they might reason about the area of a triangle by visualizing its relationship to a corresponding rectangle or other corresponding parallelogram. In addition to studying physical models of these geometric shapes, they should also develop and use mental images. Students at this age are ready to mentally manipulate shapes, and they can benefit from experiences that challenge them and that can also be verified physically. For example, "Draw a star in the upper right-hand corner of a piece of paper. If you flip the paper horizontally and then turn it 180°, where will the star be?"

Much of the work students do with three-dimensional shapes involves visualization. By representing three-dimensional shapes in two dimensions and constructing three-dimensional shapes from two-dimensional representations, students learn about the characteristics of shapes. For example, in order to determine if the two-dimensional shape in figure 5.15 is a net that can be folded into a cube, students need to pay attention to the number, shape, and relative positions of its faces.

Students should become experienced in using a variety of representations for three-dimensional shapes, for example, making a freehand drawing of a cylinder or cone or constructing a building out of cubes from a set of views (i.e., front, top, and side) like those shown in figure 5.16.

Technology affords additional opportunities for students to expand their spatial reasoning ability. Software such as Logo enables students to draw objects with specified attributes and to test and modify the results. Computer games such as Tetris (Pajithov 1996) can help develop spatial orientation and eye-hand coordination. Dynamic geometry software provides an environment in which students can explore relationships and make and test conjectures.

Students should have the opportunity to apply geometric ideas and relationships to other areas of mathematics, to other disciplines, and to problems that arise from their everyday experiences. There are many ways to make these connections. For example, measurement and geometry are closely linked, as illustrated in the problem in figure 5.11, where geometric properties are used to relate the areas of two figures of different shapes. Geometric models are also important in investigating number relationships. Number lines, arrays, and many manipulatives used for modeling number concepts are geometric realizations of arithmetic relationships. In algebra, students in grades 3–5 often work with geometric problems to explore patterns and functions (see, for example, the "tower of cubes" problem in fig. 5.5).

In addition to its utility in exploring and understanding other areas of mathematics, geometry is closely associated with other subjects, such as art, science, and social studies. For example, students' work on symmetry can enhance their creation and appreciation of art, and their work on coordinate geometry is related to the maps they create or use in their study of the world. The study of geometry promotes a deeper understanding of many aspects of mathematics, improves students' abstract reasoning, and highlights relationships between mathematics and the sciences.

Fig. **5.15.**

A task relating a two-dimensional shape to a three-dimensional shape

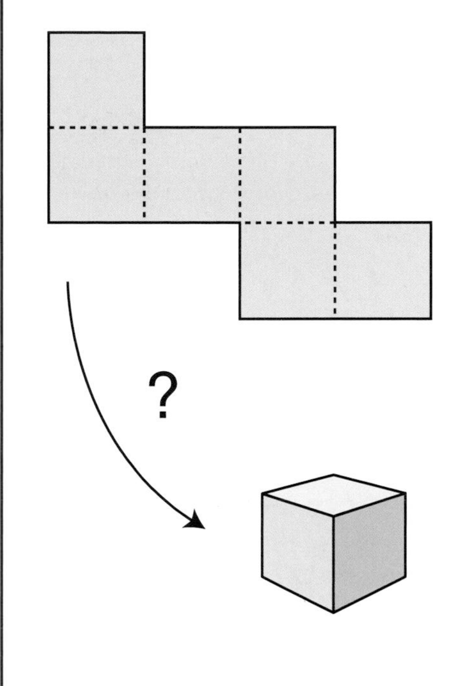

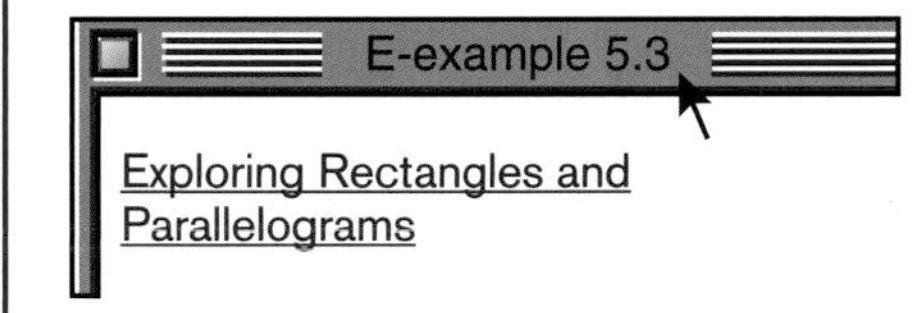

Fig. **5.16.**

Views of a three-dimensional object (Adapted from Battista and Clements 1995, p. 61)

Make a building out of ten cubes by looking at the three pictures of it below.

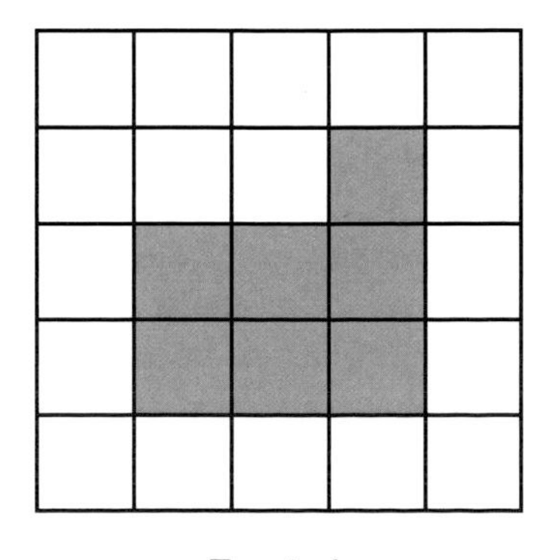

Front view

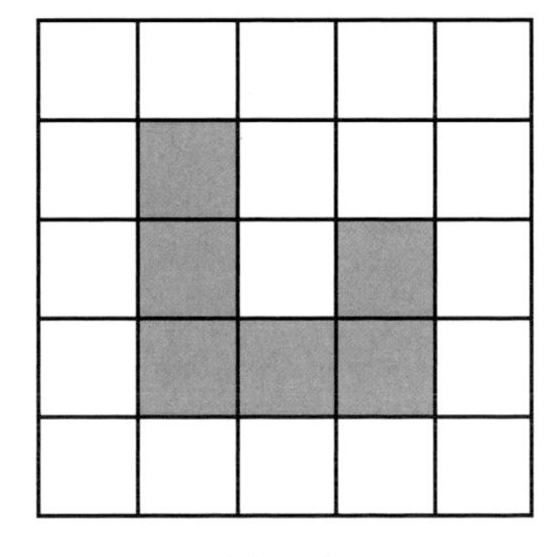

Top view

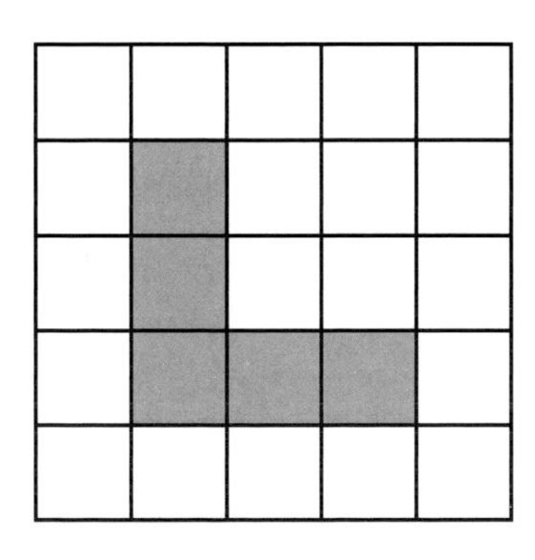

Right side view

Measurement Standard for Grades 3–5

Instructional programs from prekindergarten through grade 12 should enable all students to—

Expectations

In grades 3–5 all students should–

Standard	Expectations
Understand measurable attributes of objects and the units, systems, and processes of measurement	• understand such attributes as length, area, weight, volume, and size of angle and select the appropriate type of unit for measuring each attribute; • understand the need for measuring with standard units and become familiar with standard units in the customary and metric systems; • carry out simple unit conversions, such as from centimeters to meters, within a system of measurement; • understand that measurements are approximations and understand how differences in units affect precision; • explore what happens to measurements of a two-dimensional shape such as its perimeter and area when the shape is changed in some way.
Apply appropriate techniques, tools, and formulas to determine measurements	• develop strategies for estimating the perimeters, areas, and volumes of irregular shapes; • select and apply appropriate standard units and tools to measure length, area, volume, weight, time, temperature, and the size of angles; • select and use benchmarks to estimate measurements; • develop, understand, and use formulas to find the area of rectangles and related triangles and parallelograms; • develop strategies to determine the surface areas and volumes of rectangular solids.

Measurement

Measurement is a process that students in grades 3–5 use every day as they explore questions related to their school or home environment. For example, how much catsup is used in the school cafeteria each day? What is the distance from my house to the school? What is the range of heights of players on the basketball team? Such questions require students to use concepts and tools of measurement to collect data and to describe and quantify their world. In grades 3–5, measurement helps connect ideas within areas of mathematics and between mathematics and other disciplines. It can serve as a context to help students understand important mathematical concepts such as fractions, geometric shapes, and ways of describing data.

Prior to grade 3, students should have begun to develop an understanding of what it means to measure an object, that is, identifying an attribute to be measured, choosing an appropriate unit, and comparing that unit to the object being measured. They should have had many experiences with measuring length and should also have explored ways to measure liquid volume, weight, and time. In grades 3–5, students should deepen and expand their understanding and use of measurement. For example, they should measure other attributes such as area and angle. They need to begin paying closer attention to the degree of accuracy when measuring and use a wider variety of measurement tools. They should also begin to develop and use formulas for the measurement of certain attributes, such as area.

Measurement helps connect ideas within areas of mathematics and between mathematics and other disciplines.

In learning about measurement and learning how to measure, students should be actively involved, drawing on familiar and accessible contexts. For example, students in grades 3–5 should measure objects and space in their classroom or use maps to determine locations and distances around their community. They should determine an appropriate unit and use it to measure the area of their classroom's floor, estimate the time it takes to do various tasks, and measure and represent change in the size of attributes, such as their height.

Understand measurable attributes of objects and the units, systems, and processes of measurement

Students in grades 3–5 should measure the attributes of a variety of physical objects and extend their work to measuring more complex attributes, including area, volume, and angle. They will learn that length measurements in particular contexts are given specific names, such as perimeter, width, height, circumference, and distance. They can begin to establish some benchmarks by which to estimate or judge the size of objects. For example, they learn that a "square corner" is called a *right angle* and establish this as a benchmark for estimating the size of other angles.

Students in grades 3–5 should be able to recognize the need to select units appropriate to the attribute being measured. Different kinds of units are needed for measuring area than for measuring length. At first they might use convenient nonstandard units such as lima beans to estimate area and then come to recognize the need for a standard unit such as a unit square. Likewise, the need for a standard three-dimensional unit to measure volume grows out of initial experiences filling containers with items such as rice or packing pieces. As students find that there are spaces between the units, that the units are not easy to count, or that the units are not of a uniform size, they will appreciate the need for a standard unit.

In these grades, more emphasis should be placed on the standard units that are used to communicate in the United States (the customary units) and around the world (the metric system). Students should become familiar with the common units in these systems and establish mental images or benchmarks for judging and comparing size. For example, they may know that a paper clip weighs about a gram, the width of their forefinger is about a centimeter, or the distance from their elbow to their fingertip is about a foot.

Students in grades 3–5 should encounter the notion that measurements in the real world are approximate.

Students should gain facility in expressing measurements in equivalent forms. They use their knowledge of relationships between units and their understanding of multiplicative situations to make conversions, such as expressing 150 centimeters as 1.5 meters or 3 feet as 36 inches. Since students in the United States encounter two systems of measurement, they should also have convenient referents for comparing units in different systems—for example, 2 centimeters is a little less than an inch, a quart is a little less than a liter, a kilogram is about two pounds. However, they do not need to make formal conversions between the two systems at this level.

Students in grades 3–5 should encounter the notion that measurements in the real world are approximate, in part because of the instruments used and because of human error in reading the scales of these instruments. For example, figure 5.17 describes a measurement task and summarizes results typical of what groups of students obtain. Such an exercise provides a context in which the teacher can raise, and the class can consider, the idea of measurement as an estimation process.

Each pair of students will find slightly different measurements, even though they are measuring the same object using the same kind of measurement tools. The teacher should ask students to discuss the factors that may lead to different measurements. Students' responses will vary according to their experience, but by grade 5 they should recognize factors that affect precision. These include the limitations of the measurement tool, how precisely students read the scale on the measuring

Measure and compare:

Work in pairs and use your rulers to measure the items indicated on the chart. Record your measurements for each object on the chart.

	Objects		
	Height of Teacher's Desk	Circumference of Clockface	Length of Classroom
Jo and Rustin	70 cm	92 cm	8.0 m
Whitney and Beth	68cm	96 cm	7.5m
Ben and Anna	61 cm	91 cm	8.2 m

Fig. **5.17.**

A measurement task and typical student results

instrument (was the scale marked and read in centimeters or millimeters?), and the students' perceived need for accuracy. The discussion might lead to considering the importance of measuring precisely in certain contexts. For instance, carpenters often measure twice and use special instruments in order to minimize the waste of materials, but an estimate might be quite adequate in other instances (e.g., the scout troop hiked about 2.5 miles).

Students in grades 3–5 should explore how measurements are affected when one attribute to be measured is held constant and the other is changed. For example, consider the area of four tiles joined along adjacent sides (see fig. 5.18). The area of each tile is a square unit. When joined, the area of the resulting polygon is always four square units, but the perimeter varies from eight to ten units, depending on how the tiles are arranged. Or suppose students are given twenty toothpicks with which to build a rectangle. How many different rectangles are possible if all twenty toothpicks are used? This activity provides an opportunity to discuss the relationship of area to perimeter. It also highlights the importance of organizing solutions systematically.

Fig. **5.18.**

Polygons with the same area and different perimeters

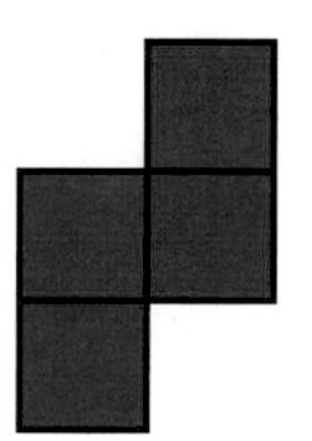

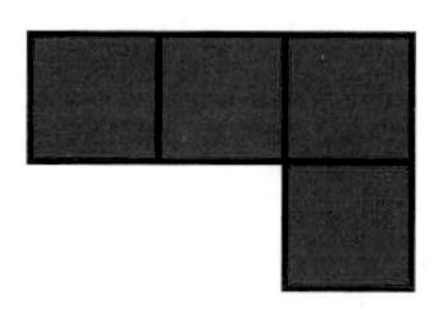

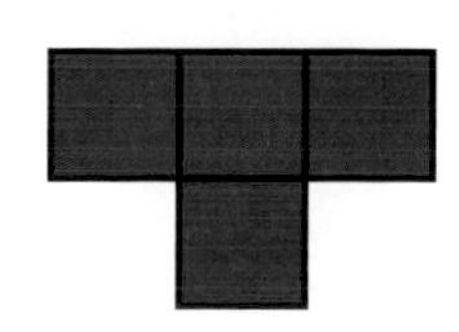

Apply appropriate techniques, tools, and formulas to determine measurements

In grades 3–5, an expanded number of tools and range of measurement techniques should be available to students. When using conventional tools such as rulers and tape measures for measuring length, students will need instruction to learn to use these tools properly. For example, they will need to recognize and understand the markings on a ruler, including where the "0," or beginning point, is located. When standard measurement tools are difficult to use in a particular situation, they must learn to adapt their tools or invent techniques that will work. In the earlier example (fig. 5.17) measuring the circumference of a clock face with a rigid ruler presented a particular challenge. Using string or some other flexible object to outline the clock face and then measuring the string would have been a good strategy. Students should be challenged to develop measurement techniques as needed in order to measure complex figures or objects. For example, they might measure the area of an irregular polygon or a leaf by covering it with transparent grid paper and counting units or by breaking it apart into regular shapes that they can measure.

Students in grades 3–5 should develop strategies to estimate measurements. For example, to estimate the length of the classroom, they might estimate the length of one floor tile and then count the number of tiles across the room and multiply the length by the number of tiles. Another strategy for estimating measurements is to compare the item to be measured against some benchmark. For example, a student might estimate the teacher's height by noting that it is about one and a quarter times the student's own height. This particular strategy highlights the use of multiplicative reasoning, an important indication of advancing understanding.

Strategies for estimating measurements are varied and often depend on the particular situation. By sharing strategies, students can compare and evaluate different approaches. Students also need experience in judging what degree of accuracy is required in a given situation and whether an underestimate or overestimate is more desirable. For example, in estimating the time needed to get up in the morning, eat breakfast, and walk or drive to school, an overestimate makes sense. However, an underestimate of the time needed to cook vegetables on the grill might be considered appropriate, since more time can always be added to the cooking process but not taken away from it.

As students have opportunities to look for patterns in the results of their measurements, they recognize that their methods for measuring the area and volume of particular objects can be generalized as formulas. For example, the table in figure 5.19 is typical of what groups of third graders might produce when using a transparent grid to deter-

Fig. **5.19.**

Measuring the areas of a set of rectangles using a transparent grid

Rectangle	Length (cm)	Width (cm)	Area (cm^2)
A	5	2	10
B	4	3	12
C	1	6	6
D	2	3	6
E	4	4	16

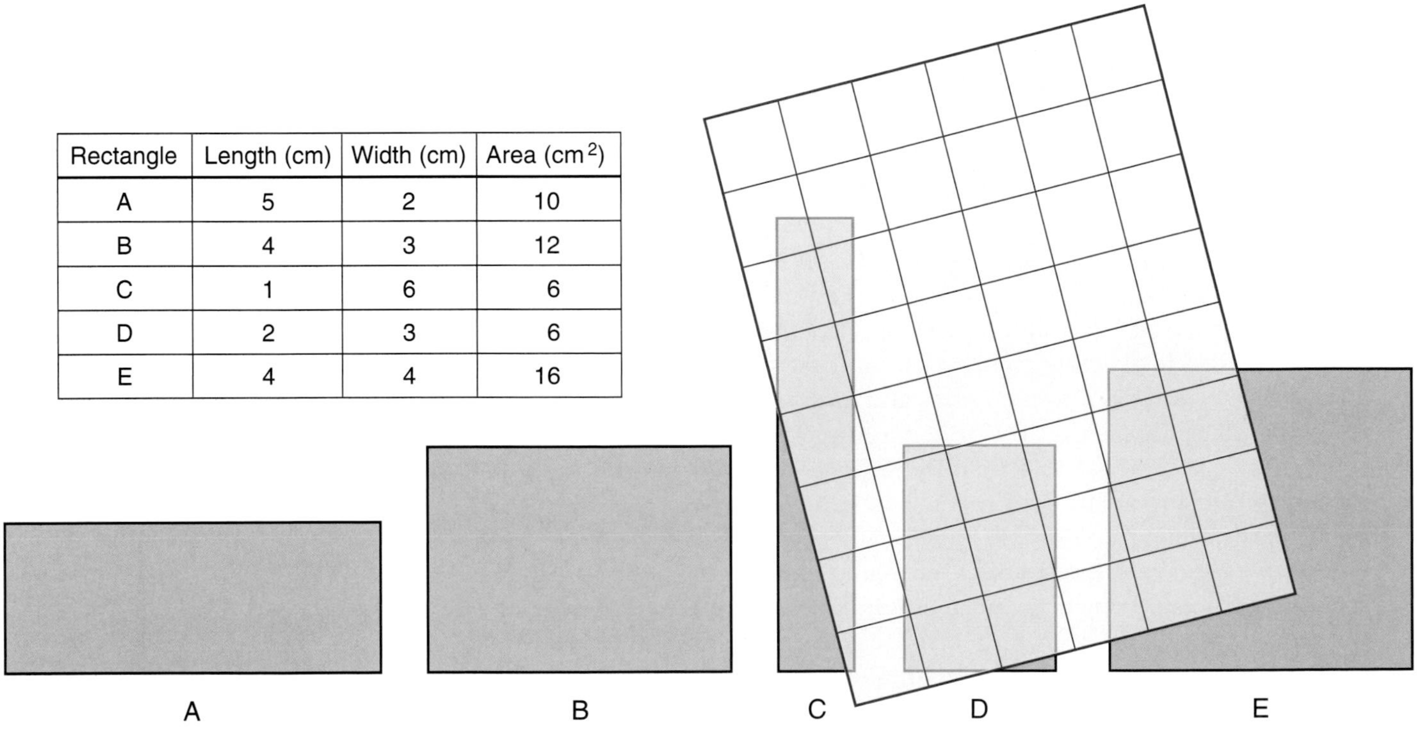

A B C D E

mine the areas of a set of rectangles. As they begin generating the table, they realize that counting all the squares is not necessary once the length (L) and width (W) of the rectangle are determined with the grid. They test their conjecture that Area = $L \times W$, and it appears to work for each rectangle in the set. Later, their teacher challenges them to think about whether and why their formula will work for big rectangles as well as small ones.

Students in grades 3–5 should develop strategies for determining surface area and volume on the basis of concrete experiences. They should measure various rectangular solids using objects such as tiles and cubes, organize the information, look for patterns, and then make generalizations. For example, the "tower of cubes" problem in figure 5.5 highlights the kind of activity that builds from concrete experiences and leads to generalizations, including the development of general formulas for measuring surface area and volume. These concrete experiences are essential in helping students understand the relationship between the measurement of an object and the succinct formula that produces the measurement.

Data Analysis and Probability

Standard for Grades 3–5

Instructional programs from prekindergarten through grade 12 should enable all students to—

Expectations

In grades 3–5 all students should–

Standard	Expectations
Formulate questions that can be addressed with data and collect, organize, and display relevant data to answer them	• design investigations to address a question and consider how data-collection methods affect the nature of the data set; • collect data using observations, surveys, and experiments; • represent data using tables and graphs such as line plots, bar graphs, and line graphs; • recognize the differences in representing categorical and numerical data.
Select and use appropriate statistical methods to analyze data	• describe the shape and important features of a set of data and compare related data sets, with an emphasis on how the data are distributed; • use measures of center, focusing on the median, and understand what each does and does not indicate about the data set; • compare different representations of the same data and evaluate how well each representation shows important aspects of the data.
Develop and evaluate inferences and predictions that are based on data	• propose and justify conclusions and predictions that are based on data and design studies to further investigate the conclusions or predictions.
Understand and apply basic concepts of probability	• describe events as likely or unlikely and discuss the degree of likelihood using such words as *certain*, *equally likely*, and *impossible*; • predict the probability of outcomes of simple experiments and test the predictions; • understand that the measure of the likelihood of an event can be represented by a number from 0 to 1.

Data Analysis and Probability

In prekindergarten through grade 2, students will have learned that data can give them information about aspects of their world. They should know how to organize and represent data sets and be able to notice individual aspects of the data—where their own data are on the graph, for instance, or what value occurs most frequently in the data set. In grades 3–5, students should move toward seeing a set of data as a whole, describing its shape, and using statistical characteristics of the data such as range and measures of center to compare data sets. Much of this work emphasizes the comparison of related data sets. As students learn to describe the similarities and differences between data sets, they will have an opportunity to develop clear descriptions of the data and to formulate conclusions and arguments based on the data. They should consider how the data sets they collect are samples from larger populations and should learn how to use language and symbols to describe simple situations involving probability.

Investigations involving data should happen frequently during grades 3–5. These can range from quick class surveys to projects that take several days. Frequent work with brief surveys (How many brothers and sisters do people in our class have? What's the farthest you have ever been from home?) can acquaint students with particular aspects of collecting, representing, summarizing, comparing, and interpreting data. More extended projects can engage students in a cycle of data analysis—formulating questions, collecting and representing the data, and considering whether their data are giving them the information they need to answer their question. Students in these grades are also becoming more aware of the world beyond themselves and are ready to address some questions that have the potential to influence decisions. For example, one class that studied playground injuries at their school gathered evidence that led to the conclusion that the bars on one piece of playground equipment were too large for the hands of most students below third grade. This finding resulted in a new policy for playground safety.

Investigations involving data should happen frequently during grades 3–5.

Formulate questions that can be addressed with data and collect, organize, and display relevant data to answer them

At these grade levels, students should pose questions about themselves and their environment, issues in their school or community, and content they are studying in different subject areas: How do fourth graders spend their time after school? Do automobiles stop at the stop signs in our neighborhood? How can the amount of water used for common daily activities be decreased? Once a question is posed, students can develop a plan to collect information to address the question. They may collect their own data, use data already collected by their school or town, or use other existing data sets such as the census or weather data accessible on the Internet to examine particular questions. If students collect their own data, they need to decide whether it is appropriate to conduct a survey or to use observations or measurements. As part of their plan, they often need to refine their question and to consider aspects of data collection such as how to word questions, whom to ask, what and when

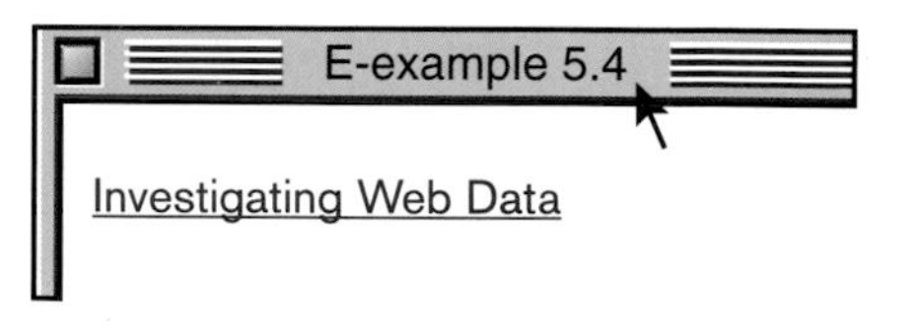

to observe, what and how to measure, and how to record their data. When they use existing data, they still need to consider and evaluate the ways in which the data were collected.

Students should become familiar with a variety of representations such as tables, line plots, bar graphs, and line graphs by creating them, watching their teacher create them, and observing those representations found in their environment (e.g., in newspapers, on cereal boxes, etc.). In order to select and interpret appropriate representations, students in grades 3–5 need to understand the nature of different kinds of data: categorical data (data that can be categorized, such as types of lunch foods) and numerical data (data that can be ordered numerically, such as heights of students in a class). Students should examine classifications of categorical data that produce different views. For example, in a study of which cafeteria foods are eaten and which are thrown out, different classifications of the types of foods may highlight different aspects of the data.

As students construct graphs of ordered numerical data, teachers need to help them understand what the values along the horizontal and vertical axes represent. Using experience with a variety of graphs, teachers should make sure that students encounter and discuss issues such as why the scale on the horizontal axis needs to include values that are not in the data set and how to represent zero on a graph. Students should also use computer software that helps them organize and represent their data, including graphing software and spreadsheets. Spreadsheets allow students to organize and order a large set of data and create a variety of graphs (see fig. 5.20).

Fig. **5.20.**

Spreadsheet with weather data

	A	B	C	D
1	Daily Precipitation and Temperatures for San Francisco, California			
2				
3		Precipitation (inches)	Temperature (°F)	
4	Date		Hi	Low
5	1/1	0.01	58	48
6	1/2	0.88	60	51
7	1/3	0.43	58	50
8	1/4	0.25	56	44
9	1/5	0	51	40
10	1/6	0.25	54	40
11	1/7	0.09	50	47
12	1/8	0	51	47

When students are ready to present their data to an audience, they need to consider aspects of their representations that will help people understand them: the type of representation they choose, the scales used in a graph, and headings and titles. Comparing different representations helps students learn to evaluate how well important aspects of the data are shown.

Select and use appropriate statistical methods to analyze data

In prekindergarten through grade 2, students are often most interested in individual pieces of data, especially their own, or which value is "the most" on a graph. A reasonable objective for upper elementary and middle-grades students is that they begin to regard a set of data as a whole that can be described as a set and compared to other data sets (Konold forthcoming). As students examine a set of ordered numerical data, teachers should help them learn to pay attention to important characteristics of the data set: where data are concentrated or clumped, values for which there are no data, or data points that appear to have unusual values. For example, in figure 5.21 consider the line plot of the heights of fast-growing plants grown in a fourth-grade classroom (adapted from Clement et al. [1997, p. 10]). Students describing these data might mention that the shortest plant measures about 14 centimeters and the tallest plant about 41 centimeters; most of the data are concentrated from 20 to 23 centimeters; and the plant that grew to a height of 41 centimeters is very unusual (an outlier), far removed from the rest of the data. As teachers guide students to focus on the shape of the data and how the data are spread across the range of values, the students should learn statistical terms such as *range* and *outlier* that help them describe the set of data.

Fig. **5.21.**

Plant height data from a fourth-grade class

Plant Height Data	
Height (in cm)	Number of Plants
15	I
20	IIII
22	II
23	𝍸 II
40	I

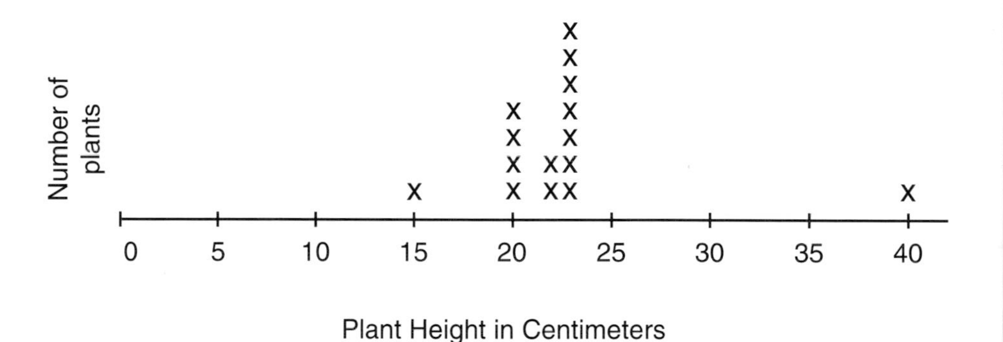

Much of students' work with data in grades 3–5 should involve comparing related data sets. Noting the similarities and differences between two data sets requires students to become more precise in their descriptions of the data. In this context, students gradually develop the idea of a "typical," or average, value. Building on their informal understanding of "the most" and "the middle," students can learn about three measures of center—mode, median, and, informally, the mean. Students need to learn more than simply how to identify the mode or median in a data set. They need to build an understanding of what, for example, the median tells them about the data, and they need to see this value in the context of other characteristics of the data. Figure 5.22 shows the results of plant growth in a third-grade classroom (adapted from Clement et al. [1997, p. 10]). Students should compare the two sets of data from the fourth- and third-grade classrooms. They may note that the median of the fourth-grade data is 23 centimeters and the median of the third-grade data is 28 centimeters. This comparison provides information

Plant Height Data	
Height (in cm)	Number of Plants
9	/
14	\
17	ǀ
22	ǀ
23	ǀ
25	/
26	ǀ
27	ǀ/\
28	ǀ
29	/ǀ
30	/
31	ǀǀ/
32	\
33	ǀǀ
35	\
39	/
40	\

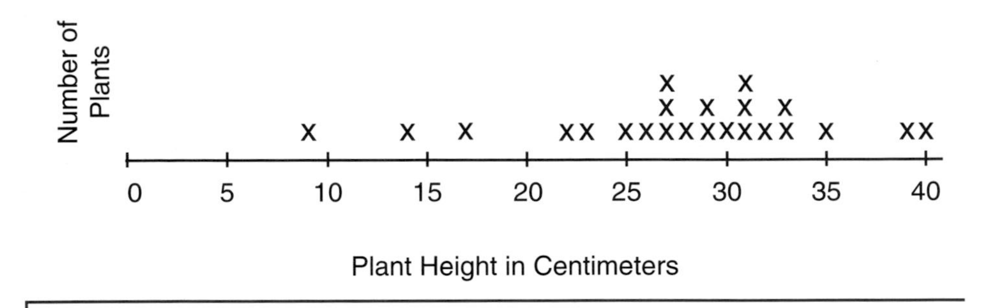

Fig. **5.22.**

Plant height data from a third-grade class

that, overall, the set of third-grade plants grew taller than the set of fourth-grade plants. But it is also important to look at the distributions of the data, which tell an even more dramatic story: Although the ranges of the two data sets are about the same, most of the third graders' plants grew taller than all but a few of the fourth graders' plants.

In grade 5, once students are experienced using the mode and median as part of their data descriptions, they can begin to conceptually explore the role of the mean as a balance point for the data set, using small data sets. The idea of a mean value—what it is, what information it gives about the data, and how it must be interpreted in the context of other characteristics of the data—is a complex one, which will continue to be developed in later grades.

Develop and evaluate inferences and predictions that are based on data

Data can be used for developing arguments that are based on evidence and for continued problem posing. As students discuss data gathered to address a particular question, they should begin to distinguish between what the data show and what might account for the results. For example, a fourth-grade class investigating the sleep patterns of first graders and fifth graders found that first graders were heavier sleepers than fifth graders, as shown in the graphs in figure 5.23 (Russell, Schifter, and Bastable 1999). They had predicted that first graders would be lighter sleepers and were surprised by their results. After describing their data, they developed a hypothesis: First graders have a higher activity level because they play outside more, and this higher activity level leads to deeper sleep. They realized they would need to collect data about a typical day for first and fifth graders in order to investigate their hypothesis. This example demonstrates how students can be encouraged to develop conjectures, show how these are based on the data, consider alternative explanations, and design further studies to examine their conjectures.

With appropriate experiences, students should begin to understand that many data sets are samples of larger populations. They can look at several samples drawn from the same population, such as different classrooms in their school, or compare statistics about their own sample to known parameters for a larger population, for example, how the median family size for their class compares with the median family size reported for their town. They can think about the issues that affect the representativeness of a sample—how well it represents the population

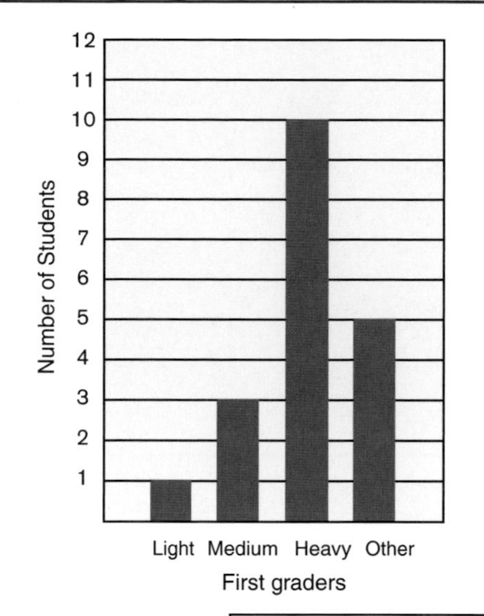

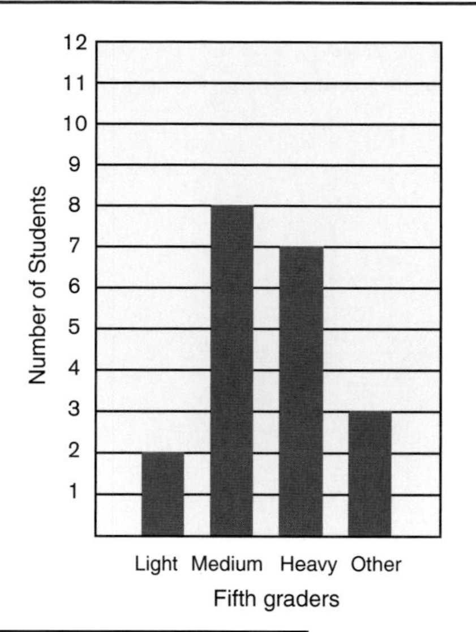

Type of Sleeper	
Light	Wakes up to the slightest noise
Medium	Wakes up to louder noises
Heavy	Sleeps through the night without waking up
Other	None of the previous three

(Russell, Schifter, and Bastable 1999)

Fig. **5.23.**

A student investigation of sleeping habits

from which it is drawn—and begin to notice how samples from the same population can vary.

Understand and apply basic concepts of probability

Students in grades 3–5 should begin to learn about probability as a measurement of the likelihood of events. In previous grades, they will have begun to describe events as certain, likely, or impossible, but now they can begin to learn how to quantify likelihood. For instance, what is the likelihood of seeing a commercial when you turn on the television? To estimate this probability, students could collect data about the number of minutes of commercials in an hour.

Students should also explore probability through experiments that have only a few outcomes, such as using game spinners with certain portions shaded and considering how likely it is that the spinner will land on a particular color. They should come to understand and use 0 to represent the probability of an impossible event and 1 to represent the probability of a certain event, and they should use common fractions to represent the probability of events that are neither certain nor impossible. Through these experiences, students encounter the idea that although they cannot determine an individual outcome, such as which color the spinner will land on next, they can predict the frequency of various outcomes.

Problem Solving Standard for Grades 3–5

Instructional programs from prekindergarten through grade 12 should enable all students to—

Build new mathematical knowledge through problem solving

Solve problems that arise in mathematics and in other contexts

Apply and adapt a variety of appropriate strategies to solve problems

Monitor and reflect on the process of mathematical problem solving

Problem solving is the cornerstone of school mathematics. Without the ability to solve problems, the usefulness and power of mathematical ideas, knowledge, and skills are severely limited. Students who can efficiently and accurately multiply but who cannot identify situations that call for multiplication are not well prepared. Students who can both develop *and* carry out a plan to solve a mathematical problem are exhibiting knowledge that is much deeper and more useful than simply carrying out a computation. Unless students can solve problems, the facts, concepts, and procedures they know are of little use. The goal of school mathematics should be for all students to become increasingly able and willing to engage with and solve problems.

Problem solving is also important because it can serve as a vehicle for learning new mathematical ideas and skills (Schroeder and Lester 1989). A problem-centered approach to teaching mathematics uses interesting and well-selected problems to launch mathematical lessons and engage students. In this way, new ideas, techniques, and mathematical relationships emerge and become the focus of discussion. Good problems can inspire the exploration of important mathematical ideas, nurture persistence, and reinforce the need to understand and use various strategies, mathematical properties, and relationships.

What should problem solving look like in grades 3–5?

Students in grades 3–5 should have frequent experiences with problems that interest, challenge, and engage them in thinking about important mathematics. Problem solving is not a distinct topic, but a process that should permeate the study of mathematics and provide a context in which concepts and skills are learned. For instance, in the following hypothetical example, a teacher poses these questions to her students:

> If you roll two number cubes (both with the numbers 1–6 on their faces) and subtract the smaller number from the larger or subtract one number from the other if they are the same, what are the possible outcomes? If you did this twenty times and created a chart and line plot of the results, what do you think the line plot would look like? Is one particular difference more likely than any other differences?

Initially, the students predict that they will roll as many of one difference as of another. As they begin rolling the cubes and making a list of the differences, some are surprised that the numbers in their lists range only from 0 to 5. They organize their results in a chart and continue to mark the differences they roll (see fig. 5.24). After the students have worked for a few minutes, the teacher calls for a class discussion and asks the students to summarize their results and reflect on their predictions. Some notice that they are getting only a few 0's and 5's but many 1's and 2's. This prompts the class to generate a list of rolls that produce each difference. Others list combinations that produce a difference of 2 and find many possibilities. The teacher helps students express this probability and questions them about the likelihood of rolling other differences, such as 0, 3, and 5.

Fig. **5.24.**

A chart of the frequency of the differences between the numbers on the faces of two dice rolled simultaneously

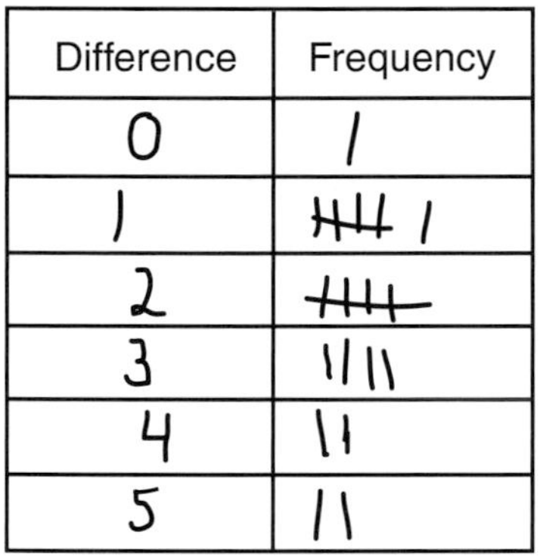

Difference	Frequency
0	𝍷
1	𝍸 𝍷
2	𝍸
3	𝍷𝍷𝍷𝍷
4	𝍷𝍷
5	𝍷𝍷𝍷

The questions posed in this episode were "problems" for the students in that the answers were not immediately obvious. They had to generate and organize information and then evaluate and explain the results. The teacher was able to introduce notions of probability such as predicting and describing the likelihood of an event, and the problem was accessible and engaging for every student. It also provided a context for encouraging students to formulate a new set of questions. For example: Could we create a table that would make it easy to compute the probabilities of each value? Suppose we use a set of number cubes with the numbers 4–9 on the faces. How will the results be similar? How will they be different? What if we change the rules to allow for negative numbers?

Good problems and problem-solving tasks encourage reflection and communication and can emerge from the students' environment or from purely mathematical contexts. They generally serve multiple purposes, such as challenging students to develop and apply strategies, introducing them to new concepts, and providing a context for using skills. They should lead somewhere, mathematically. In the following episode drawn from an unpublished classroom experience, a fourth-grade teacher asked students to work on the following task:

> Show all the rectangular regions you can make using 24 tiles (1-inch squares). You need to use all the tiles. Count and keep a record of the area and perimeter of each rectangle and then look for and describe any relationships you notice.

When the students were ready to discuss their results, the teacher asked if anyone had a rectangle with a length of 1, of 2, of 3, and so on, and modeled a way to organize the information (see fig. 5.25).

The teacher asked if anyone had tried to form a rectangle of length 5 and, if so, what had happened. The students were encouraged work with partners to make observations about the information in the chart and their rectangular models. They noticed that the numbers in the first two columns of any row could be multiplied to get 24 (the area). The teacher noted their observation by writing "$L \times W = 24$" and used the term *factors of 24* as another way, in addition to length and width, to describe the numbers in the first two columns. Some students noticed that as the numbers for one dimension increased, those for the other dimension decreased. Still others noted that the perimeters were always even. One student asked if the rectangles at the bottom of the chart were the same as the ones at the top, just turned different ways. This observation prompted the teacher to remind the students that they had talked about this idea as a property of multiplication—the commutative property—and as congruence of figures.

The teacher then asked the students to describe the rectangles with the greatest and smallest perimeters. They pointed out that the long "skinny" rectangles had greater perimeters than the "fatter" rectangles. The teacher modeled this by taking the 1-unit-by-24-unit rectangle of perimeter 50, splitting it in half, and connecting the halves to form the 2-unit-by-12-unit rectangle (see fig. 5.26). As she moved the tiles, she explained that some tile edges on the outside boundary of the skinny rectangle were moved to the inside of the wider rectangle. Because there were fewer edges on the outside, the perimeter of the rectangle decreased.

Fig. **5.25.**

The dimensions of the rectangular regions made with 24 one-inch square tiles

Length (L) (units)	Width (W) (units)	Area (A) (sq. units)	Perimeter (P) (units)
1	24	24	50
2	12	24	28
3	8	24	22
4	6	24	20
6	4	24	20
8	3	24	22
12	2	24	28
24	1	24	50

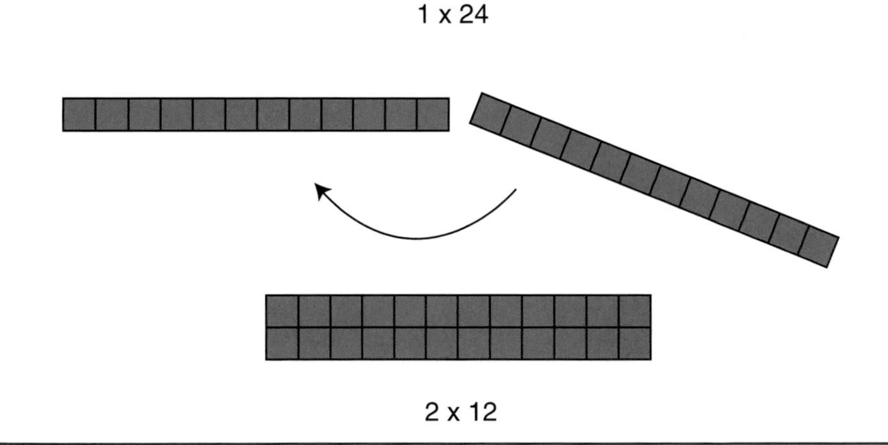

Fig. **5.26.**

Forming a 2×12 rectangle from a 1×24 rectangle

The "24 tiles" problem provides opportunities for students to consider the relationship between area and perimeter, to model the commutative property of multiplication, to use particular vocabulary (*factor* and *multiple*), to record data in an organized way, and to review basic number combinations. It reinforces the relationship $L \times W = A$. It also allows the teacher to help students with different needs focus on different aspects of the problem—building all the rectangles, organizing the data, looking for patterns, or making and justifying conjectures.

Reflecting on different ways of thinking about and representing a problem solution allows comparisons of strategies and consideration of different representations. For example, students might be asked to find several ways to determine the number of dots on the boundary of the square in figure 5.27 and then to represent their solutions as equations (Burns and Mclaughlin 1990).

Fig. **5.27.**

The "dot square" problem

Students will likely see different patterns. Several possibilities are shown in figure 5.28. The teacher should ask each student to relate the drawings to the numbers in their equations. When several different strategies have been presented, the teacher can ask students to examine the various ways of solving the problem and to notice how they are alike and how they are different. This problem offers a natural way to introduce the concept and term *equivalent expressions.*

In addition to developing and using a variety of strategies, students also need to learn how to ask questions that extend problems. In this way, they can be encouraged to follow up on their genuine curiosity about mathematical ideas. For example, the teacher might ask students to create a problem similar to the "dot square" problem or to extend it in some way: If there were a total of 76 dots, how many would be on each side of the square? Could a square be formed with a total of 75 dots? Students could also work with extensions involving dots on the perimeter of other regular polygons. By extending problems and asking different questions, students become problem posers as well as problem solvers.

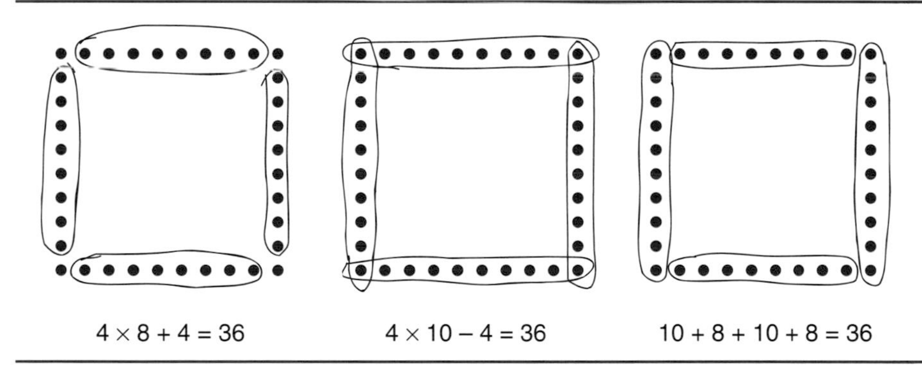

Fig. **5.28.**

Several possible solutions to the "dot square" problem

What should be the teacher's role in developing problem solving in grades 3 through 5?

Teachers can help students become problem solvers by selecting rich and appropriate problems, orchestrating their use, and assessing students' understanding and use of strategies. Students are more likely to develop confidence and self-assurance as problem solvers in classrooms where they play a role in establishing the classroom norms and where everyone's ideas are respected and valued. These attitudes are essential if students are expected to make sense of mathematics and to take intellectual risks by raising questions, formulating conjectures, and offering mathematical arguments. Since good problems challenge students to think, students will often struggle to arrive at solutions. It is the teacher's responsibility to know when students need assistance

and when they are able to continue working productively without help. It is essential that students have time to explore problems. Giving help too soon can deprive them of the opportunity to make mathematical discoveries. Students need to know that a challenging problem will take some time and that perseverance is an important aspect of the problem-solving process and of doing mathematics.

As students share their solutions with classmates, teachers can help them probe various aspects of their strategies. Explanations that are simply procedural descriptions or summaries should give way to mathematical arguments. In this upper elementary class, a teacher questioned two students as they described how they divided nine brownies equally among eight people (Kazemi 1998, pp. 411–12):

Sarah: The first four we cut them in half. (*Jasmine divides squares in half on an overhead transparency.*)

Ms. Carter: Now as you explain, could you explain why you did it in half?

Sarah: Because when you put it in half, it becomes four … four … eight halves.

Ms. Carter: Eight halves. What does that mean if there are eight halves?

Sarah: Then each person gets a half.

Ms. Carter: Okay, that each person gets a half. (*Jasmine labels halves 1 through 8 for each of the eight people.*)

Sarah: Then there were five boxes [brownies] left. We put them in eighths.

Ms. Carter: Okay, so they divided them into eighths. Could you tell us why you chose eighths?

Sarah: It's easiest. Because then everyone will get … each person will get a half and (*addresses Jasmine*) … how many eighths?

Jasmine: (*Quietly*) Five-eighths.

Ms. Carter: I didn't know why you did it in eighths. That's the reason. I just wanted to know why you chose eighths.

Perseverance is an important aspect of the problem-solving process.

Jasmine: We did eighths because then if we did eighths, each person would get each eighth, I mean one-eighth out of each brownie.

Ms. Carter: Okay, one-eighth out of each brownie. Can you just, you don't have to number, but just show us what you mean by that? I heard the words, but…

Jasmine: (*Shades in one-eighth of each of the five brownies that were divided into eighths.*) Person one would get this … (*points to one-eighth*)

Ms. Carter: Oh, out of each brownie.

Sarah: Out of each brownie, one person will get one-eighth.

Ms. Carter: One-eighth. Okay. So how much then did they get if they got their fair share?

Jasmine & Sarah: They got a half and five-eighths.

Ms. Carter: Do you want to write that down at the top, so I can see what you did? (*Jasmine writes 1/2 + 1/8 + 1/8 + 1/8 + 1/8 + 1/8 at the top of the overhead transparency.*)

In this discussion, the teacher pressed students to give reasons for their decisions and actions: What does it mean if there are eight halves? Could you tell us why you chose eighths? Can you show us what you mean by that? She was not satisfied with a simple summary of the steps but instead expected the students to give verbal justifications all along the way and to connect those justifications with both numbers and representations. This particular pair of students used a strategy that was different from that of other students. Although it was not the most efficient strategy, it did reveal that these students could solve a problem they had not encountered before and that they could explain and represent their thinking.

Listening to discussions, the teacher is able to assess students' understanding. In the conversation about sharing brownies, the teacher asked students to justify their responses in order to gain information about their conceptual knowledge. For any assessment of problem solving, teachers must look beyond the answer to the reasoning behind the solution. This evidence can be found in written and oral explanations, drawings, and models. Reflecting on these assessment data, teachers can choose directions for future instruction that fit with their mathematical goals.

Reasoning and Proof STANDARD for Grades 3–5

Instructional programs from prekindergarten through grade 12 should enable all students to—

Recognize reasoning and proof as fundamental aspects of mathematics

Make and investigate mathematical conjectures

Develop and evaluate mathematical arguments and proofs

Select and use various types of reasoning and methods of proof

During grades 3–5, students should be involved in an important transition in their mathematical reasoning. Many students begin this grade band believing that something is true because it has occurred before, because they have seen several examples of it, or because their experience to date seems to confirm it. During these grades, formulating conjectures and assessing them on the basis of evidence should become the norm. Students should learn that several examples are not sufficient to establish the truth of a conjecture and that counterexamples can be used to disprove a conjecture. They should learn that by considering a range of examples, they can reason about the general properties and relationships they find.

Much of the work in these grades should be focused on reasoning about mathematical relationships, such as the structure of a pattern, the similarities and differences between two classes of shapes, or the overall shape of the data represented on a line plot. Students should move from considering *individual* mathematical objects—this triangle, this number, this data point—to thinking about *classes* of objects—all triangles, all numbers that are multiples of 4, a whole set of data. Further, they should be developing descriptions and mathematical statements about relationships between these classes of objects, and they can begin to understand the role of definition in mathematics.

Mathematical reasoning develops in classrooms where students are encouraged to put forth their own ideas for examination. Teachers and students should be open to questions, reactions, and elaborations from others in the classroom. Students need to explain and justify their thinking and learn how to detect fallacies and critique others' thinking. They need to have ample opportunity to apply their reasoning skills and justify their thinking in mathematics discussions. They will need time, many varied and rich experiences, and guidance to develop the ability to construct valid arguments and to evaluate the arguments of others. There is clear evidence that in classrooms where reasoning is emphasized, students do engage in reasoning and, in the process, learn what constitutes acceptable mathematical explanation (Lampert 1990; Yackel and Cobb 1994, 1996).

What should reasoning and proof look like in grades 3 through 5?

In grades 3–5, students should reason about the relationships that apply to the numbers, shapes, or operations they are studying. They need to define the relationship, analyze why it is true, and determine to what group of mathematical objects (numbers, shapes, and operations) it can be applied. Consider the following episode drawn from unpublished classroom observation notes:

> In Ms. Taylor's third-grade class, students were having a discussion of how to compute 4×8. One student, Matt, explained, "I thought of 2×8, that's 16, then you just double it." The teacher asked several students to restate the idea and then asked the class, "Do you think Matt's way of multiplying by 4—by doubling then doubling again—works with problems other than 4×8?" When the response from students was quite mixed, she asked them to try some problems like this themselves before gathering again to discuss Matt's method.

This example shows a teacher taking advantage of an opportunity to engage students in mathematical reasoning. By asking the question "Do you think that always works?" she moved the discussion from the specific problem to a consideration of a general characteristic of multiplication problems—that a factor in a multiplication expression can itself be factored and then the new factors can be multiplied in any order.

> After students had worked on several problems and had discussed with a partner why "doubling then doubling again" was a strategy for multiplying by 4, the teacher reconvened the class for further discussion. Student responses to whether Matt's strategy would always work showed a wide range of thinking:
>
> *Carol:* Because if you have 2 times 8 and 4 times 8, you're doubling the answer. It works every time.
>
> *Malia:* It has to be doubled because you're doing the same thing over again. It's like you did 2 times 8 is 16 and then you did 2 times 8 is 16 again, so it has to be 32.
>
> *Steven:* What you're doing is counting by 8s, so you're counting ahead, you're skipping some of the 8s. You're doing another two of them, so it's like doubling them up.
>
> *Matt:* I tried to see if it would work with triples, so I did 2 times 8 and 6 times 8, and it worked. You times it by 3 and the answer is tripled.

These students' explanations are tied to the specific example, but there is evidence that some students are constructing arguments that may lead to more-general conclusions. Carol is satisfied that "it works every time" but does not have an argument that is based on the structure of multiplication. Malia refers to breaking up one of the factors in the problem into two parts, multiplying the other number by both parts, and then adding the results—the distributive property of multiplication over addition. Steven's explanation is based on modeling multiplication as skip-counting, and Matt takes his original idea further by testing whether multiplying by 6 is the same as multiplying by 2 then by 3. Although none of these third graders' arguments is stated in a way that is complete or general, they are beginning to see what it means to develop and test conjectures about mathematical relationships.

Mathematical reasoning develops in classrooms where students are encouraged to put forth their own ideas for examination.

Fig. **5.29.**

A student's solution to the problem 74×6 involves the distributive property.

74 x 2 = 148
74 x 4 = 296
148 + 296 = 444

Following this discussion, the teacher sent students off to work on a set of multiplication problems. Their work on the problems gave evidence that some of them were applying aspects of the reasoning discussed in the class session. For example, Katherine computed 74×6 by first computing 74×2 and writing the product. Then she doubled the answer to get the solution to 74×4 and added the two products together to get the solution for 74×6 (see fig. 5.29). She was using thinking similar to Malia's, which seemed to involve the distributive property.

During grades 3–5, students should move toward reasoning that depends on relationships and properties. Students need to be challenged with questions such as, What if I gave you twenty more problems like this to do—would they all work the same way? How do you know? Through comparing solutions and questioning one another's reasoning, they can begin to learn to describe relationships that hold across many instances and to develop and defend arguments about why those relationships can be generalized and to what cases they apply (Maher and Martino 1996).

At these grades, students need experiences in learning about what constitutes a convincing argument (Hanna and Yackel forthcoming). For example, in this episode drawn from unpublished classroom observation notes, a third-grade class explored the following problem (adapted from Tierney and Berle-Carman [1995, p. 22]).

> Start with two identical rectangular regions—each the same size. Cut each of the two rectangles in half as shown in figure 5.30. Compare one of the smaller rectangles to one of the right triangles; do they have the same area or does one have a larger area than the other?

Fig. **5.30.**

A rectangle cut into halves in two different ways

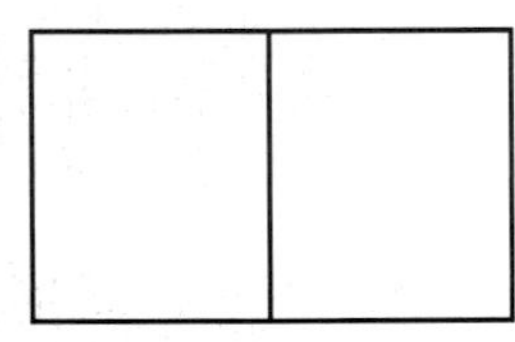

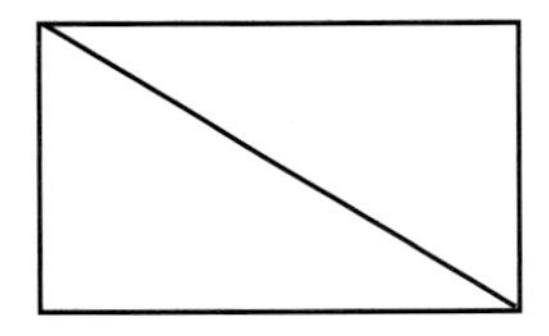

Initially, the students tried to solve the problem by just looking at the figure. For example, they reasoned:

"The triangle is bigger because it goes way up."

"I think they're the same because the triangle's taller, but the rectangle's longer."

As the students worked on this problem, some were convinced that they could decide if the areas were equal (or not) by whether or not they could cut the triangle into a set of haphazard pieces and fit them on the rectangle so that they cover the space (see fig. 5.31a). Others thought about how to organize the cutting and pasting by, for example, cutting the triangle into two pieces to make it into a rectangle that matches the other rectangles (see fig. 5.31b).

Still others developed ways to reason about the relationships in the figure without cutting and pasting. For example: "We folded each paper in half and each paper was the same size to begin with, so the half that's a rectangle is the same as the half that's a triangle."

Fig. **5.31.**

Students' attempts to demonstrate that a triangle has the same area as a rectangle

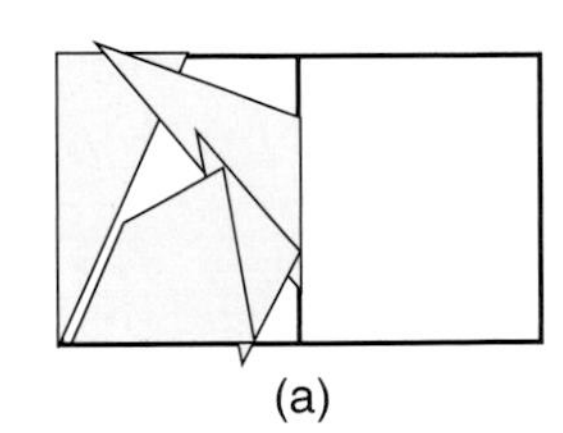

(a)

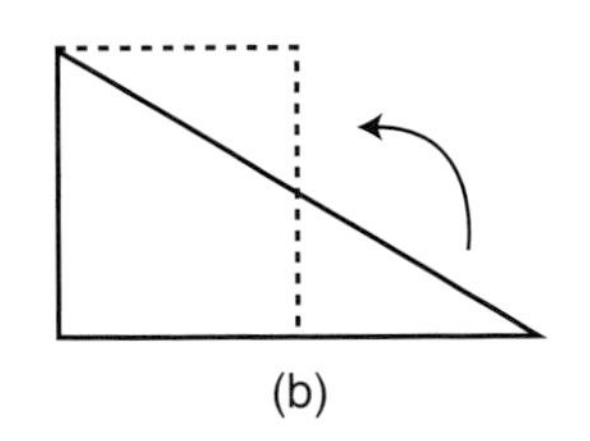

(b)

At this grade level, many students are just beginning to develop an idea about what constitutes a convincing argument. The first solution—cutting and pasting in a disorganized way—does not make use of the properties of the two shapes and therefore may not convince the student doing the cutting and pasting or other students that the areas are equal. The second solution takes into consideration geometric relationships between the particular triangle and rectangle and therefore may be more convincing. The third solution is the beginning of a logical analysis of the relationships among the shapes—that halves of equal areas must be equal to each other.

Students in grades 3–5 should frequently make conjectures about mathematical relationships, investigate those conjectures, and make mathematical arguments that are based on their work. They need to know that posing conjectures and trying to justify them is an expected part of students' mathematical activity. Justification will have a range of meanings for students in grades 3–5, but as they progress through these grades and have more experiences with making mathematical arguments, they should increasingly base their arguments on an analysis of properties, structures, and relationships.

Posing conjectures and trying to justify them is an expected part of students' mathematical activity.

Sometimes students' conjectures about mathematical properties and relationships will turn out to be wrong. Part of mathematical reasoning is examining and trying to understand why something that looks and seems as if it might be true is not and to begin to use counterexamples in this context. Coming up with ideas that turn out *not* to be true is part of the endeavor. These "wrong" ideas often are opportunities for important mathematical discussions and discoveries. For example, a student might propose that if both the numerator and denominator of a fraction are larger than the numerator and denominator, respectively, of another fraction, then the first fraction must be larger. This rule works in comparing 3/4 with 1/2 or 6/4 with 2/3. However, when thinking about this conjecture more carefully, students will find counterexamples—for example, 3/4 is not larger than 2/2 and 2/6 is smaller than 1/2.

What should be the teacher's role in developing reasoning and proof in grades 3 through 5?

In order for mathematical experiences such as those described in this section to happen frequently, the teacher must establish the expectation that the class as a mathematical community is continually developing, testing, and applying conjectures about mathematical relationships. In the episode in Ms. Taylor's third-grade classroom, where students explored the effects of multiplying by 2 and by 2 again, the teacher looked for an opportunity to go beyond finding the solution to an individual problem to focus on more-general mathematical structures and relationships. In this way, she helped her students recognize reasoning as a central part of mathematical activity.

Part of the teacher's role in making reasoning central is to make all students responsible both for articulating their own reasoning and for working hard to understand the reasoning of others, as shown in the following episode, drawn from unpublished classroom observation notes.

> In a fourth-grade classroom, the students were ordering fractions. To begin this activity, the teacher had asked them to identify fractions that are more than 1/2 and less than 1. After the students talked in pairs, the teacher asked how they were choosing their fractions:
>
> *Patrize:* We were talking about how you could get it, and if you make the top number, the numerator, higher than a half of the denominator, but you don't make it the same as the denominator like 5/5 'cause then it will be a whole.

> *Teacher:* It sounds like you have a conjecture. Can someone else explain it?
>
> *Justin:* Like if you have 3/4, half of 4 is 2, so you want the number higher than 2 but not 4.

By routinely questioning students in this way, the teacher is establishing the expectation that students listen carefully to one another's ideas and try to understand them.

Teachers should look for opportunities for students to revise, expand, and update generalizations they have made.

The teacher should continually remind students of conjectures and mathematical arguments that they have developed as part of the shared classroom experience and that can be applied to further work. Teachers should look for opportunities for students to revise, expand, and update generalizations they have made as they develop new mathematical skills and knowledge. Matt's idea about tripling in Ms. Taylor's third-grade class could provide the basis for students to reason about a larger class of problems. Even students who seem to have developed a clear argument about a mathematical relationship need to be questioned and challenged when they are ready to encounter new aspects of the relationship. For example, a class of third graders had spent a great deal of time working with arrays in their study of multiplication. As a group, they were very sure that multiplication was commutative, and they could demonstrate this property using an area model. In the fourth grade, they began encountering larger numbers; when the teacher noticed that some students were using commutativity, she asked the class what they knew about it. At first they seemed certain that multiplication is commutative in all cases, but when she pressed, "But would it work for any numbers? How about 43 279 times 6 892?" they lost their confidence. They could no longer use physical models to show commutativity with such large numbers, and they needed further work to develop mental images and mathematical arguments based on what they had learned from the physical models. It is likely that these students will also need to revisit commutativity when they study computation with fractions and decimals.

The teacher will also have to make decisions about which conjectures are mathematically significant for students to pursue. To do this, the teacher must take into account the skills, needs, and understandings of the students and the mathematical goals for the class.

Communication Standard for Grades 3–5

Instructional programs from prekindergarten through grade 12 should enable all students to—

Organize and consolidate their mathematical thinking through communication

Communicate their mathematical thinking coherently and clearly to peers, teachers, and others

Analyze and evaluate the mathematical thinking and strategies of others

Use the language of mathematics to express mathematical ideas precisely

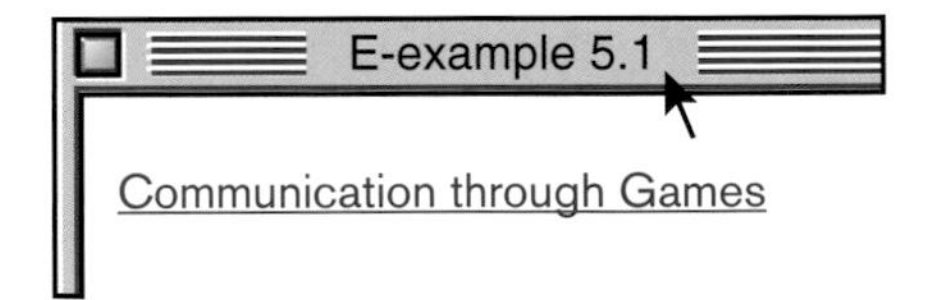

The ability to read, write, listen, think, and communicate about problems will develop and deepen students' understanding of mathematics. In grades 3–5, students should use communication as a tool for understanding and generating solution strategies. Their writing should be more coherent than in earlier grades, and their increasing mathematical vocabulary can be used along with everyday language to explain concepts. Depending on the purpose for writing, such as taking notes or writing to explain an answer, students' descriptions of problem-solving strategies and reasoning should become more detailed and coherent.

In grades 3–5, students should become more adept at learning from, and working with, others. Their communication can consist not only of conversations between student and teacher or one student and another student but also of students listening to a number of peers and joining group discussions in order to clarify, question, and extend conjectures. In classroom discussions, students should become the audience for one another's comments. This involves speaking to one another in order to convince or question peers. The discourse should not be a goal in itself but rather should be focused on making sense of mathematical ideas and using them effectively in modeling and solving problems. The value of mathematical discussions is determined by whether students are learning as they participate in them (Lampert and Cobb forthcoming).

What should communication look like in grades 3 through 5?

In a grades 3–5 classroom, communication should include sharing thinking, asking questions, and explaining and justifying ideas. It should be well integrated in the classroom environment. Students should be encouraged to express and write about their mathematical conjectures, questions, and solutions. For example, after preparatory work in decimals, a fifth-grade teacher engaged her students in the following problem in order to help them think about and develop methods for adding decimals (episode adapted from Schifter, Bastable, and Russell [1999, pp. 114–20]).

> Pretend you are a jeweler. Sometimes people come in to get rings resized. When you cut down a ring to make it smaller, you keep the small portion

of gold in exchange for the work you have done. Recently you have collected these amounts:

1.14 g .089 g .3 g

Now you have a repair job to do for which you need some gold. You are wondering if you have enough. Work together with your group to figure out how much gold you have collected. Be prepared to show the class your solution. (P. 114)

In this activity, the teacher presented the students with a problem-solving situation. Although they had worked with representing decimals, they had not discussed adding them. As was customary in the class, the students were expected to talk with their peers to solve the problem and to share their results and thinking with the class. The students used communication as a natural and essential part of the problem-solving process. As the groups worked, the teacher circulated among the students:

Nikki: We could line the numbers up on the right like you do with other numbers.

Ned: Maybe we should line up the decimals, but I don't know why we would do that.

Teacher: I think you're suggesting that you might line this problem up differently from the way you line up whole-number addition. Is that right?

Ned: (*Nods.*)

Teacher: Why do you line whole numbers up the way you do? What's the reason for it?

Ned: I don't know. It's just the way you do it. That's how we learned to do it.

Malik: I think it would help if we drew a picture, like of the [base-ten] blocks. Maybe we could figure it out then. (P. 114)

The teacher moved to another group where the students had represented their problem as shown in figure 5.32.

Teacher: What happened to the decimal numbers?

Jaron: We just decided to drop the decimals and add the numbers like usual. That way we could line them up on the right and add. We left the zero in there, but you can just leave it out since it doesn't mean anything.

Teacher: Do you all agree?

Johanna & Jerry: Yes.

Teacher: Are you saying, then, that if you start out with 1 and 14 hundredths grams of gold and some other little bits that it adds up to 206 grams of gold? (P. 115)

Listening carefully to the discussions, the teacher rephrased Ned's suggestion in order to make sure she had accurately captured his thinking, to help him focus on the important mathematical concepts, and to guide him in considering how this problem is related to those more

Fig. **5.32.**

Jaron's group's incorrect solution to 1.14 g + .089 g + .3 g

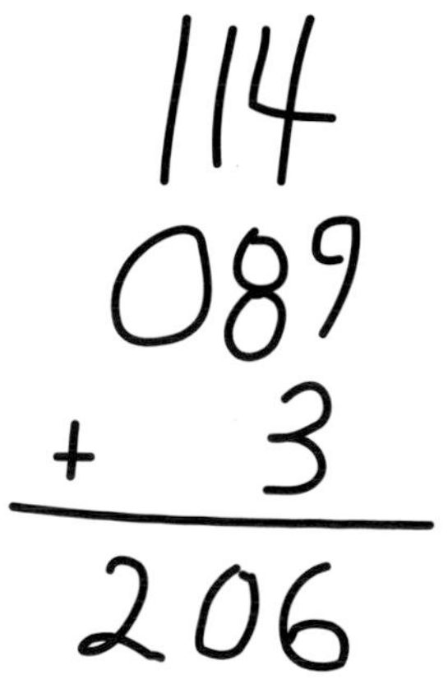

```
  114
  089
+   3
-----
  206
```

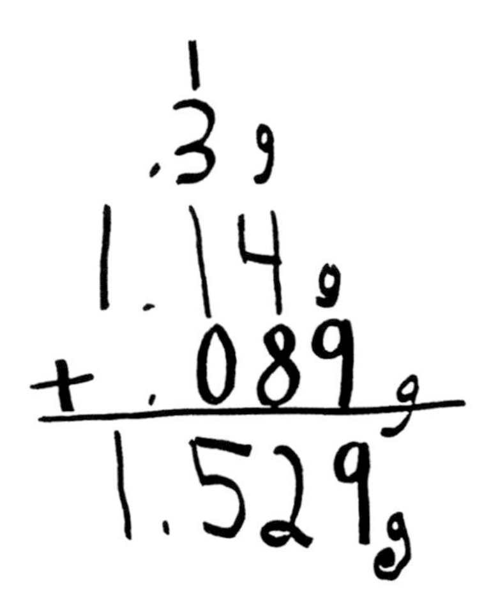

Fig. **5.33.**

Rob's group's solution to 1.14 g + .089 g + .3 g

familiar to him. Ned's response gave her important assessment information about whether he understood his method for adding whole numbers. Although he was able to use an algorithm to add whole numbers, he lacked an understanding of the concepts behind the procedure and therefore was unsure if or how it could be used or adapted for this new purpose.

In talking with Jaron's group, the teacher asked a question that led students to think about the reasonableness of their response by considering it in relation to its real-world context. The realization that their response didn't make sense caused the students to revisit the problem. In this particular instance, the teacher chose to let students work through their confusion. The teacher's decisions about what to say or not say, what to ask or not ask, were based on her observations of the students and their conversations. For example, What strategies were they using? Were misconceptions being challenged? Her goal was to nudge the students to reflect on their answer and to do further mathematical reasoning.

> After the groups finished their work, the class as a whole had a discussion. Rob reported that the students in his group represented the problem as shown in figure 5.33 (p. 116).
>
> Ned immediately asked why they had decided to line up the numbers that way, and Rob responded that the group thought they needed to line up the tenths with the tenths and the hundredths with the hundredths to "make it come out right." Jaron speculated that it was possible to drop the zero in .089, since "it doesn't stand for anything." Teresa jumped in the conversation by stating, "You can't just drop that zero. It has to be there or you get 89 hundredths instead of 89 thousandths, and they're not the same at all." Malik continued to push for a model, but he was stumped. "If I had the flats be one whole, then the rods are tenths and the units are hundredths, but I don't know how to draw the thousandths except as dots. Then I can't really tell what's going on." Another student, Ben, suggested that the block be one whole, so a flat could be tenths, the rod could be hundredths, and the unit could be thousandths. He and several other students drew and presented a picture to illustrate their thinking (see fig. 5.34).

Fig. **5.34.**

Ben's group's solution to 1.14 g + .089 g + .3 g

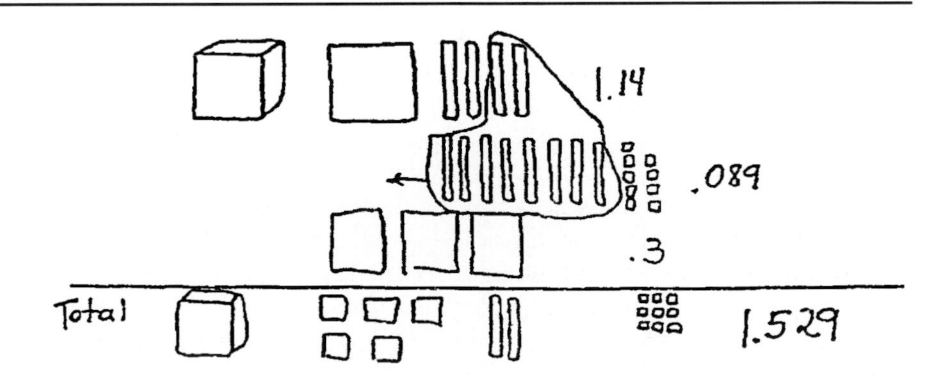

> Malik paid close attention during Ben's explanation, nodding that he understood. Teresa was also interested in the explanation, noting the significance of how zero was represented. After this presentation, the

students wrote in their journals, explaining what they thought was a correct procedure for adding the numbers. Many mentioned that the demonstration had made it clear that tenths had to be added to tenths and hundredths to hundredths for the right answer. Some made up new problems and made drawings of the base-ten model.

Because discussion of thinking was a regular occurrence in this classroom, students were comfortable describing their thinking, even if their ideas were different from the ideas of their peers. Besides focusing on their own thinking, students also attempted to understand the thinking of others and in some cases to relate it to their own. Ned, who earlier had been unable to articulate why he lined up whole numbers in a particular way when he added, questioned Rob about why his group had lined up the numbers the way they did. Ned was taking responsibility for his learning by asking questions about a concept that wasn't quite clear to him. Ben thought about Malik's dilemma and came up with a solution that became clear to Malik.

The use of models and pictures provides a further opportunity for understanding and conversation. Having a concrete referent helps students develop understandings that are clearer and more easily shared (Hiebert et al. 1997). The talk that preceded, accompanied, and followed Ben's presentation gave meaning to the base-ten model. Malik had been "stuck" by viewing the model in one way until Ben showed him another way to look at it.

Throughout the lesson, the interactions among students were necessary in helping them make sense of what they were doing. Because there was time to talk, write, model, and draw pictures, as well as occasions for work in small groups, large groups, and as individuals, students who worked best in different ways all had opportunities to learn.

Teachers may need to explicitly discuss students' effective and ineffective communication strategies.

What should be the teacher's role in developing communication in grades 3 through 5?

With appropriate support and a classroom environment where communication about mathematics is expected, teachers can work to build the capacity of students to think, reason, solve complex problems, and communicate mathematically. This involves creating classroom environments in which intellectual risks and sense making are expected. Teachers must also routinely provide students with rich problems centered on the important mathematical ideas in the curriculum so that students are working with situations worthy of their conversation and thought. In daily lessons, teachers must make on-the-spot decisions about which points of the mathematical conversation to pick up on and which to let go, and when to let students struggle with an issue and when to give direction. For example, the teacher in the episode above chose to let one group of students struggle with the fact that their answer was unreasonable. Teachers must refine their listening, questioning, and paraphrasing techniques, both to direct the flow of mathematical learning and to provide models for student dialogue. Well-posed questions can simultaneously elicit, extend, and challenge students' thinking and at the same time give the teacher an opportunity to assess the students' understanding.

Periodically, teachers may need to explicitly discuss students' effective and ineffective communication strategies. Teachers can model questioning and explaining, for example, and then point out and explain those techniques to their students. They can also highlight examples of good communication among students. ("I noticed that Karen and Malia disagreed on an answer. They not only explained their reasoning to each other very carefully, but they listened to one another. Each understood the other's reasoning. It was hard, but eventually, they realized that one way made more sense than the other.")

Teachers need to help students learn to ask questions when they disagree or do not understand a classmate's reasoning. It is important that students understand that the focus is not on who is right or wrong but rather on whether an answer makes sense and can be justified. Students need to learn that mathematical arguments are logical and connected to mathematical relationships. When making a concept or strategy clear to a peer, the student-explainer is forced to reexamine and thus deepen his or her mathematical understanding. In settings where communication strategies are taught, modeled, and expected, students will eventually begin to adopt listening, paraphrasing, and questioning techniques in their own mathematical conversations.

Teachers must help students acquire mathematical language to describe objects and relationships. For example, as students use informal language such as "the corner-to-corner lines" to describe the diagonals of a rectangle, the teacher should point out the mathematical term given to these lines. Specialized vocabulary is much more meaningful if it is introduced in an appropriate context. Teachers in grades 3–5 should look for, and take advantage of, such opportunities to introduce mathematical terms. In this way, words such as *equation*, *variable*, *perpendicular*, *product*, and *factor* should become part of students' normal vocabulary.

Teachers also need to provide students with assistance in writing about mathematical concepts. They should expect students' writing to be correct, complete, coherent, and clear. Especially in the beginning,

Teachers must help students acquire mathematical language.

teachers need to send writing back for revision. Students will also need opportunities to check the clarity of their work with peers. Initially, when they have difficulty knowing what to write about in mathematics class, the teacher might ask them to use words, drawings, and symbols to explain a particular mathematical idea. For example, students could write about how they know that 1/2 is greater than 2/5 and show at least three different ways to justify this conclusion. To help students write about their reasoning processes, the teacher can pose a problem-solving activity and later ask, "What have you done so far to solve this problem, what decisions did you make, and why did you make those decisions?" As students respond, the teacher can explain, "This is exactly what I'd like you to tell me in your writing."

Having students compare and analyze different pieces of their work is another way to convey expectations and help them understand what complete and incomplete responses look like. For example, students were asked to use pictures and words to explain their thinking for the following question (Kouba, Zawojewski, and Strutchens 1997, p. 119):

> José ate 1/2 of a pizza.
>
> Ella ate 1/2 of another pizza.
>
> José said that he ate more pizza than Ella, but Ella said they both ate the same amount. Use words and pictures to show that José could be right.

Students' responses reflected different levels of understanding (see the examples in fig. 5.35). The first student assumed that each pizza was the same size. Although the student used words and drawings in the response, the answer was correct only if the units were the same, an assumption that cannot be made from the statement of the problem. The second student suggested by the drawing that the size of 1/2 depends on the size of the unit. The teacher might ask this student to explain his or her thinking. The third solution, including written words and drawings, was correct and complete in that it communicated why José could be correct. Discussion of various student responses, especially as mathematical concepts and problems become more complex, is an effective way to help students continue to improve their ability to communicate.

Fig. **5.35.**

Students' responses to the "pizza" problem (Dossey, Mullis, and Jones 1993)

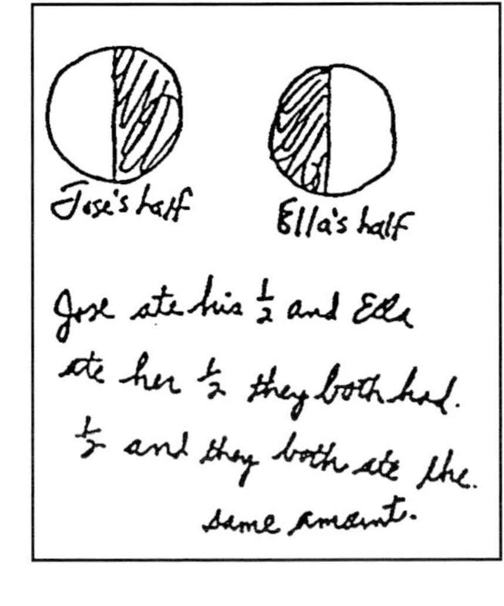

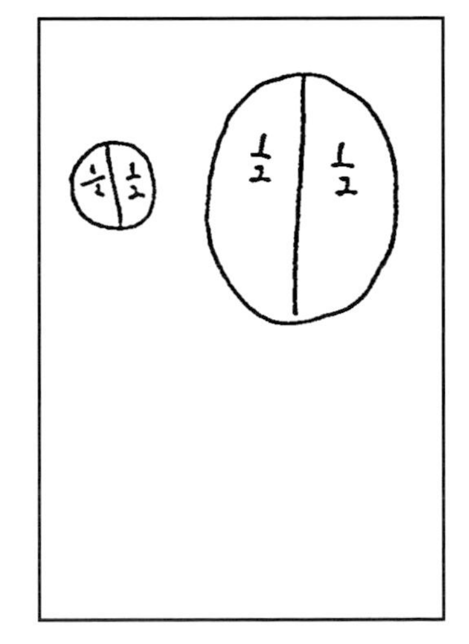

Connections STANDARD for Grades 3–5

Instructional programs from prekindergarten through grade 12 should enable all students to—

Recognize and use connections among mathematical ideas

Understand how mathematical ideas interconnect and build on one another to produce a coherent whole

Recognize and apply mathematics in contexts outside of mathematics

Students in grades 3–5 study a considerable amount of new mathematical content, and their ability to understand and manage these new ideas will rest, in part, on how well the ideas are connected. Connecting mathematical ideas includes linking new ideas to related ideas considered previously. These connections help students see mathematics as a unified body of knowledge rather than as a set of complex and disjoint concepts, procedures, and processes.

What should connections look like in grades 3 through 5?

Two big ideas that recur throughout the study of mathematics in grades 3–5 were elaborated on in the introduction at the beginning of the chapter: equivalence and multiplicative reasoning. Each should receive major emphasis at this level, in part because each is connected to so many topics studied in grades 3–5. For example, students learn that a fraction has an equivalent decimal representation, that the area of a right triangle is equal to half of the area of a related rectangle, that 150 centimeters is the same as 1.5 meters, and that the likelihood of getting heads when flipping a coin is the same as the likelihood of rolling an even number on a number cube. Some equivalences are not obvious to students and thus prompt further exploration to understand "why." As equivalence continues to emerge in the study of different mathematical content areas, it fosters a sense of unity and connectedness in the study of mathematics. Likewise, as students solve problems as diverse as counting the possible combinations of shirts and shorts in a wardrobe and measuring the area of a rectangle, they begin to see and use a similar multiplicative structure in both situations. Their work in developing computational algorithms highlights properties of multiplication that they can model geometrically, reason about, and express in general terms. Thus, multiplicative structures connect ideas from number, algebra, and geometry. Equivalence and multiplicative reasoning help students see that mathematics is not a set of isolated topics but rather a web of closely connected ideas.

Real-world contexts provide opportunities for students to connect what they are learning to their own environment. Students' experiences at home, at school, and in their community provide contexts for worthwhile mathematical tasks. For example, ideas of position and direction such as those used in walking from one place to another can be used to develop the geometric idea of using coordinates to describe a location. In a fourth-grade class, students could make a map on a coordinate system of the various routes they use to walk to school. With the map, they could determine and compare the distances traveled. Everyday experiences can also be the source of data. In a fifth-grade classroom, students may want to investigate questions about after-school activities. How many students participate in such activities? What are the activities? How frequently do they participate? Is the level of participation consistent across the year? Is there a way to describe the class on the basis of their activities? Encouraging students to ask questions and to use mathematical approaches to find answers helps them see the value of mathematics and also motivates them to study new mathematical ideas.

There are connections within mathematics, and mathematics is also connected to, and used within, other disciplines. Building on these connections provides opportunities to enrich the learning in both areas. For example, in a social studies unit, a fifth-grade class might discuss the population and area of selected states. They can investigate which states are most and least crowded. By using almanacs, Web-based databases, and maps, they can collect data and construct charts to summarize the information. Once the information is collected, they will need to determine how to consider both area and population in order to judge crowdedness. Such discussions could lead to an informal consideration of population density and land use.

In grades 3–5, students should be developing the important processes needed for scientific inquiry and for mathematical problem solving—inferring, measuring, communicating, classifying, and predicting. The kinds of investigations that enable students to build these processes often include significant mathematics as well as science. It is important that teachers stimulate discussion about both the mathematics and the science ideas that emerge from the investigations, whether they occur in a science lesson or a mathematics lesson. For example, students might study the evaporation of liquid from an open container. How does the volume of liquid in the container change over time? From which type of jar does 100 cubic centimeters of water evaporate faster—one with a large opening or one with a small opening? Figure 5.36 (Goldberg 1997, p. 2) shows the results of an experiment to examine this question. The table shows the volume of water in each jar over a five-day period. Is there a pattern in the data? If so, what are some ways to describe the pattern? How many days will it take for all the water in jar 1 to evaporate? A discussion about *why* the water evaporates faster from a wider container and what might happen if certain conditions are altered—for example, if a fan is left blowing on the containers—integrates concepts of both mathematics and science.

The development of mathematical ideas and the use of mathematics in other disciplines are intertwined. At times, new ideas develop in a purely mathematical context and are applied to other situations. At other times, new mathematics arises out of situations in other disciplines or in real-world contexts. Mathematical investigations that are drawn strictly from the realm of mathematics are also appropriate and important. The value of a mathematical task is not dependent on whether it has a real-world context but rather on whether it addresses important mathematics, is intellectually engaging, and is solvable using

The value of a mathematical task is not dependent on whether it has a real-world context.

Time (in days)	Jar 1 (volume in cc)	Jar 2 (volume in cc)
0	100	100
2	91	84
5	80	80
7	74	66
9	67	50

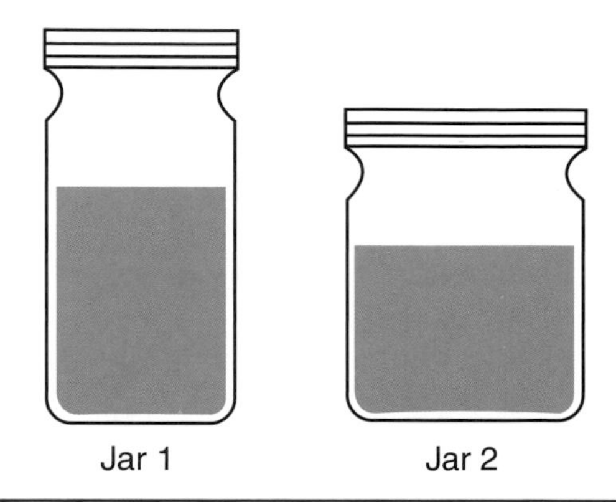

Fig. **5.36.**
Results of an evaporation experiment

tools the learner has or can draw on. The use of similar mathematics within different contexts gives students an appreciation of the power of mathematics and its generality. As stated in a National Research Council report (1996, p. 105):

> Students at all grade levels and in every domain of science should have the opportunity to use scientific inquiry and develop the ability to think and act in ways associated with inquiry, including asking questions, planning and conducting investigations, using appropriate tools and techniques to gather data, thinking critically and logically about relationships between evidence and explanations, constructing and analyzing alternative explanations, and communicating scientific arguments.

What should be the teacher's role in developing connections in grades 3 through 5?

Teachers should select tasks that help students explore and develop increasingly sophisticated mathematical ideas. They should ask questions that encourage and challenge students to explain new ideas and develop new strategies based on mathematics they already know. For example, asking students to describe two ways they can estimate the cost of twelve notebooks can prompt different strategies. Figure 5.37 illustrates two strategies that might emerge—a rounding strategy and another strategy based on proportionality, a new idea that will receive considerable attention in later grades.

Fig. **5.37.**

Two estimation strategies—one using rounding and the other based on proportionality

Teachers should help students explore and describe mathematical connections and ensure that they see mathematical ideas in a variety of contexts and models. For example, as students explain their strategies for estimating the cost of the twelve notebooks, the teacher should point out how the second strategy relates to multiplication and how it can be modeled using a fractional representation (e.g., 2 for $1 means 12 for $6, or 2/1 = 12/6).

Teachers should encourage students to look for mathematical ideas throughout the school day. For example, geometry can play an important role in art, data should have a prominent role in social studies discussions, and communication and problem solving should be integrated with language arts. Scientific contexts can be especially productive for exploring and using mathematics. Mathematics and science have a long history of close ties, and many mathematical notions arose from scientific problems. Teachers should build on everyday experiences to encourage the study of mathematical ideas through systematic, quantitative investigations of phenomena that students can experience directly. These may include applications as varied as studying the relationship between the arm span and height of students, investigating the strength of a particular brand of paper towel, or studying the volume and surface area of different cereal boxes.

At times, opportunities for mathematical investigations arise spontaneously in class. For example, after a fifth-grade class spent some time learning about environmental issues, a question arose as to whether the water fountain was an efficient way of getting water to students. The class formulated a plan to respond to the question. This included estimating how much water the fountain released during a "typical" turn at the fountain and how much water was actually consumed. In a situation like this, the teacher plays an important role in helping students understand and think about the scientific and mathematical topics that this investigation evokes.

At times, opportunities for mathematical investigations arise spontaneously in class.

Building on connections can make mathematics a challenging, engaging, and exciting domain of study.

Although the teacher's role includes being alert and responsive to unexpected opportunities, it is also important that teachers plan ahead to integrate mathematics into other subject areas and experiences that students will have during the year. Consider, for example, the following episode, adapted from Russell, Schifter, and Bastable (1999).

> Ms. Watson's fourth grade runs a snack shop for two weeks every school year to pay for a trip to meet the class's pen pals in a neighboring state. Since the students run the whole project, from planning what to sell to recording sales and reordering stock, Ms. Watson uses this project as an opportunity for students to develop and use mathematical ideas. It is clear that a great deal of estimation and calculation takes place naturally as part of the project: projecting what will be needed for the trip, making change, keeping records of expenses, calculating income, and so forth. This year Ms. Watson decided to extend some of the ideas her students had encountered about collecting and describing data through their work on this project.
>
> At the beginning of the project, she gave the class a list of twenty-one items, available at a local warehouse club, that she and the principal had approved as possible sale items. The students needed to decide which of these products they would sell and how they would allocate the $100 provided for their start-up costs to buy certain quantities of those products. They had limited time to make these decisions, and the class engaged in a lively discussion about how best to find out which of the snack items were most popular among the students in the school. Some students insisted that they would need to survey all classes in order to get "the correct information." If they surveyed only some students, this group contended, then "we won't give everyone a chance, so we won't know about something that maybe only one person likes." Others argued that surveying one or two classes at each grade level would provide enough of an idea of what students across the grades like and would result in a set of data they could collect and organize more efficiently. As they talked, the teacher reminded them of the purpose of their survey: "Will our business fail if we don't have everyone's favorite?" The class eventually decided to survey one class at each grade. Even the students who had worried that a sample would not give them complete information had become convinced that this procedure would give them *enough* information to make good choices about which snacks to buy.
>
> The students went on to design their survey—which raised new issues—and to collect, organize, and use the data to develop their budget. Once they had their data, another intense discussion ensued about how to use the information to guide their choices on how to stock their snack shop. They eventually chose to buy the two top choices in each category (they had classified the snacks into four categories), and since that didn't use up their budget, they ordered additional quantities of the overall top two snacks.

Ms. Watson used this realistic context to help her students see how decisions about designing data investigations are tied to the purpose or the problem being addressed. The real restrictions of time and resources made it natural for the students to consider how a sample can be selected to represent a population, and they were able to interpret their data in light of the decisions they needed to make.

Ultimately, connections within mathematics, connections between mathematics and everyday experience, and connections between mathematics and other disciplines can support learning. Building on the connections can also make mathematics a challenging, engaging, and exciting domain of study.

Representation Standard for Grades 3–5

Instructional programs from prekindergarten through grade 12 should enable all students to—

Create and use representations to organize, record, and communicate mathematical ideas

Select, apply, and translate among mathematical representations to solve problems

Use representations to model and interpret physical, social, and mathematical phenomena

In grades 3–5, students need to develop and use a variety of representations of mathematical ideas to model problem situations, to investigate mathematical relationships, and to justify or disprove conjectures. They should use informal representations, such as drawings, to highlight various features of problems; they should use physical models to represent and understand ideas such as multiplication and place value. They should also learn to use equations, charts, and graphs to model and solve problems. These representations serve as tools for thinking about and solving problems. They also help students communicate their thinking to others. Students in these grades will use both external models—ones that they can build, change, and inspect—as well as mental images.

What should representation look like in grades 3–5?

Students in grades 3–5 should continue to develop the habit of representing problems and ideas to support and extend their reasoning. Such representations help to portray, clarify, or extend a mathematical idea by focusing on essential features. Students represent ideas when they create a table of data about weather patterns, when they describe in words or with a picture the important features of an object such as a cylinder, or when they translate aspects of a problem into an equation. Good representations fulfill a dual role: they are tools for thinking and instruments for communicating. Consider the following problem:

> What happens to the area of a rectangle if the lengths of its sides are doubled?

Students who represent the problem in some way are more likely to see important relationships than those who consider the problem without a representation. One student's initial response to the problem was that the new rectangle would be twice the size of the first rectangle. Her thinking might have stopped there, but another student questioned her answer, prompting her to think more deeply. She decided she needed a picture to help her think about the problem. Her drawing (see fig. 5.38) helped her consider the complexity of the problem more carefully and showed her that the new rectangle is not only bigger but that it is four times bigger than the original rectangle. It was also a way to show her answer and to justify it to others.

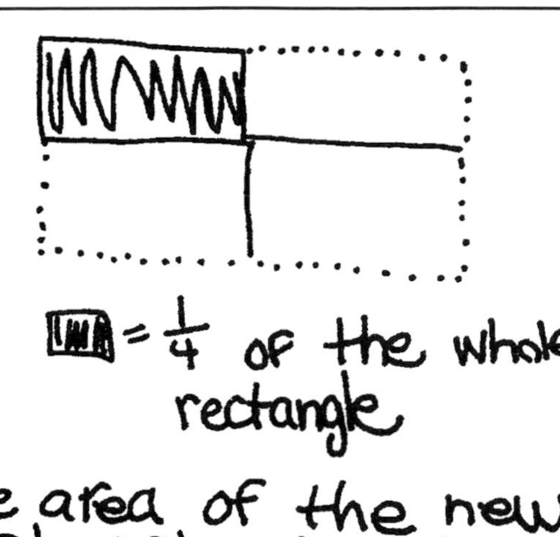

Fig. **5.38.**

A student's representation of the results of doubling the lengths of the sides of a rectangle

Students will have learned about, and begun to use, many symbolic and graphical representations (e.g., numerals, equals sign, and bar graphs) in the primary grades. In grades 3–5, students should create representations that are more detailed and accurate than is expected in the primary grades. Their repertoire of symbols, tools, and conventional notation should expand and be clearly connected to concepts as they are explored. For example, in representing algebraic and numerical relationships, students should become comfortable using equations and understanding the equals sign as a balance point in the equation. Many students who have only seen equations with an arithmetic expression on the left side of the equation and a call for the numerical answer on the right side, such as $6 \times 30 = \square$, don't understand that equations may have several symbols on each side, as in $2 \times 5 \times 6 = 3 \times 4 \times 5$.

Students in grades 3–5 should also become familiar with technological tools such as dynamic geometry software and spreadsheets. They should learn to set up a simple spreadsheet (see fig. 5.39) and use it to pose and solve problems, examine data, and investigate patterns. For example, a fourth-grade class could keep track of the daily temperature and other features of the weather for the whole year and consider questions such as these: What month is coldest? What would we tell a visitor to expect for weather in October? After two months, they might find that they are having difficulty managing and ordering the quantity of data they have collected. By entering the data in a spreadsheet, they can easily see and select the data they want, compare certain columns, or graph particular aspects of the data. They can conveniently find the median temperature for February or calculate the total amount of rainfall for April. In January, if the class notices that temperature alone is no longer giving them enough information, they can add a column for wind chill to get a more accurate summary of the weather they are experiencing.

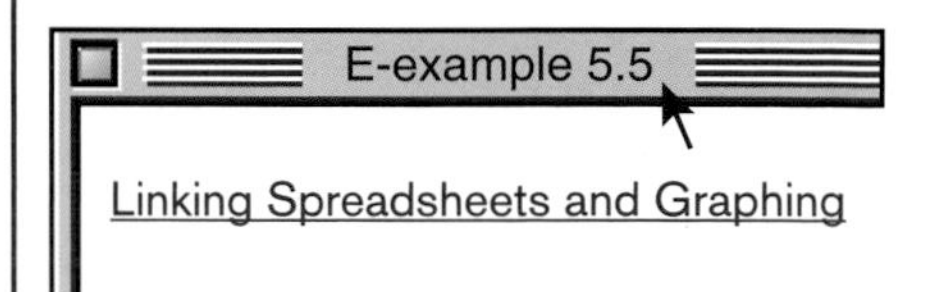

Learning to interpret, use, and construct useful representations needs careful and deliberate attention in the classroom. Teaching *forms* of representation (e.g., graphs or equations) as ends in themselves is not productive. Rather, representations should be portrayed as useful tools

Fig. **5.39.**

A simple spreadsheet can be used to organize and examine data, pose and solve problems, and investigate patterns.

	A	B	C	D	E
1			JANUARY		
2					
3	Date	High Temperature	Precipitation	Rain or Snow?	Wind Chill
4					
5	1-Jan	22			20
6	2-Jan	20	5.0"	snow	20
7	3-Jan	23	3.5"	snow	23
8	4-Jan	24			20
9	5-Jan	14			10
10	6-Jan	5			-5
11	7-Jan	24			20
12	8-Jan	8			0
13	9-Jan	6			-10
14	10-Jan	9			-5
15	11-Jan	16			10
16	12-Jan	20			20
17	13-Jan	32			32

for building understanding, for communicating information, and for demonstrating reasoning (Greeno and Hall 1997). Students should become flexible in choosing and creating representations—standard or nonstandard, physical models or mental images—that fit the purpose at hand. They should also have many opportunities to consider the advantages and limitations of the various representations they use.

What should be the teacher's role in developing representation in grades 3–5?

Learning to record or represent thinking in an organized way, both in solving a problem and in sharing a solution, is an acquired skill for many students. Teachers can and should emphasize the importance of representing mathematical ideas in a variety of ways. Modeling this process as they work through a problem with the class is one way to stimulate students to use and analyze representations. Talking through why some representations are more effective than others in a particular situation gives prominence to the process and helps students critique aspects of their representations. Teachers can strategically choose student representations that will be fruitful for the whole class to discuss. For example, consider the following question, which a third-grade class might explore:

> Are there more even or odd products in the multiplication table shown in figure 5.41? Explain why.

Students may initially generate many examples to formulate an answer, as illustrated in figure 5.40. Other students may use the multiplication table to organize their work, as illustrated in figure 5.41. Organizing the work in this way highlights patterns that support students in thinking more systematically about the problem.

Each representation reveals a different way of thinking about the problem. Giving attention to the different methods as well as to the different representations will help students see the power of viewing a problem from different perspectives. Observing how different students select and use representations also gives the teacher assessment information about what aspects of the problem they notice and how they reason about the patterns and regularities revealed in their representations.

Fig. **5.40.**

An exploration of odd and even numbers in the multiplication table

$4 \times 3 = 12$ even
$2 \times 5 = 10$ ~~odd~~ even
$3 \times 7 = 21$ odd

x	1	2	3	4	5	6	7	8	9
1	1	2	3	4	5	6	7	8	9
2	2	4	6	8	10	12	14	16	18
3	3	6	9	12	15	18	21	24	27
4	4	8	12	16	20	24	28	32	36
5	5	10	15	20	25	30	35	40	45
6	6	12	18	24	30	36	42	48	54
7	7	14	21	28	35	42	49	56	63
8	8	16	24	32	40	48	56	64	72
9	9	18	27	36	45	54	63	72	81

Fig. **5.41.**

Using a multiplication table to solve an "odd and even numbers" problem

As students discuss their ideas and begin to develop conjectures based on representations of the problem, the teacher might want to represent the students' thinking in other ways in order to support and extend their ideas. For example, when students notice that an even number multiplied by an even number always produces an even result, the teacher might record this idea as "even × even = even." This representation serves as a summary of the students' thinking. It suggests a way to record the generalization and may prompt students to look for other generalizations of the same type.

Some students will need explicit help in representing problems. Although in the rectangle problem (fig. 5.38), the student quickly decided on a representation that was effective in showing the important relationships, many students need support in constructing pictures, graphs, tables, and other representations. If they have many opportunities for using, developing, comparing, and analyzing a variety of representations, students will become competent in selecting what they need for a particular problem.

As students work with a variety of representations, teachers need to observe carefully how they understand and use them. Representations do not "show" the mathematics to the students. Rather, the students need to work with each representation extensively in many contexts as well as move between representations in order to understand how they can use a representation to model mathematical ideas and relationships.

By listening carefully to students' ideas and helping them select and organize representations that will show their thinking, teachers can help students develop the inclination and skills to model problems effectively, to clarify their own understanding of a problem, and to use representations to communicate effectively with one another.

Middle-grades students are drawn toward mathematics if they find both challenge and support in the mathematics classroom.

Ambitious expectations in algebra and geometry

stretch the middle-grades program beyond a preoccupation with number.

CHAPTER 6

Standards for Grades 6–8

Middle-grades students should see mathematics as an exciting, useful, and creative field of study. As they enter adolescence, students experience physical, emotional, and intellectual changes that mark the middle grades as a significant transition point in their lives. During this time, many students will solidify conceptions about themselves as learners of mathematics—about their competence, their attitude, and their interest and motivation. These conceptions will influence how they approach the study of mathematics in later years, which will in turn influence their life opportunities. Middle-grades students are drawn toward mathematics if they find both challenge and support in the mathematics classroom. Students acquire an appreciation for, and develop an understanding of, mathematical ideas if they have frequent encounters with interesting, challenging problems.

In the middle-grades mathematics classroom, young adolescents should regularly engage in thoughtful activity tied to their emerging capabilities of finding and imposing structure, conjecturing and verifying, thinking hypothetically, comprehending cause and effect, and abstracting and generalizing. In these grades, each student follows his or her own developmental timetable. Some mature early, and others late. Some progress rapidly, others more slowly. Thus, every middle-grades teacher faces the challenge of dealing with many aspects of diversity. Yet students also display some commonalities. For example, young adolescents are almost universally sensitive to the influence of their peers. The differences in intellectual development and emotional maturity and the sensitivity of individuals to peer-group perceptions make it especially important for teachers to create classroom environments in which clearly established norms support the learning of mathematics by everyone.

An ambitious, focused mathematics program for *all* students in the middle grades is proposed in these Standards. Ambitious expectations are identified in algebra and geometry that would stretch the middle-grades program beyond a preoccupation with number. In recent years, the possibility and necessity of students' gaining facility in algebraic thinking have been widely

Students should learn significant amounts of algebra and geometry and see them as interconnected.

recognized. Accordingly, these Standards propose a significant amount of algebra for the middle grades. In addition, there is a need for increased attention to geometry in these grades. Facility in geometric thinking is essential to success in the later study of mathematics and also in many situations that arise outside the mathematics classroom. Moreover, geometry is typically the area in which U.S. students perform most poorly on domestic and international assessments of mathematics proficiency. Therefore, significantly more geometry is recommended in these Standards for the middle grades than has been the norm. The recommendations are ambitious—they call for students to learn many topics in algebra and geometry and also in other content areas. To guard against fragmentation of the curriculum, therefore, middle-grades mathematics curriculum and instruction must also be focused and integrated.

Specific foci are identified in several content areas. For example, in number and operations, these Standards propose that students develop a deep understanding of rational-number concepts, become proficient in rational-number computation and estimation, and learn to think flexibly about relationships among fractions, decimals, and percents. This facility with rational numbers should be developed through experience with many problems involving a range of topics, such as area, volume, relative frequency, and probability. In algebra, the focus is on proficiency in recognizing and working effectively with linear relationships and their corresponding representations in tables, graphs, and equations; such proficiency includes competence in solving linear equations. Students can develop the desired algebraic facility through problems and contexts that involve linear and nonlinear relationships. Appropriate problem contexts can be found in many areas of the curriculum, such as using scatterplots and approximate lines of fit to give meaning to the concept of slope or noting that the relationship between the side lengths and the perimeters of similar figures is linear, whereas the relationship between the side lengths and the areas of similar figures is nonlinear.

Curricular focus and integration are also evident in the proposed emphasis on proportionality as an integrative theme in the middle-grades mathematics program. Facility with proportionality develops through work in many areas of the curriculum, including ratio and proportion, percent, similarity, scaling, linear equations, slope, relative-frequency histograms, and probability. The understanding of proportionality should also emerge through problem solving and reasoning, and it is important in connecting mathematical topics and in connecting mathematics and other domains such as science and art.

In the recommendations for middle-grades mathematics outlined here, students will learn significant amounts of algebra and geometry throughout grades 6, 7, and 8. Moreover, they will see algebra and geometry as interconnected with each other and with other content areas in the curriculum. They will have experience with both the geometric representation of algebraic ideas, such as visual models of algebraic identities, and the algebraic representation of geometric ideas, such as equations for lines represented on coordinate grids. They will see the value of interpreting both algebraically and geometrically such important mathematical ideas as the slope of a line and the Pythagorean relationship. They also will relate algebraic and geometric ideas to other topics—for example, when they reason about percents using visual models or equations or when they represent an approximate line of fit for a scatterplot both geometrically and algebraically. Students can gain a deeper understanding of proportionality if it develops along with foundational algebraic ideas such as linear relationships and geometric ideas such as similarity.

Students' understanding of foundational algebraic and geometric ideas should be developed through extended experience over all three years in the middle grades and across a broad range of mathematics content, including statistics, number, and measurement. How these ideas are packaged into courses and what names are given to the resulting arrangement are far less important than ensuring that students have opportunities to see and understand the connections among related ideas. This approach is a challenging alternative to the practice of offering a select group of middle-grades students a one-year course that focuses narrowly on algebra or geometry. All middle-grades students will benefit from a rich and integrated treatment of mathematics content. Instruction that segregates the content of algebra or geometry from that of other areas is educationally unwise and mathematically counterproductive.

Principles and Standards for School Mathematics proposes an ambitious and rich experience for middle-grades students that both prepares them to use mathematics effectively to deal with quantitative situations in their lives outside school and lays a solid foundation for their study of mathematics in high school. Students are expected to learn serious, substantive mathematics in classrooms in which the emphasis is on thoughtful engagement and meaningful learning.

Students are expected to learn serious, substantive mathematics with an emphasis on thoughtful engagement and meaningful learning.

For those who make decisions about the design and organization of middle-grades mathematics education, it would be insufficient simply to announce new and more-ambitious goals like those suggested here. School system leaders need to commit to and support steady, long-term improvement and capacity building to accomplish such goals. The capacity of schools and middle-grades teachers to provide the kind of mathematics education envisioned needs to be built. Special attention must be given to the preparation and ongoing professional support of teachers in the middle grades. Teachers need to develop a sound knowledge of mathematical ideas and excellent pedagogical practices and become aware of current research on students' mathematics learning. Professional development is especially important in the middle grades because so little attention has been given in most states and provinces to the special preparation that may be required for mathematics teachers at these grade levels. Many such teachers hold elementary school generalist certification, which typically involves little specific preparation in mathematics. Yet teachers in the middle grades need to know much more mathematics than is required in most elementary school teacher-certification programs. Some middle-grades mathematics teachers hold secondary school mathematics-specialist certification. But middle-grades teachers need to know much more about adolescent development, pedagogical alternatives, and interdisciplinary approaches to teaching than most secondary school teacher-certification programs require. In order to accomplish the ambitious goals for the middle grades that are presented here, special teacher-preparation programs must be developed.

Number and Operations STANDARD for Grades 6–8

Instructional programs from prekindergarten through grade 12 should enable all students to—

Expectations

In grades 6–8 all students should–

Standard	Expectations
Understand numbers, ways of representing numbers, relationships among numbers, and number systems	• work flexibly with fractions, decimals, and percents to solve problems; • compare and order fractions, decimals, and percents efficiently and find their approximate locations on a number line; • develop meaning for percents greater than 100 and less than 1; • understand and use ratios and proportions to represent quantitative relationships; • develop an understanding of large numbers and recognize and appropriately use exponential, scientific, and calculator notation; • use factors, multiples, prime factorization, and relatively prime numbers to solve problems; • develop meaning for integers and represent and compare quantities with them.
Understand meanings of operations and how they relate to one another	• understand the meaning and effects of arithmetic operations with fractions, decimals, and integers; • use the associative and commutative properties of addition and multiplication and the distributive property of multiplication over addition to simplify computations with integers, fractions, and decimals; • understand and use the inverse relationships of addition and subtraction, multiplication and division, and squaring and finding square roots to simplify computations and solve problems.
Compute fluently and make reasonable estimates	• select appropriate methods and tools for computing with fractions and decimals from among mental computation, estimation, calculators or computers, and paper and pencil, depending on the situation, and apply the selected methods; • develop and analyze algorithms for computing with fractions, decimals, and integers and develop fluency in their use; • develop and use strategies to estimate the results of rational-number computations and judge the reasonableness of the results; • develop, analyze, and explain methods for solving problems involving proportions, such as scaling and finding equivalent ratios.

Number and Operations

In grades 6–8, students should deepen their understanding of fractions, decimals, percents, and integers, and they should become proficient in using them to solve problems. By solving problems that require multiplicative comparisons (e.g., "How many times as many?" or "How many per?"), students will gain extensive experience with ratios, rates, and percents, which helps form a solid foundation for their understanding of, and facility with, proportionality. The study of rational numbers in the middle grades should build on students' prior knowledge of whole-number concepts and skills and their encounters with fractions, decimals, and percents in lower grades and in everyday life. Students' facility with rational numbers and proportionality can be developed in concert with their study of many topics in the middle-grades curriculum. For example, students can use fractions and decimals to report measurements, to compare survey responses from samples of unequal size, to express probabilities, to indicate scale factors for similarity, and to represent constant rate of change in a problem or slope in a graph of a linear function.

Understand numbers, ways of representing numbers, relationships among numbers, and number systems

In the middle grades, students should become facile in working with fractions, decimals, and percents. Teachers can help students deepen their understanding of rational numbers by presenting problems, such as those in figure 6.1 that call for flexible thinking. For more discussion of useful representations for rational numbers, see the "Representation" section of this chapter.

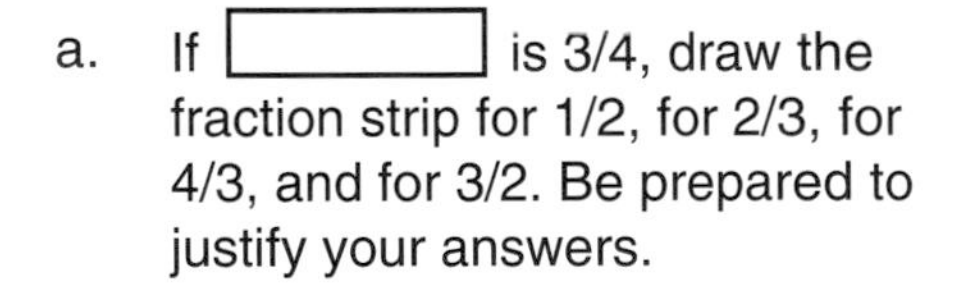

a. If [rectangle] is 3/4, draw the fraction strip for 1/2, for 2/3, for 4/3, and for 3/2. Be prepared to justify your answers.

b.

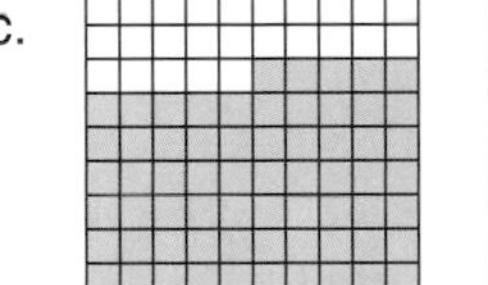

Using the points you are given on the number line above, locate 1/2, 2 1/2, and 1/4. Be prepared to justify your answers.

c.

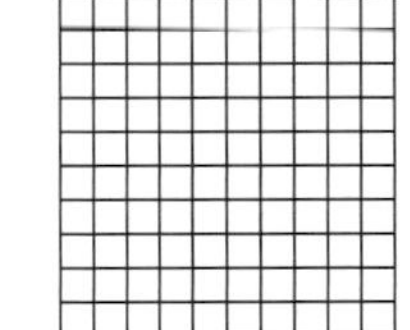

Use the drawing above to justify in as many different ways as you can that 75% of the square is equal to 3/4 of the square. You may reposition the shaded squares if you wish.

Fig. **6.1.**

Problems that require students to think flexibly about rational numbers

At the heart of flexibility in working with rational numbers is a solid understanding of different representations for fractions, decimals, and percents. In grades 3–5, students should have learned to generate and recognize equivalent forms of fractions, decimals, and percents, at least in some simple cases. In the middle grades, students should build on and extend this experience to become facile in using fractions, decimals, and percents meaningfully. Students can develop a deep understanding of rational numbers through experiences with a variety of models, such as fraction strips, number lines, 10 × 10 grids, area models, and objects. These models offer students concrete representations of abstract ideas and support students' meaningful use of representations and their flexible movement among them to solve problems.

As they solve problems in context, students also can consider the advantages and disadvantages of various representations of quantities. For example, students should understand not only that 15/100, 3/20, 0.15, and 15 percent are all representations of the same number but also that these representations may not be equally suitable to use in a particular context. For example, it is typical to represent a sales discount as 15%, the probability of winning a game as 3/20, a fraction of a dollar in writing a check as 15/100, and the amount of the 5 percent tax added to a purchase of $2.98 as $0.15.

In the middle grades, students should expand their repertoire of meanings, representations, and uses for nonnegative rational numbers. They should recognize and use fractions not only in the ways they have in lower grades—as measures, quantities, parts of a whole, locations on a number line, and indicated divisions—but also in new ways. For example, they should encounter problems involving ratios (e.g., 3 adult chaperones for every 8 students), rates (e.g., scoring a soccer goal on 3 of every 8 penalty kicks), and operators (e.g., multiplying by 3/8 means generating a number that is 3/8 of the original number). In the middle grades, students also need to deepen their understanding of decimal numbers and extend the range of numbers and tasks with which they work. The foundation of students' work with decimal numbers must be an understanding of whole numbers and place value. In grades 3–5, students should have learned to think of decimal numbers as a natural extension of the base-ten place-value system to represent quantities less than 1. In grades 6–8, they should also understand decimals as fractions whose denominators are powers of 10. The absence of a solid conceptual foundation can greatly hinder students. Without a solid conceptual foundation, students often think about decimal numbers incorrectly; they may, for example, think that 3.75 is larger than 3.8 because 75 is more than 8 (Resnick et al. 1989). Students also need to interpret decimal numbers as they appear on calculator screens, where they may be truncated or rounded.

Fig. **6.2.**

A student's reasoning about the sizes of rational numbers

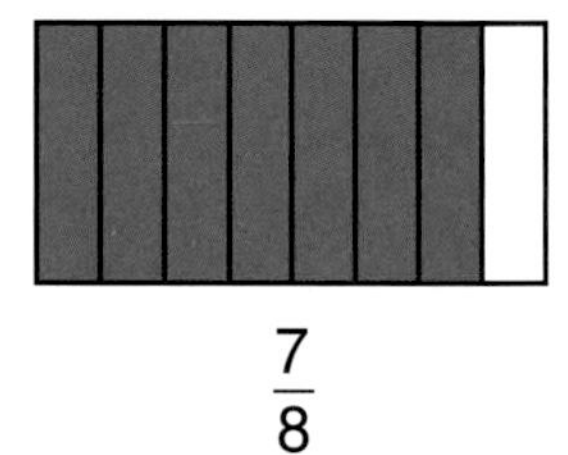

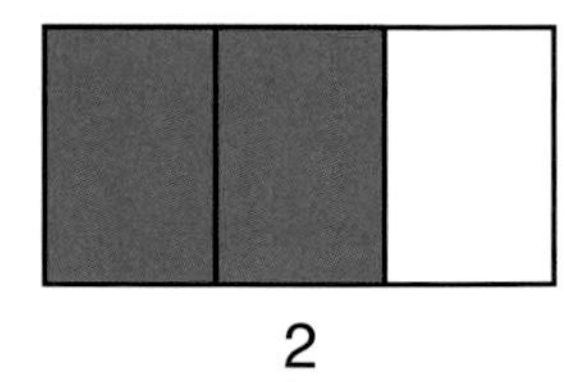

The 7/8 portion is one piece less than a whole, and so is 2/3. But the missing piece for 7/8 is smaller than the missing piece for 2/3. So 7/8 is bigger than 2/3.

In the lower grades, students should have had experience in comparing fractions between 0 and 1 in relation to such benchmarks as 0, 1/4, 1/2, 3/4, and 1. In the middle grades, students should extend this experience to tasks in which they order or compare fractions, which many students find difficult. For example, fewer than one-third of the thirteen-year-old U.S. students tested in the National Assessment of Educational Progress (NAEP) in 1988 correctly chose the largest number from 3/4, 9/16, 5/8, and 2/3 (Kouba, Carpenter, and Swafford 1989). Students' difficulties with comparison of fractions have also been documented in more recent NAEP administrations (Kouba, Zawojewski, and Strutchens 1997). Visual images of fractions as fraction strips should help many students think flexibly in comparing fractions. As shown in figure 6.2, a student might conclude that 7/8 is greater than 2/3 because each fraction is exactly "one piece" smaller than 1 and the missing 1/8 piece is smaller than the missing 1/3 piece. Students may also be helped by thinking about the relative locations of fractions and decimals on a number line.

Percents, which can be thought about in ways that combine aspects of both fractions and decimals, offer students another useful form of rational number. Percents are particularly useful when comparing fractional parts of sets or numbers of unequal size, and they are also frequently encountered in problem-solving situations that arise in everyday life. As with fractions and decimals, conceptual difficulties need to be carefully addressed in instruction. In particular, percents less than 1 percent and greater than 100 percent are often challenging, and most students are likely to benefit from frequent encounters with problems involving percents of these magnitudes in order to develop a solid understanding.

Attention to developing flexibility in working with rational numbers contributes to students' understanding of, and facility with, proportionality. Facility with proportionality involves much more than setting two ratios equal and solving for a missing term. It involves recognizing quantities that are related proportionally and using numbers, tables, graphs, and equations to think about the quantities and their relationship. Proportionality is an important integrative thread that connects many of the mathematics topics studied in grades 6–8. Students encounter proportionality when they study linear functions of the form $y = kx$, when they consider the distance between points on a map drawn to scale and the actual distance between the corresponding locations in the world, when they use the relationship between the circumference of a circle and its diameter, and when they reason about data from a relative-frequency histogram.

Proportionality connects many of the mathematics topics studied in grades 6–8.

In the middle grades, students should continue to work with whole numbers in a variety of problem-solving settings. They should develop a sense of the magnitude of very large numbers (millions and billions) and become proficient at reading and representing them. For example, they should recognize and represent 2 300 000 000 as 2.3×10^9 in scientific notation and also as 2.3 billion. Contexts in which large numbers arise naturally are found in other school subjects as well as in everyday life. For example, a newspaper headline may proclaim, "Clean-Up Costs from Oil Spill Exceed $2 Billion!" or a science textbook may indicate that the number of red blood cells in the human body is about 1.9×10^{13}. Students also need to understand various forms of notation and recognize, for instance, that the number 2.5×10^{11} might appear on a calculator as 2.5E11 or 2.5 11, depending on the make and model of the machine. Students' experiences in working with very large numbers and in using the idea of orders of magnitude will also help build their facility with proportionality.

Students can also work with whole numbers in their study of number theory. Tasks, such as the following, involving factors, multiples, prime numbers, and divisibility, can afford opportunities for problem solving and reasoning.

1. Explain why the sum of the digits of any multiple of 3 is itself divisible by 3.
2. A number of the form *abcabc* always has several prime-number factors. Which prime numbers are always factors of a number of this form? Why?

Middle-grades students should also work with integers. In lower grades, students may have connected negative integers in appropriate ways to informal knowledge derived from everyday experiences, such as below-zero winter temperatures or lost yards on football plays. In the middle grades, students should extend these initial understandings of integers. Positive and negative integers should be seen as useful for noting relative changes or values. Students can also appreciate the utility of negative integers when they work with equations whose solution requires them, such as $2x + 7 = 1$.

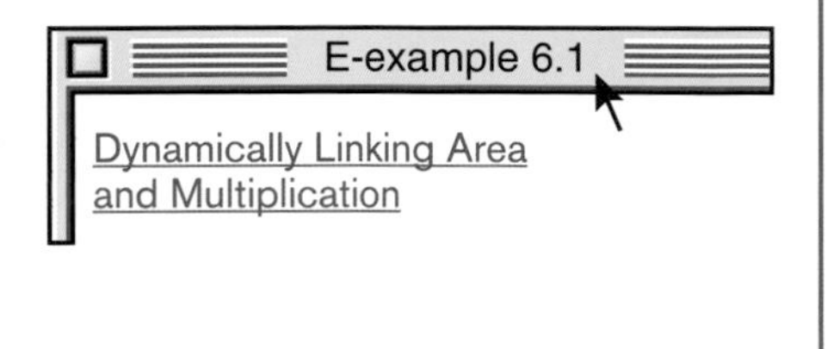

Understand meanings of operations and how they relate to one another

In the middle grades, students should continue to refine their understandings of addition, subtraction, multiplication, and division as they use these operations with fractions, decimals, percents, and integers. Teachers need to be attentive to conceptual obstacles that many students encounter as they make the transition from operations with whole numbers.

Multiplying and dividing fractions and decimals can be challenging for many students because of problems that are primarily conceptual rather than procedural. From their experience with whole numbers, many students appear to develop a belief that "multiplication makes bigger and division makes smaller." When students solve problems in which they need to decide whether to multiply or divide fractions or decimals, this belief has negative consequences that have been well researched (Greer 1992). Also, a mistaken expectation about the magnitude of a computational result is likely to interfere with students' making sense of multiplication and division of fractions or decimals (Graeber and Tanenhaus 1993). Teachers should check to see if their students harbor this misconception and then take steps to build their understanding.

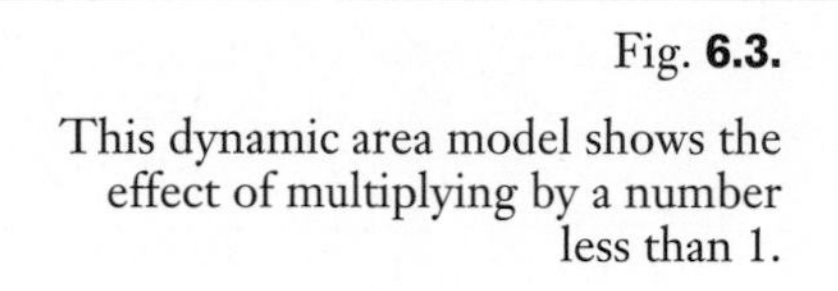

Fig. **6.3.**

This dynamic area model shows the effect of multiplying by a number less than 1.

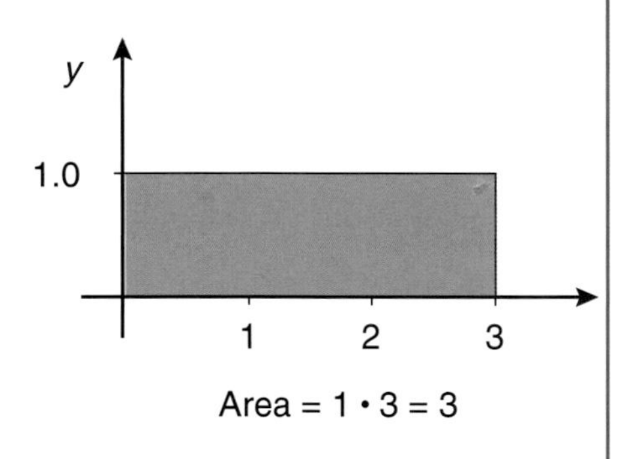

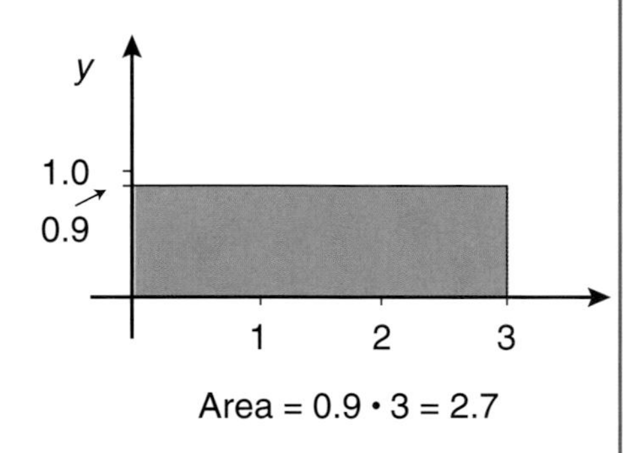

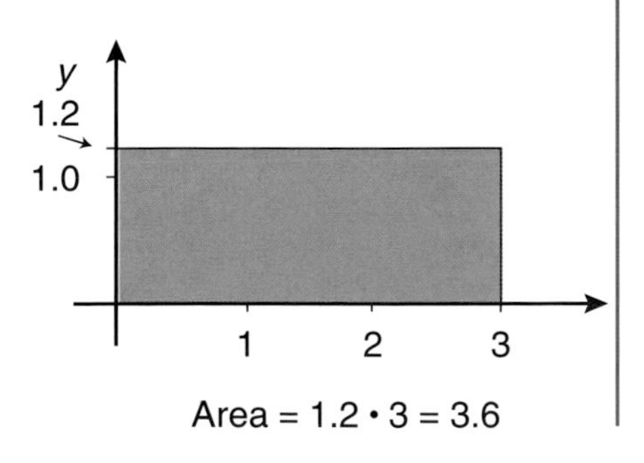

Figure 6.3 illustrates how students might use dynamic geometry software to examine how the product $3 \cdot y$ is affected by the magnitude of y for nonnegative y-values. In this illustration, the product is represented as the area of a $3 \times y$ rectangle. As a student changes the value of y by dragging a point along the vertical axis, the area of the rectangle changes. Referring to the area of the 3×1 rectangle, students can see that the area of the $3 \times y$ rectangle is smaller when y is less than 1 and larger when y is greater than 1. That is, in contrast to the expectation that multiplication makes bigger, multiplying 3 by a number smaller than 1 results in a product that is less than 3.

Teachers can help students extend their understanding of addition and subtraction with whole numbers to decimals by building on a solid understanding of place value. Students should be able to compute $1.4 + 0.67$ by applying their knowledge about $140 + 67$ and their understanding of the magnitude of the numbers involved in the computation. Without such a foundation, students may operate with decimal numbers inappropriately by, say, placing the decimal point in the wrong place after multiplying or dividing. Teachers can also help students add and subtract fractions correctly by helping them develop meaning for numerator, denominator, and equivalence and by encouraging them to use benchmarks and estimation (see fig. 6.4). Students who have a solid conceptual foundation in fractions should be less prone to committing computational errors than students who do not have such a foundation.

The division of fractions has traditionally been quite vexing for students. Although "invert and multiply" has been a staple of conventional mathematics instruction and although it seems to be a simple way to remember how to divide fractions, students have for a long time had difficulty doing so. Some students forget which number is to be inverted, and others are confused about when it is appropriate to apply the procedure. A common way of formally justifying the "invert and multiply" procedure is to use sophisticated arguments involving the manipulation of algebraic rational expressions—arguments beyond the reach of many middle-grades students. This process can seem very remote and mysterious to many students. Lacking an understanding of the underlying rationale, many students are therefore unable to repair their errors and clear up their confusions about division of fractions on their own. An alternative approach involves helping students understand the division of fractions by building on what they know about the division of whole numbers. If students understand the meaning of division as repeated subtraction, they can recognize that 24 ÷ 6 can be interpreted as "How many sets of 6 are there in a set of 24?" This view of division can also be applied to fractions, as seen in figure 6.5. To solve this problem, students can visualize repeatedly cutting off 3/4 yard of ribbon. The 5 yards of ribbon would provide enough for 6 complete bows, with a remainder of 2/4, or 1/2, yard of ribbon, which is enough for only 2/3 of a bow. Carefully sequenced experiences with problems such as these can help students build an understanding of division of fractions.

Fig. **6.4.**

Using benchmarks to estimate the results of a fraction computation

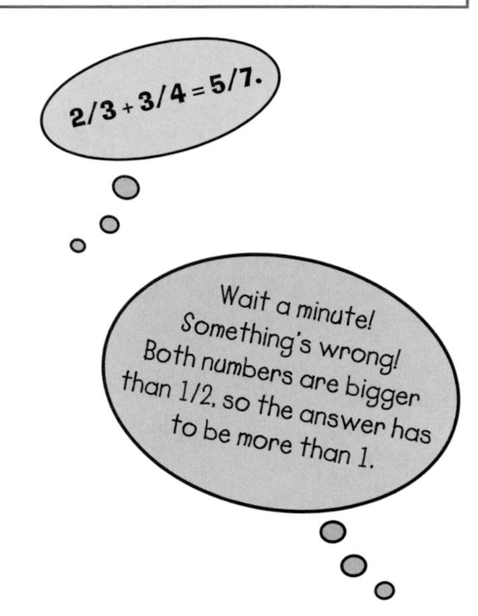

If 5 yards of ribbon are cut into pieces that are each 3/4 yard long to make bows, how many bows can be made?

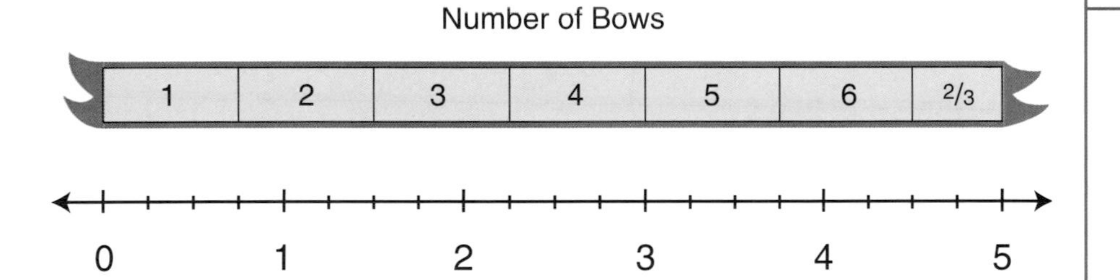

Fig. **6.5.**

Using the idea of division as repeated subtraction to solve a problem involving fractions

Students' understanding of operations with fractions, decimals, and integers can also be enhanced as they examine the validity and utility of properties of operations, such as the commutative and associative properties of addition and multiplication, with which they are familiar from their experiences with whole numbers. These properties can be used to simplify many computations with fractions, for example, 3 × (4/5 × 2/3) can be expressed as (3 × 2/3) × 4/5, which makes the calculation easier. The familiar distributive properties for whole-number operations can also be applied to fractions, decimals, and integers. Students already know that 3 × 26 can be computed by decomposing 26 and using the distributive property of multiplication over addition to get (3 × 20 + 3 × 6); in a similar fashion, they can compute 3 × 2 1/2 by expressing it as (3 × 2 + 3 × 1/2).

From earlier work with whole numbers, students should be familiar with the inverse relationship between the operation pairs of addition-subtraction and multiplication-division. In the middle grades, they can continue to apply this relationship as they work with fractions, decimals, and integers. In the middle grades, students should also add another pair to their repertoire of inverse operations—squaring and taking square roots. In grades 6–8, students frequently encounter squares and square roots when they use the Pythagorean relationship. They can use the inverse relationship to determine the approximate location of square roots between whole numbers on a number line. Figure 6.6 illustrates this reasoning for $\sqrt{27}$ and $\sqrt{99}$.

Fig. **6.6.**
Locating square roots on a number line

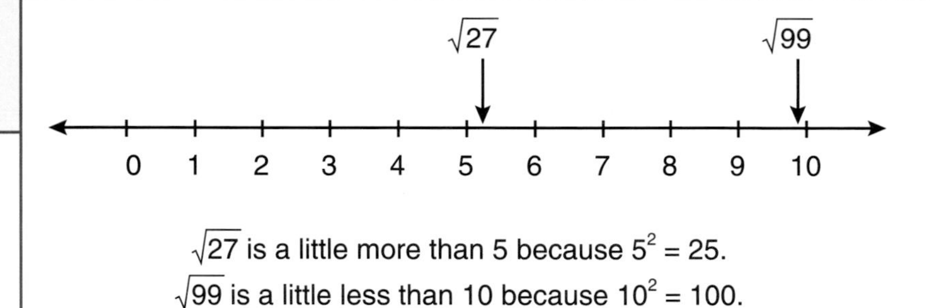

In grades 6–8, students should acquire computational fluency—the ability to compute efficiently and accurately—with fractions, decimals, and integers.

Compute fluently and make reasonable estimates

In grades 6–8, students should acquire computational fluency—the ability to compute efficiently and accurately—with fractions, decimals, and integers. Teachers should help students learn how to decide when an exact answer or an estimate would be more appropriate, how to choose the computational method that would be best to use, and how to evaluate the reasonableness of answers to computations. Most calculations should arise as students solve problems in context. Students should consider the features of the problem and the likely use of an answer to a calculation in deciding whether an exact answer or an estimate is needed, and then select an appropriate mode of calculation from among mental calculation, paper-and-pencil methods, or calculator use. For example, the cost of 1 1/4 pounds of cheese at $2.40 a pound can be found mentally, whereas the cost of 1.37 pounds of cheese at $2.95 a pound might be estimated, although a calculator would probably be the preferred tool if an exact answer were needed. Students should regularly analyze the answers to their calculations to evaluate their reasonableness.

Through teacher-orchestrated discussions of problems in context, students can develop useful methods to compute with fractions, decimals, percents, and integers in ways that make sense. Students' understanding of computation can be enhanced by developing their own methods and sharing them with one another, explaining why their methods work and are reasonable to use, and then comparing their methods with the algorithms traditionally taught in school. In this way, students can appreciate the power and efficiency of the traditional algorithms and also connect them to student-invented methods that may sometimes be less powerful or efficient but are often easier to understand.

Students should also develop and adapt procedures for mental calculation and computational estimation with fractions, decimals, and integers. Mental computation and estimation are also useful in many calculations involving percents. Because these methods often require flexibility in moving from one representation to another, they are useful in deepening students' understanding of rational numbers and helping them think flexibly about these numbers.

Instruction in solving proportions should include methods that have a strong intuitive basis. The so-called cross-multiplication method can be developed meaningfully if it arises naturally in students' work, but it can also have unfortunate side effects when students do not adequately understand when the method is appropriate to use. Other approaches to solving proportions are often more intuitive and also quite powerful. For example, when trying to decide which is the better buy—12 tickets for $15.00 or 20 tickets for $23.00—students might choose to use a scaling strategy (finding the cost for a common number of tickets) or a unit-rate strategy (finding the cost for one ticket). (See fig. 6.7.)

Fig. **6.7.**

Two approaches to solving a problem involving proportions

Which is the better buy—12 tickets for $15.00 or 20 tickets for $23.00?

Scaling Strategy

12 tickets for $15.00 → 60 tickets for $75.00.
20 tickets for $23.00 → 60 tickets for $69.00.

Unit-Rate Strategy

$15.00 for 12 tickets → $1.25 for 1 ticket.
$23.00 for 20 tickets → $1.15 for 1 ticket.

So 20 tickets for $23.00 is the better buy.

As different ways to think about proportions are considered and discussed, teachers should help students recognize when and how various ways of reasoning about proportions might be appropriate to solve problems. Further discussion about ways to approach a contextualized problem involving a proportional relationship is found in the "Connections" section of this chapter.

Algebra Standard for Grades 6–8

Instructional programs from prekindergarten through grade 12 should enable all students to—

Expectations

In grades 6–8 all students should–

Standard	Expectations
Understand patterns, relations, and functions	• represent, analyze, and generalize a variety of patterns with tables, graphs, words, and, when possible, symbolic rules; • relate and compare different forms of representation for a relationship; • identify functions as linear or nonlinear and contrast their properties from tables, graphs, or equations.
Represent and analyze mathematical situations and structures using algebraic symbols	• develop an initial conceptual understanding of different uses of variables; • explore relationships between symbolic expressions and graphs of lines, paying particular attention to the meaning of intercept and slope; • use symbolic algebra to represent situations and to solve problems, especially those that involve linear relationships; • recognize and generate equivalent forms for simple algebraic expressions and solve linear equations.
Use mathematical models to represent and understand quantitative relationships	• model and solve contextualized problems using various representations, such as graphs, tables, and equations.
Analyze change in various contexts	• use graphs to analyze the nature of changes in quantities in linear relationships.

Algebra

Students in the middle grades should learn algebra both as a set of concepts and competencies tied to the representation of quantitative relationships and as a style of mathematical thinking for formalizing patterns, functions, and generalizations. In the middle grades, students should work more frequently with algebraic symbols than in lower grades. It is essential that they become comfortable in relating symbolic expressions containing variables to verbal, tabular, and graphical representations of numerical and quantitative relationships. Students should develop an initial understanding of several different meanings and uses of variables through representing quantities in a variety of problem situations. They should connect their experiences with linear functions to their developing understandings of proportionality, and they should learn to distinguish linear relationships from nonlinear ones. In the middle grades, students should also learn to recognize and generate equivalent expressions, solve linear equations, and use simple formulas. Whenever possible, the teaching and learning of algebra can and should be integrated with other topics in the curriculum.

Understand patterns, relations, and functions

The study of patterns and relationships in the middle grades should focus on patterns that relate to linear functions, which arise when there is a constant rate of change. Students should solve problems in which they use tables, graphs, words, and symbolic expressions to represent and examine functions and patterns of change. For example, consider the following problem:

> Charles saw advertisements for two cellular telephone companies. Keep-in-Touch offers phone service for a basic fee of $20.00 a month plus $0.10 for each minute used. ChitChat has no monthly basic fee but charges $0.45 a minute. Both companies use technology that allows them to charge for the exact amount of time used; they do not "round up" the time to the nearest minute, as many of their competitors do. Compare these two companies' charges for the time used each month.

Students might begin by making a table, picking convenient numbers of minutes, and finding the corresponding costs for the two companies, as shown in figure 6.8a. Using a graphing calculator, students might then plot the points as ordered pairs (minutes, cost) on the coordinate plane, obtaining a graph for each of the two companies (see fig. 6.8b). Some students might describe the pattern in each graph verbally: "Keep-in-Touch costs $20.00 and then $0.10 more per minute." Others might write an equation to represent the cost (y) in dollars in terms of the number of minutes (x), such as $y = 20.00 + 0.10x$.

Before the students solve the problem, a teacher might ask them to use their table and graph to focus on important basic issues regarding

Fig. **6.8.**

Students can compare the charges for two telephone companies by making a table (a) and by representing the charges on a graphing calculator (b).

No. of minutes	0	10	20	30	40	50	60
Keep-in-Touch	$20.00	$21.00	$22.00	$23.00	$24.00	$25.00	$26.00
ChitChat	$0.00	$4.50	$9.00	$13.50	$18.00	$22.50	$27.00

(a)

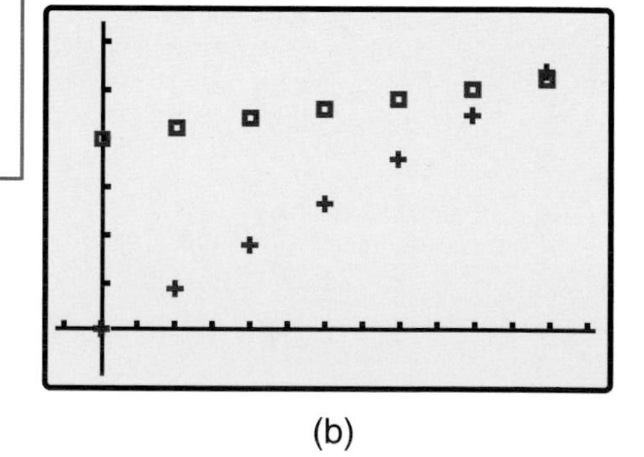

(b)

Students should develop a general understanding of, and facility with, slope and y-intercept and their manifestations in tables, graphs, and equations.

the relationships they represent. By asking, "How much would each company charge for 25 minutes? For 100 minutes?" the teacher could find out if students can interpret and extend the patterns. Since the table identifies only a small number of distinct points, a teacher could ask why it is legitimate to connect the points on the graph to make a line. Students might also be asked why one graph (for ChitChat) includes the origin but the other (for Keep-in-Touch) does not (see fig. 6.8b). Most students will recognize that the ChitChat graph includes the origin because there is no charge if no calls are made but the Keep-in-Touch graph includes (0, 20) because the company charges $20.00 even if the telephone is not used.

Many students will naturally seek a formula to express these patterns, but questions such as the following would be a good catalyst for others: How can you find the cost for any number of minutes for the Keep-in-Touch plan? For the ChitChat plan? What aspects of the stated price schedule are indicated in the graph? How? Students are likely to observe the constant difference between both the successive entries in the table and the coordinates of the points for each company along a straight line. They may explain the pattern underlying the function by saying, "Whenever you talk for one more minute, you pay $0.10 more (or $0.45 more), so the points go up the same amount each time." Others might say that a straight line is reasonable because each company charges a constant amount for each minute. Teachers should encourage students to explain their observations in their own words. Their explanations will provide the teacher with important insights into the students' thinking, particularly how well they recognize and represent linear relationships.

A solution to the stated problem requires comparing data from the two companies. A teacher might want to ask additional questions about this comparison: Which company is cheaper if you use the telephone infrequently? If you use it frequently? If you cannot spend more than $50.00 in a month but you want to talk for as many minutes as possible, which company would be the better choice? Considering questions such as these can lay the groundwork for a pivotal question: Is there a number of minutes that costs the same for both companies? Such questions could give rise to many observations. For example, most students will notice in their table that something important happens between 50 and 60 minutes, namely, using ChitChat becomes more expensive than using Keep-in-Touch. From the graph, some students may observe that this shift occurs at about 57 minutes: Keep-in-Touch is the cheaper company when a customer uses more than 57 minutes in a month. Experiences such as this can lay a foundation for solving systems of simultaneous equations.

The problem could also easily be extended or adapted in ways that would draw students' attention to important characteristics of the line graph for each company's charges. For example, to draw attention to the y-intercept, students could be asked to use a graphing calculator to examine how the graph would be affected if Keep-in-Touch increased or decreased its basic fee or if ChitChat decided to begin charging a basic fee. Students' attention could be drawn to the slope by asking them to consider the steepness of the lines using a question such as, What happens to the graph for Keep-in-Touch if the company increases its cost per minute from $0.10 to $0.15? Through experiences such as these, students should develop a general understanding of, and facility with, slope and y-intercept and their manifestations in tables, graphs, and equations.

The problem could also easily be extended to nonlinear relationships if, for instance, the companies did not charge proportionally for portions of minutes used. If they rounded to the nearest minute, then the cost for each company would be graphed as a step function rather than a linear function. In another variation, a nonlinear pricing scheme for a third company could be introduced.

Another important topic for class discussion is comparing and contrasting the merits of graphical, tabular, and symbolic representations in this example. A teacher might ask, "Which helps us see better the point at which the two companies switch position and Keep-in-Touch becomes the more economical—a table or a graph?" "Is it easier to see the rate per minute from the graph or from the equation?" or "How can you determine the rate per minute from the table?" Through discussion, students can identify the strengths and the limitations of different forms of representation. Graphs give a picture of a relationship and allow the quick recognition of linearity when change is constant. Algebraic equations typically offer compact, easily interpreted descriptions of relationships between variables.

Represent and analyze mathematical situations and structures using algebraic symbols

Working with variables and equations is an important part of the middle-grades curriculum. Students' understanding of variable should go far beyond simply recognizing that letters can be used to stand for unknown numbers in equations (Schoenfeld and Arcavi 1988). The following equations illustrate several uses of variable encountered in the middle grades:

$$27 = 4x + 3$$
$$1 = t(1/t)$$
$$A = LW$$
$$y = 3x$$

The role of variable as "place holder" is illustrated in the first equation: x is simply taking the place of a specific number that can be found by solving the equation. The use of variable in denoting a generalized arithmetic pattern is shown in the second equation; it represents an identity when t takes on any real value except 0. The third equation is a formula, with A, L, and W representing the area, length, and width, respectively, of a rectangle. The third and fourth equations offer examples of covariation: in the fourth equation, as x takes on different values, y also varies.

Students can identify the strengths and the limitations of different forms of representation.

Most students will need extensive experience in interpreting relationships among quantities in a variety of problem contexts before they can work meaningfully with variables and symbolic expressions. An understanding of the meanings and uses of variables develops gradually as students create and use symbolic expressions and relate them to verbal, tabular, and graphical representations. Relationships among quantities can often be expressed symbolically in more than one way, providing opportunities for students to examine the equivalence of various algebraic expressions. Fairly simple equivalences can be involved: the cost (in dollars) of using Keep-in-Touch can be expressed as $y = 0.10x + 20$, as $y = 20 + 0.10x$, as $20 + 0.10x = y$, and as $0.10x + 20 = y$. Complex symbolic expressions also can be examined, such as the equivalence of $4 + 2L + 2W$ and $(L + 2)(W + 2) - LW$ when representing the number of unit tiles to be placed along the border of a rectangular pool with length L units and width W units; see the "Representation" section of this chapter for a discussion of this example.

A problem such as the one in figure 6.9 (adapted from Educational Development Center, Inc. 1998, p. 41) could give students valuable experience in deciding whether two expressions are equivalent. A teacher might encourage students to begin solving this problem by drawing several more boxes of various sizes so they can look for a pattern. Some students will probably note that the caramels are also arranged in a rectangular pattern, which is narrower and shorter than the rectangular arrangement of chocolates. Using this observation, they might report, "To find the length and width of the caramel rectangle, take 1 off the length and 1 off the width of the chocolate rectangle. Multiply the length and width of the caramel rectangle to find the number of caramels." If L and W are the dimensions of the array of chocolates, and C is the number of caramels, then this generalization could be expressed symbolically as $C = (L - 1)(W - 1)$. Other students might find and use the number of chocolates to find the number of caramels. For example, for a 3×5 box of chocolates, they might propose starting with the 15 chocolates, then taking off 3 because "there is one less column of caramels" and then taking off 5 "because there is one less row of caramels." This could be expressed generally as $C = LW - L - W$. Although both expressions for the number of caramels are likely to seem reasonable to many students, they do not yield the same answer. For the 3×5 array, the first produces the correct answer, 8 caramels. The second gives the answer 7 caramels. Either examining a few more boxes with different dimensions or reconsidering the process represented by the second equation would confirm that the second equation needs to be corrected by adding 1 to obtain $C = LW - L - W + 1$. The algebraic equivalence of $(L - 1)(W - 1)$ and $LW - L - W + 1$ can be demonstrated in general using the distributive property of multiplication over subtraction.

Fig. **6.9.**

The Super Chocolates problem

Super Chocolates are arranged in boxes so that a caramel is placed in the center of each array of four chocolates, as shown below. The dimensions of the box tell you how many columns and how many rows of chocolates come in the box. Develop a method to find the number of caramels in any box if you know its dimensions. Explain and justify your method using words, diagrams, or expressions.

2×2

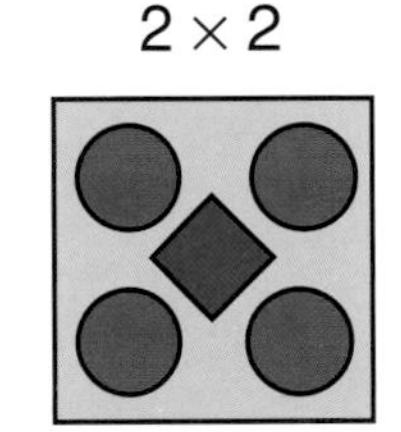

2×4

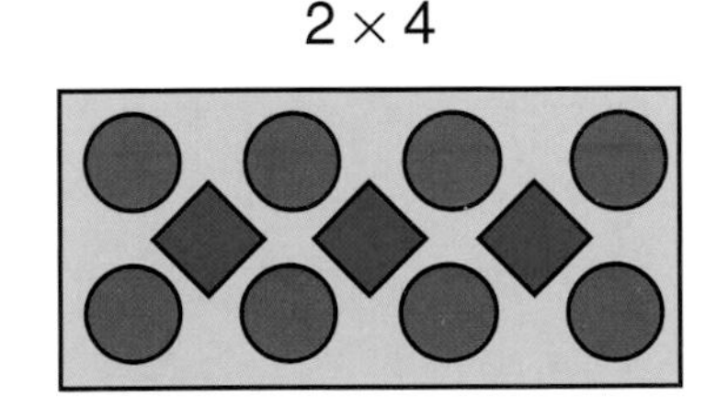

3×5

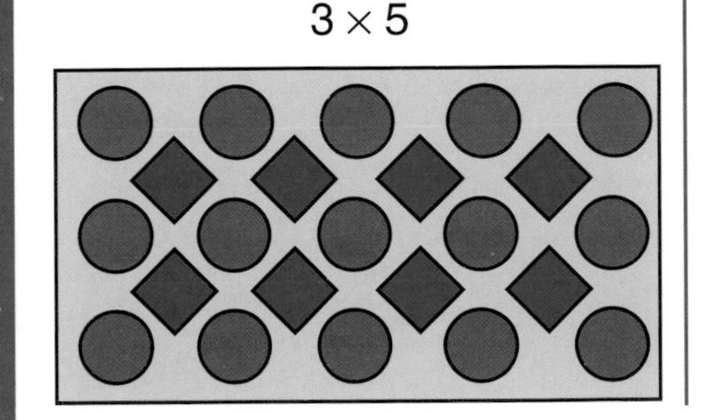

Through a variety of experiences such as these, students can learn the strengths and limitations of various methods for checking the equivalence of expressions. In some instances, the equivalence of algebraic expressions can be demonstrated geometrically; see the "Geometry" section of this chapter for a demonstration that $(a + b)^2 = a^2 + 2ab + b^2$.

Most middle-grades students will need considerable experience with linear equations before they will be comfortable and fluent in transforming or solving them. Although students will probably acquire facility with equations at different times during the middle grades, by the end of grade 8, students should be able to solve equations like $84 - 2x = 5x + 12$ for the unknown number, to recognize as identities such equations as $1 = t(1/t)$ (when t is not 0), to apply formulas such as $V = \pi r^2 h$, and to recognize that equations such as $y = -3x + 10$ represent linear functions that are satisfied by many ordered pairs (x, y). Students should be able to use equations of the form $y = mx + b$ to represent linear relationships, and they should know how the values of the slope (m) and the y-intercept (b)

affect the line. For example, in the "cellular telephone" problem discussed earlier, they should recognize that $y = 0.10x + 20$ and $y = 0.45x$ are both linear equations, that the graph of the latter will be steeper than that of the former, and that the former intersects the y-axis at (0, 20) rather than at the origin.

Students' facility with symbol manipulation can be enhanced if it is based on extensive experience with quantities in contexts through which students develop an initial understanding of the meanings and uses of variables and an ability to associate symbolic expressions with problem contexts. Fluency in manipulating symbolic expressions can be further enhanced if students understand equivalence and are facile with the order of operations and the distributive, associative, and commutative properties.

Students' facility with symbol manipulation can be enhanced if it is based on extensive experience with quantities in contexts.

Use mathematical models to represent and understand quantitative relationships

A major goal in the middle grades is to develop students' facility with using patterns and functions to represent, model, and analyze a variety of phenomena and relationships in mathematics problems or in the real world. With computers and graphing calculators to produce graphical representations and perform complex calculations, students can focus on using functions to model patterns of quantitative change. Students should have frequent experiences in modeling situations with equations of the form $y = kx$, such as relating the side lengths and the perimeters of similar shapes. Opportunities can be found in many other areas of the curriculum; for example, scatterplots and approximate lines of fit can model trends in data sets. Students also need opportunities to model relationships in everyday contexts, such as the "cellular telephone" problem.

Students also should have experience in modeling situations and relationships with nonlinear functions, such as compound-interest problems, the relationship between the length of the radius of a circle and the area of the circle, or situations like the one in figure 6.10. If students have only a few points to examine, it can be difficult to see that

Fig. **6.10.**

A problem involving a nonlinear relationship, with an associated table and graph

Consider rectangles with a fixed area of 36 square units. The width (W) of the rectangles varies in relation to the length (L) according to the formula $W = 36/L$. Make a table showing the widths for all the possible whole-number lengths for these rectangles up to $L = 36$.

Solution:

Length	1	2	3	4	5	6	7	8	9	10	11	...	36
Width	36	18	12	9	7.2	6	5.14	4.5	4	3.6	3.27	...	1

Look at the table and examine the pattern of the difference between consecutive entries for the length and the width. As the length increases by 1, the width decreases, but not at a constant rate. What do you expect the graph of the relationship between L and W to look like? Will it be a straight line? Why or why not?

Solution:

The graph is not a straight line because the rate of change is not constant. Instead the graph appears to be a curve that bends sharply downward and then becomes more level.

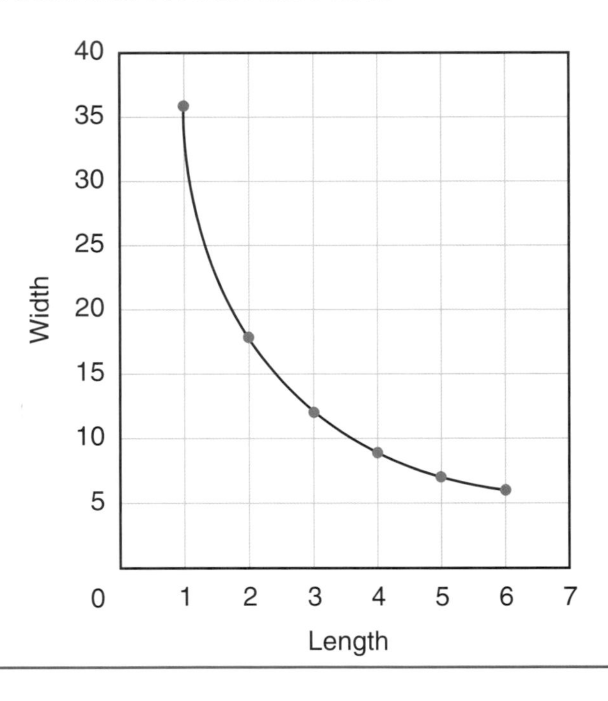

With a graphing calculator or computer graphing software, students can test some conjectures more easily than with paper-and-pencil methods.

the graph for this problem is not linear. As more points are graphed, however, the curve becomes more apparent. Students could use graphing calculators or computer graphing tools to do problems such as this.

When doing experiments or dealing with real data, students may encounter "messy data," for which a line or a curve may not be an exact fit. They will need experience with such situations and assistance from the teacher to develop their ability to find a function that fits the data well enough to be useful as a prediction tool. In their later study of statistics, students may learn sophisticated methods to determine lines of best fit for data. In the middle grades when students encounter a set of points suggesting a linear relationship, they can simply use a ruler to try several lines until they find one that appears to be a good fit and then write an equation for that line. An example of this sort of activity, related to a scatterplot of measurements, can be found in the "Data Analysis and Probability" section of this chapter. With a graphing calculator or computer graphing software, students can test some conjectures more easily than with paper-and-pencil methods.

Analyze change in various contexts

In their study of algebra, middle-grades students should encounter questions that focus on quantities that change. Recall, for example, that ChitChat charges $0.45 a minute for phone calls. The cost per minute does not change, but the total cost changes as the telephone is used. This can be seen quite readily from the two graphs in figure 6.11. The meaning of the term *flat rate* can be seen in the cost-per-minute graph, which shows points along a horizontal line at $y = 0.45$, representing a constant rate of $0.45 a minute. The total-cost graph shows points along a straight line that includes the origin and has a slope of 0.45.

Fig. **6.11**.

These two graphs represent different relationships in ChitChat's pricing scheme.

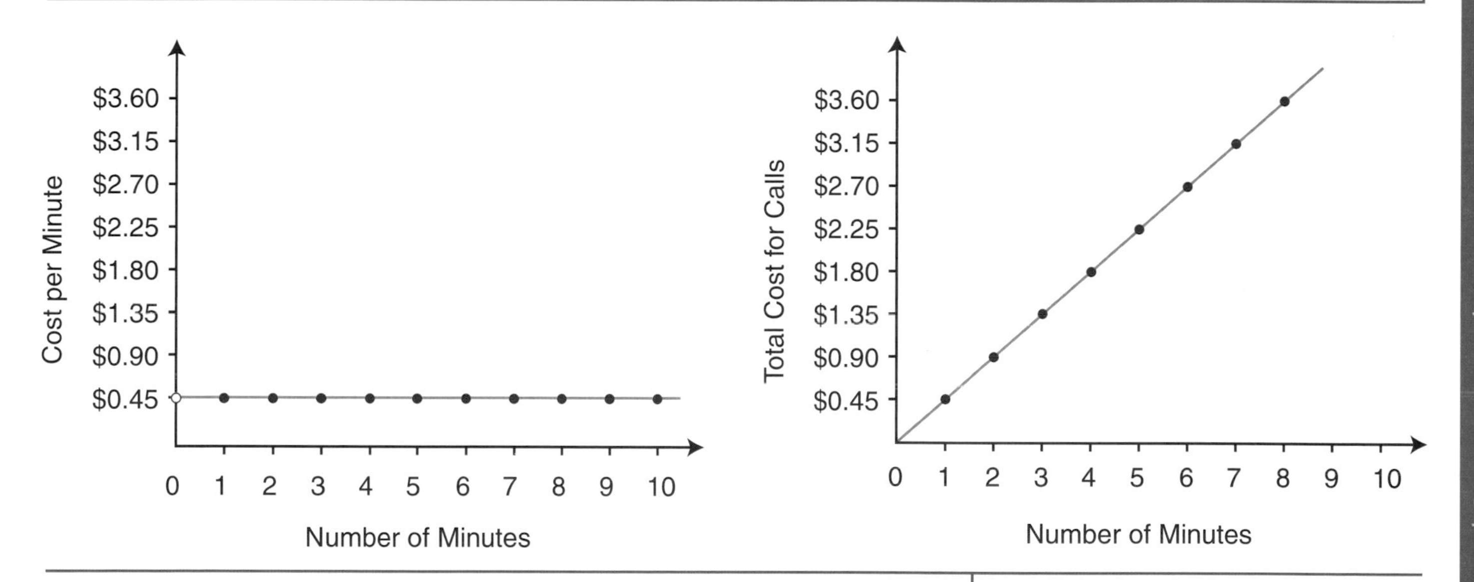

Students may be confused when they first encounter two different graphs to represent different relationships in the same situation. Teachers can assist students in understanding what relationships are represented in the two graphs by asking such questions as these: When the number of minutes is 4, what do the values of the corresponding point on each graph represent? When the number of minutes is 8? Why is the y-value of the cost-per-minute graph constant at 0.45? How much does the total-cost graph increase from 5 minutes to 6 minutes? Why? How would each graph change, if at all, if the cost per minute were changed to $0.20?

A slight modification of the problem, such as the addition of a third telephone company with a different pricing scheme, can allow the analysis of change in nonlinear relationships:

> Quik-Talk advertises monthly cellular phone service for $0.50 a minute for the first 60 minutes but only $0.10 a minute for each minute thereafter. Quik-Talk also charges for the exact amount of time used.

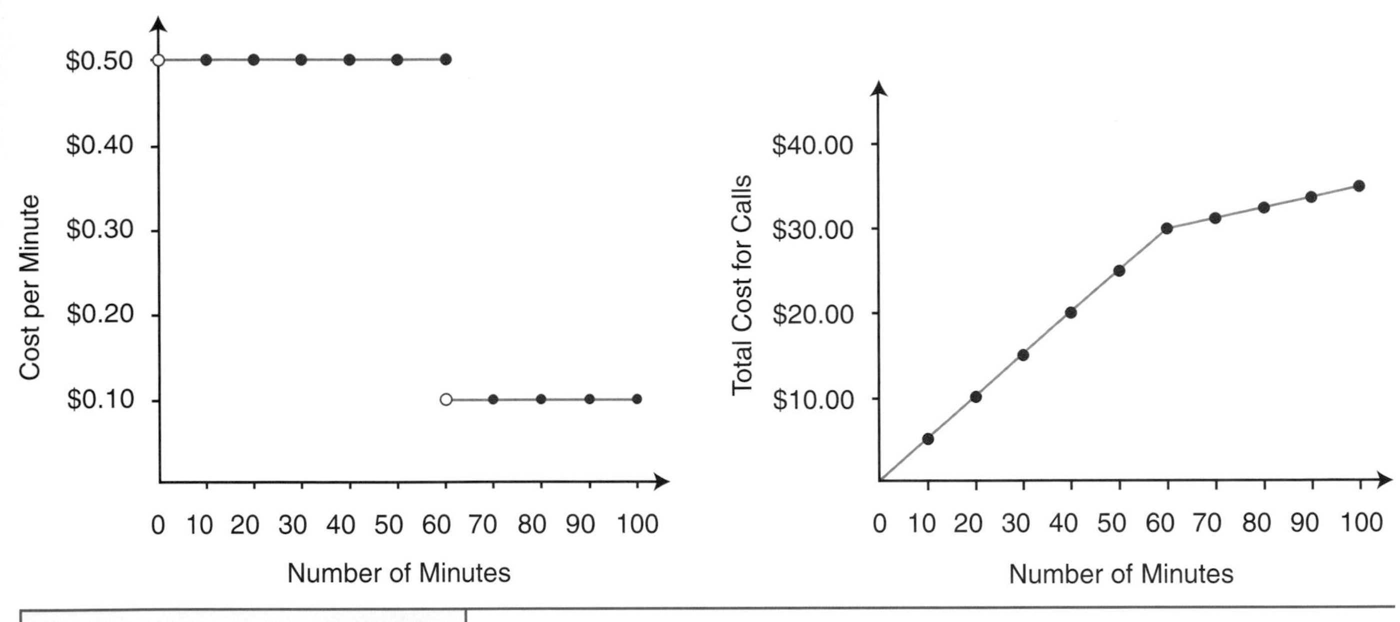

Fig. **6.12.**

These two graphs represent different relationships in Quik-Talk's pricing scheme.

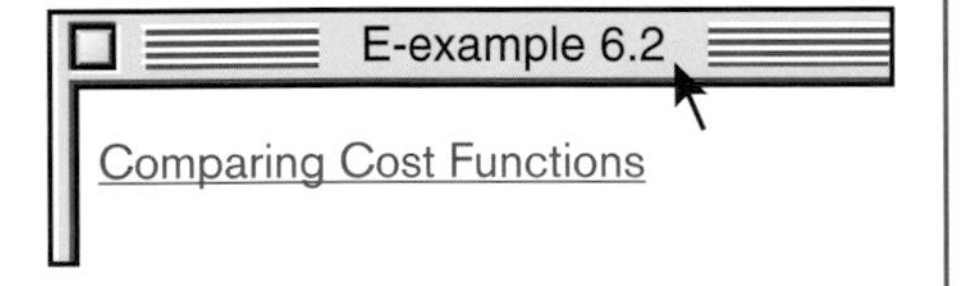

A teacher might ask students to graph the rates of change in this example. Figure 6.12 shows the cost-per-minute graph and a total-cost graph for Quik-Talk's pricing scheme. Students can answer questions about the relationships represented in the graphs: Why does the cost-per-minute graph consist of two different line segments? How can we tell from the graph that the pricing scheme changes after 60 minutes? Why is part of the total-cost function steeper than the rest of the graph?

Comparing the cost-per-minute graph to the total-cost graph for phone calls can help students develop a clearer understanding of the relationship between change (cost per minute) and accumulation (total cost of calls). These concepts are precursors to the later study of change in calculus.

Students' examination of graphs of change and graphs of accumulation can be facilitated with specially designed computer software. Such software allows students to change either the number of minutes used in one month by dragging a horizontal "slider" (see fig. 6.13) or the cost per minute by dragging a vertical slider. They can then observe the corresponding changes in the graphs and in the symbolic expression for the relationship. Technological tools can also help students examine the nature of change in many other settings. For example, students could examine distance-time relationships using computer-based laboratories, as discussed in the "Measurement" section of this chapter. Such experiences with appropriate technology, supported by careful planning by teachers and interactions with classmates, can help students develop a solid understanding of some fundamental notions of change.

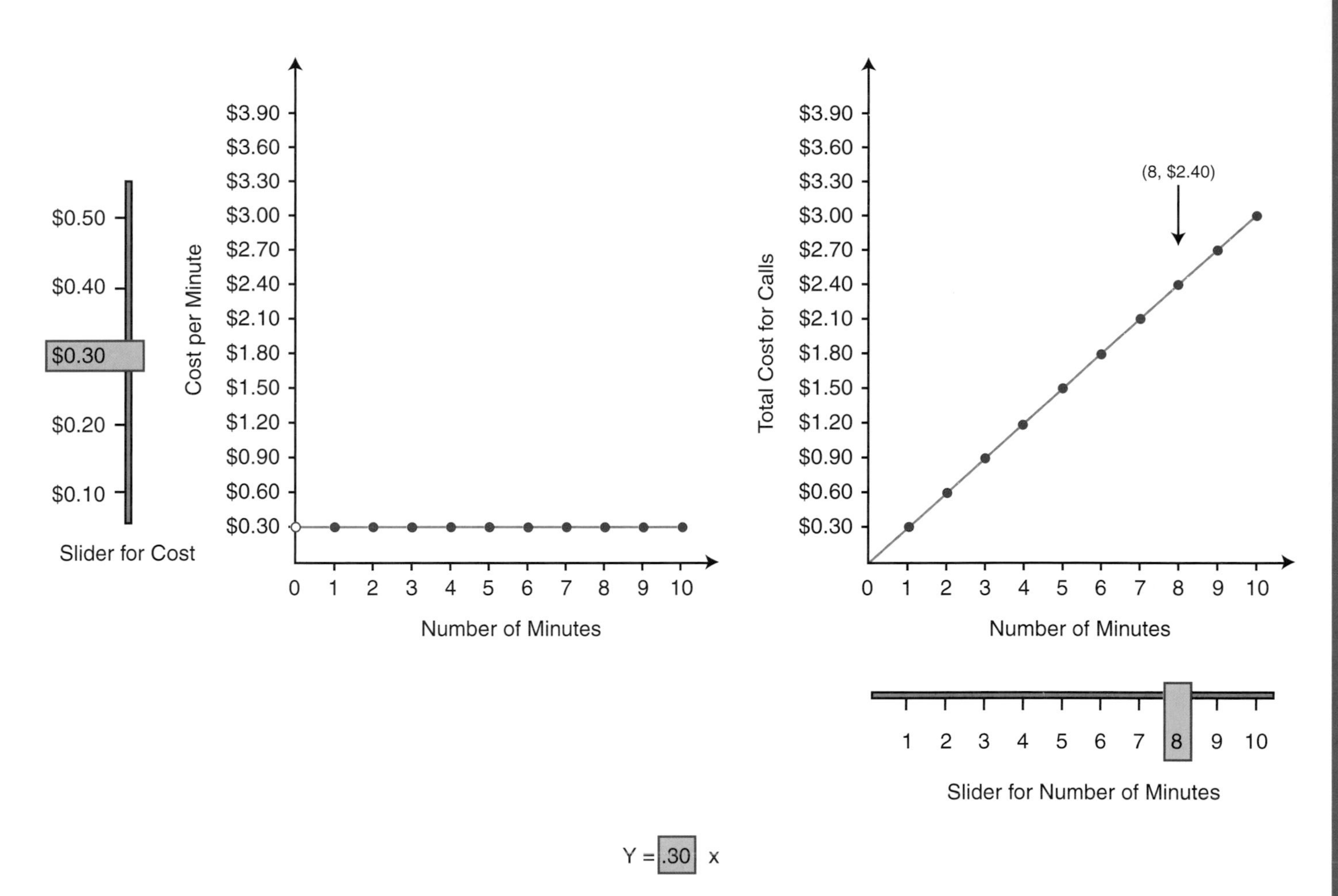

Fig. **6.13.**

Computer software can help students understand some fundamental notions of change.

Geometry Standard for Grades 6–8

Instructional programs from prekindergarten through grade 12 should enable all students to—

Expectations

In grades 6–8 all students should–

Instructional programs from prekindergarten through grade 12 should enable all students to—	In grades 6–8 all students should–
Analyze characteristics and properties of two- and three-dimensional geometric shapes and develop mathematical arguments about geometric relationships	• precisely describe, classify, and understand relationships among types of two- and three-dimensional objects using their defining properties; • understand relationships among the angles, side lengths, perimeters, areas, and volumes of similar objects; • create and critique inductive and deductive arguments concerning geometric ideas and relationships, such as congruence, similarity, and the Pythagorean relationship.
Specify locations and describe spatial relationships using coordinate geometry and other representational systems	• use coordinate geometry to represent and examine the properties of geometric shapes; • use coordinate geometry to examine special geometric shapes, such as regular polygons or those with pairs of parallel or perpendicular sides.
Apply transformations and use symmetry to analyze mathematical situations	• describe sizes, positions, and orientations of shapes under informal transformations such as flips, turns, slides, and scaling; • examine the congruence, similarity, and line or rotational symmetry of objects using transformations.
Use visualization, spatial reasoning, and geometric modeling to solve problems	• draw geometric objects with specified properties, such as side lengths or angle measures; • use two-dimensional representations of three-dimensional objects to visualize and solve problems such as those involving surface area and volume; • use visual tools such as networks to represent and solve problems; • use geometric models to represent and explain numerical and algebraic relationships; • recognize and apply geometric ideas and relationships in areas outside the mathematics classroom, such as art, science, and everyday life.

Geometry

Students should come to the study of geometry in the middle grades with informal knowledge about points, lines, planes, and a variety of two- and three-dimensional shapes; with experience in visualizing and drawing lines, angles, triangles, and other polygons; and with intuitive notions about shapes built from years of interacting with objects in their daily lives.

In middle-grades geometry programs based on these recommendations, students investigate relationships by drawing, measuring, visualizing, comparing, transforming, and classifying geometric objects. Geometry provides a rich context for the development of mathematical reasoning, including inductive and deductive reasoning, making and validating conjectures, and classifying and defining geometric objects. Many topics treated in the Measurement Standard for the middle grades are closely connected to students' study of geometry.

Analyze characteristics and properties of two- and three-dimensional geometric shapes and develop mathematical arguments about geometric relationships

Middle-grades students should explore a variety of geometric shapes and examine their characteristics. Students can conduct these explorations using materials such as geoboards, dot paper, multiple-length cardboard strips with hinges, and dynamic geometry software to create two-dimensional shapes.

Students must carefully examine the features of shapes in order to precisely define and describe fundamental shapes, such as special types of quadrilaterals, and to identify relationships among the types of shapes. A teacher might ask students to draw several parallelograms on a coordinate grid or with dynamic geometry software. Students should make and record measurements of the sides and angles to observe some of the characteristic features of each type of parallelogram. They should then generate definitions for these shapes that are correct and consistent with the commonly used ones and recognize the principal relationships among elements of these parallelograms. A Venn diagram like the one shown in figure 6.14 might be used to summarize observations that a square is a special case of a rhombus and rectangle, each of which is a special case of a parallelogram.

The teacher might also ask students to draw the diagonals of multiple examples of each shape, as shown in figure 6.15, and then measure the lengths of the diagonals and the angles they form. The results can be summarized in a table like that in figure 6.16. Students should observe that the diagonals of these parallelograms bisect each other, which they might propose as a defining characteristic of a parallelogram. Moreover, they might observe, the diagonals are perpendicular in rhombuses (including squares) but not in other parallelograms and the diagonals are of equal length in rectangles (including squares) but not in other parallelograms. These observations might suggest other defining characteristics of special quadrilaterals, for instance, that a square is a parallelogram with diagonals that are perpendicular and of equal

Fig. **6.14.**

A diagram showing the relationships among types of parallelograms

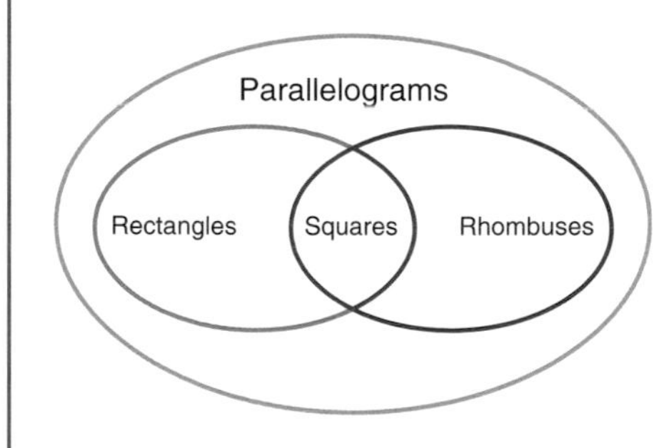

Fig. **6.15.**

Students can draw the diagonals of parallelograms to make further observations.

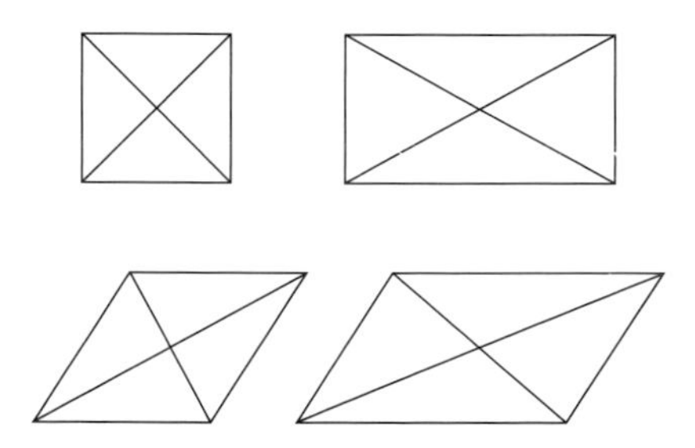

Fig. **6.16.**

A table of students' observations about the properties of the diagonals of special types of quadrilaterals

	Do the diagonals always bisect each other?	Are the diagonals always equal in length?	Are the diagonals always perpendicular?
Parallelograms	Yes	No	No
Rectangles	Yes	Yes	No
Rhombuses	Yes	No	Yes
Squares	Yes	Yes	Yes

length. Using dynamic geometry software, students could explore the adequacy of this definition by trying to generate a counterexample.

Middle-grades students also need experience in working with congruent and similar shapes. From their earlier work, students should understand that congruent shapes and angles are identical and can be "matched" by placing one atop the other. Students can begin with an intuitive notion of similarity: similar shapes have congruent angles but not necessarily congruent sides. In the middle grades, they should extend their understanding of similarity to be more precise, noting, for instance, that similar shapes "match exactly when magnified or shrunk" or that their corresponding angles are congruent and their corresponding sides are related by a scale factor.

Students can investigate congruence and similarity in many settings, including art, architecture, and everyday life. For example, observe the overlapping pairs of triangles in the design of the kite in figure 6.17. The overlapping triangles, which have been disassembled in the figure, can be shown to be similar. Students can measure the angles of the triangles in the kite and see that their corresponding angles are congruent. They can measure the lengths of the sides of the triangles and see that the differences are not constant but are instead related by a constant scale factor. With the teacher's guidance, students can thus begin to develop a more formal definition of similarity in terms of relationships among sides and angles.

Investigations into the properties of, and relationships among, similar shapes can afford students many opportunities to develop and evaluate conjectures inductively and deductively. For example, an investigation of the perimeters, areas, and side lengths of the similar and

Fig. **6.17.**

Kite formed by overlapping triangles

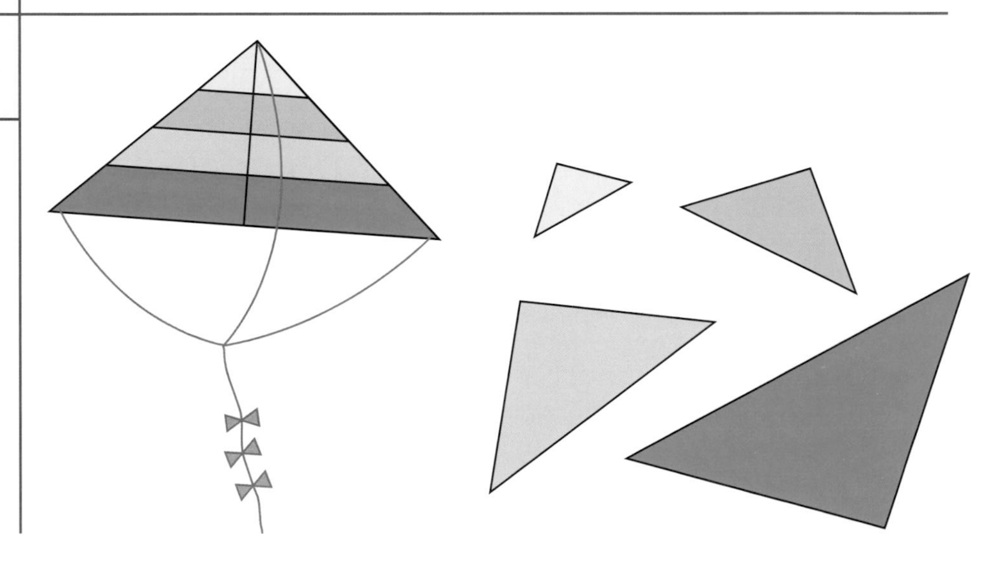

congruent triangles in the kite example could reveal relationships and lead to generalizations. Teachers might encourage students to formulate conjectures about the ratios of the side lengths, of the perimeters, and of the areas of the four similar triangles. They might conjecture that the ratio of the perimeters is the same as the scale factor relating the side lengths and that the ratio of the areas is the square of that scale factor. Then students could use dynamic geometry software to test the conjectures with other examples. Students can formulate deductive arguments about their conjectures. Communicating such reasoning accurately and clearly prepares students for creating and understanding more-formal proofs in subsequent grades.

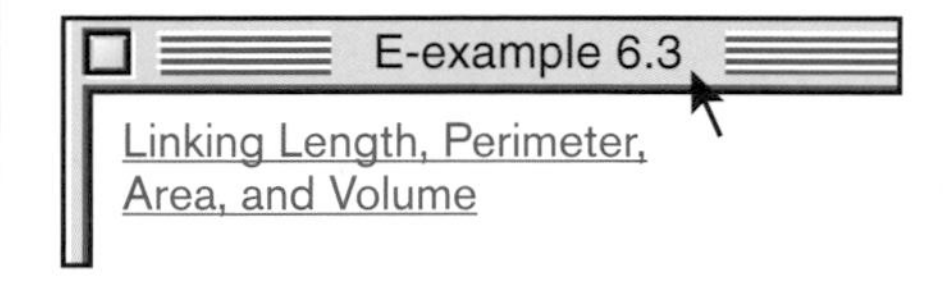

Specify locations and describe spatial relationships using coordinate geometry and other representational systems

Geometric and algebraic representations of problems can be linked using coordinate geometry. Students could draw on the coordinate plane examples of the parallelograms discussed previously, examine their characteristic features using coordinates, and then interpret their properties algebraically. Such an investigation might include finding the slopes of the lines containing the segments that compose the shapes. From many examples of these shapes, students could make important observations about the slopes of parallel lines and perpendicular lines. Figure 6.18 helps to illustrate for one specific rhombus what might be observed in general: the slopes of parallel lines (in this instance, the opposite sides of the rhombus) are equal and the slopes of perpendicular lines (in this instance, the diagonals of the rhombus) are negative reciprocals. The slopes of the diagonals are

$$\frac{19-(-5)}{11-(-5)}=\frac{24}{16}=\frac{3}{2}$$

and

$$\frac{11-3}{-3-9}=\frac{8}{-12}=-\frac{2}{3}.$$

Fig. **6.18.**

A rhombus drawn on the coordinate plane

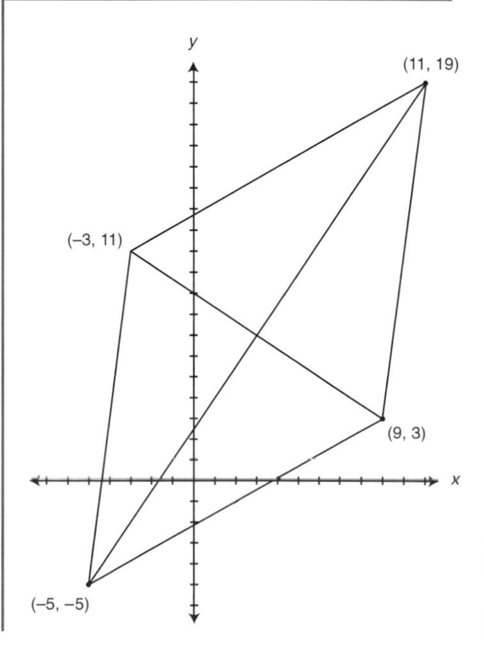

Apply transformations and use symmetry to analyze mathematical situations

Transformational geometry offers another lens through which to investigate and interpret geometric objects. To help them form images of shapes through different transformations, students can use physical objects, figures traced on tissue paper, mirrors or other reflective surfaces, figures drawn on graph paper, and dynamic geometry software. They should explore the characteristics of flips, turns, and slides and should investigate relationships among compositions of transformations. These experiences should help students develop a strong understanding of line and rotational symmetry, scaling, and properties of polygons.

From their experiences in grades 3–5, students should know that rotations, slides, and flips produce congruent shapes. By exploring the positions, side lengths, and angle measures of the original and resulting figures, middle-grades students can gain new insights into congruence. They could, for example, note that the images resulting from transformations have different positions and sometimes different orientations

Fig. **6.19.**

Using dynamic geometry software, students can explore the results of reflections (a) and rotations (b).

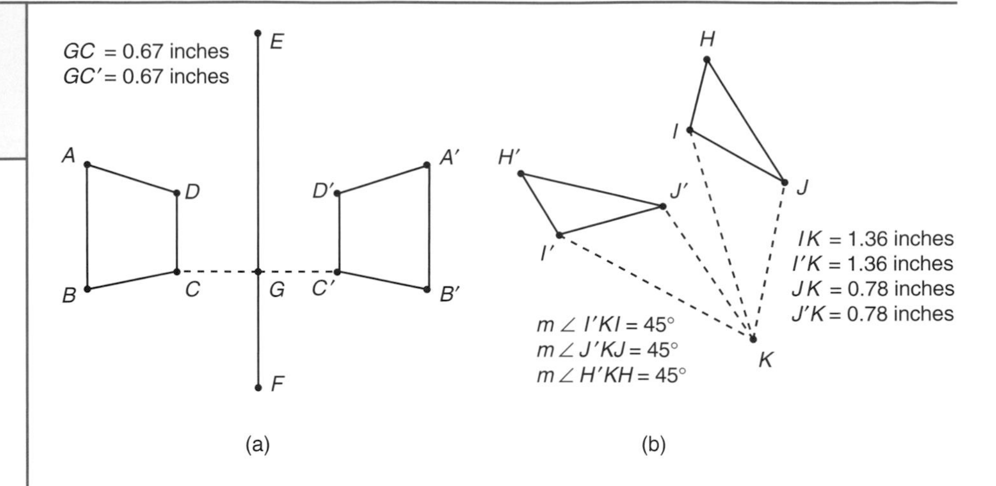

E-example 6.4

Congruence, Similarity, and Symmetry through Transformations

from those of the original figure (the preimage), although they have the same side lengths and angle measures as the original. Thus congruence does not depend on position and orientation.

Transformations can become an object of study in their own right. Teachers can ask students to visualize and describe the relationship among lines of reflection, centers of rotation, and the positions of preimages and images. Using dynamic geometry software, students might see that each point in a reflection is the same distance from the line of reflection as the corresponding point in the preimage, as shown in figure 6.19a. In a rotation, such as the one shown in figure 6.19b, students might note that the corresponding vertices in the preimage and image are the same distance from the center of rotation and that the angles formed by connecting the center of rotation to corresponding pairs of vertices are congruent.

Fig. **6.20.**

Three pairs of congruent shapes

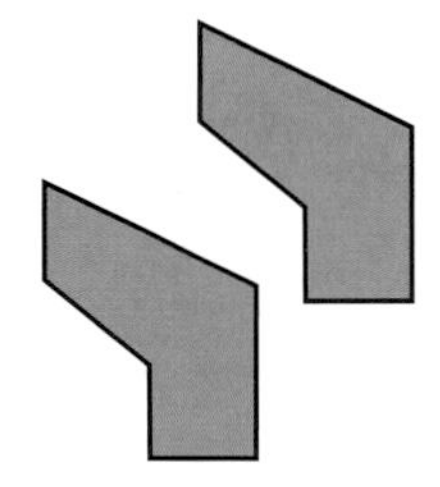

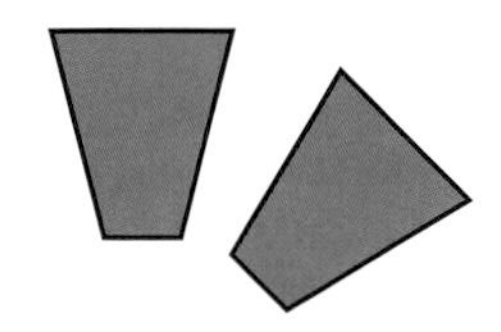

Teachers can pose additional challenges to develop students' understanding of transformations and congruence. For example, given the three pairs of congruent shapes in figure 6.20, students might be asked to identify a transformation applied to transform one shape into the other. Most students who have had extensive experience with transformations should see that the first pair appears to be related by a reflection, the second pair by a translation, and the third pair by either a reflection or a rotation. As students develop a more sophisticated understanding of transformations, they could be asked to describe the transformation more exactly, using distance, angles, and headings. The transformation for the second pair of shapes, for example, is a translation of 1.5 cm at a 45-degree angle.

Teachers may also want students to consider what happens when transformations are composed. For example, the image produced when a figure is reflected through one line and the resulting image is reflected through a different line will be either a translation of the preimage if the lines of reflection are parallel or a rotation if they intersect. To assess students' understanding of transformations, teachers can give students two congruent shapes and have them specify a transformation or a composition of transformations that will map one to the other. This can be done using dynamic geometry software.

Transformations can also be used to help students understand similarity and symmetry. Work with magnifications and contractions, called *dilations*, can support students' developing understanding of similarity. For example, dilation of a shape affects the length of each side by a constant scale factor, but it does not affect the orientation or the magnitude of the angles. In a similar manner, rotations and reflections can help students understand symmetry. Students can observe that when a figure has rotational symmetry, a rotation can be found such that the preimage (original shape) exactly matches the image but its vertices map to different vertices. Looking at line symmetry in certain classes of shapes can also lead to interesting observations. For example, isosceles trapezoids have a line of symmetry containing the midpoints of the parallel opposite sides (often called *bases*). Students can observe that the pair of sides not intersected by the line of symmetry (often called the *legs*) are congruent, as are the two corresponding pairs of angles. Students can conclude that the diagonals are the same length, since they can be reflected onto each other, and that several pairs of angles related to those diagonals are also congruent. Further exploration reveals that rectangles and squares also have a line of symmetry containing the midpoints of a pair of opposite sides (and other lines of symmetry as well) and all the resulting properties.

Students' skills in visualizing and reasoning about spatial relationships are fundamental in geometry.

Use visualization, spatial reasoning, and geometric modeling to solve problems

Students' skills in visualizing and reasoning about spatial relationships are fundamental in geometry. Some students may have difficulty finding the surface area of three-dimensional shapes using two-dimensional representations because they cannot visualize the unseen faces of the shapes. Experience with models of three-dimensional shapes and their two-dimensional "nets" is useful in such visualization (see fig. 6.25 in the "Measurement" section for an example of a net). Students also need to examine, build, compose, and decompose complex two- and three-dimensional objects, which they can do with a variety of media, including paper-and-pencil sketches, geometric models, and dynamic geometry software. Interpreting or drawing different views of buildings, such as the base floor plan and front and back views, using dot paper can be useful in developing visualization. Students should build three-dimensional objects from two-dimensional representations; draw objects from a geometric description; and write a description, including its geometric properties, for a given object.

Students can also benefit from experience with other visual models, such as networks, to use in analyzing and solving real problems, such as those concerned with efficiency. To illustrate the utility of networks, students might consider the problem and the networks given in figure 6.21 (adapted from Roberts [1997, pp. 106–7]). The teacher could ask students to determine one or several efficient routes that Caroline might use for the streets on map A, share their solutions with the class, and describe how they found them. Students should note the start-end point of each route and the number of different routes that they find. Students could then find an efficient route for map B. They should eventually conclude that no routes in map B satisfy the conditions of the problem. They should discuss why no such route can be found; the teacher might suggest that students count the number of paths attached to each node and look at where they "get stuck" in order to better

Fig. **6.21.**

Networks used to solve efficiency problems

Caroline's job is to collect money from parking meters. She wants to find an efficient route that starts and ends at the same place and travels on each street only once.

A. The streets she has to cover are shown in map A. Find and trace such a route for her.

B. A new street, shown in map B, may be added to her route. Can you find an efficient route that includes the new street?

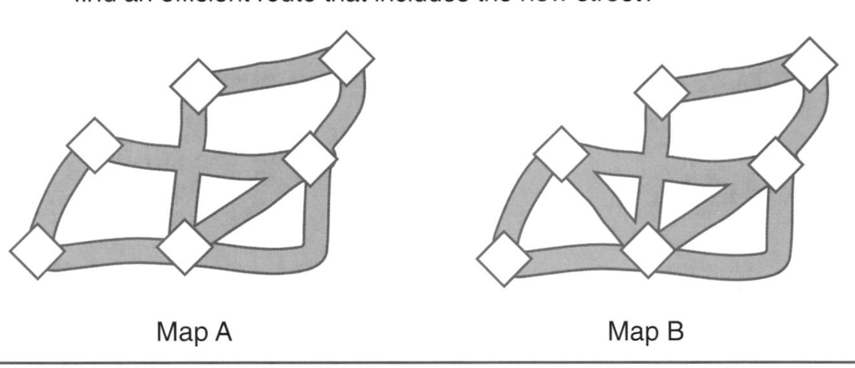

understand why they reach an impasse. To extend this investigation, students could look for efficient paths in other situations or they might change the conditions of the map B problem to find the pathway with the least backtracking. Such an investigation in the middle grades is a precursor of later work with Hamiltonian circuits, a foundation for work with sophisticated networks.

Visual demonstrations can help students analyze and explain mathematical relationships. Eighth graders should be familiar with one of the many visual demonstrations of the Pythagorean relationship—the diagram showing three squares attached to the sides of a right triangle. Students could replicate some of the other visual demonstrations of the relationship using dynamic geometry software or paper-cutting procedures, and then discuss the associated reasoning.

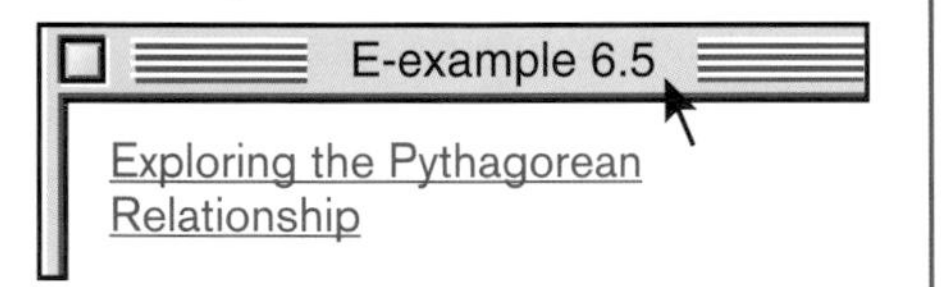

Geometric models are also useful in representing other algebraic relationships, such as identities. For example, the visual demonstrations of the identity $(a + b)^2 = a^2 + 2ab + b^2$ in figure 6.22 makes it easy to remember. A teacher might begin by asking students to draw a square with side lengths (2 + 5). Students could then partition the square as shown in fig. 6.22a, calculate the area of each section, and finally represent the total area. Students could then apply this approach to the

Fig. **6.22.**

Geometric representation demonstrating the identity $(a + b)^2 = a^2 + 2ab + b^2$

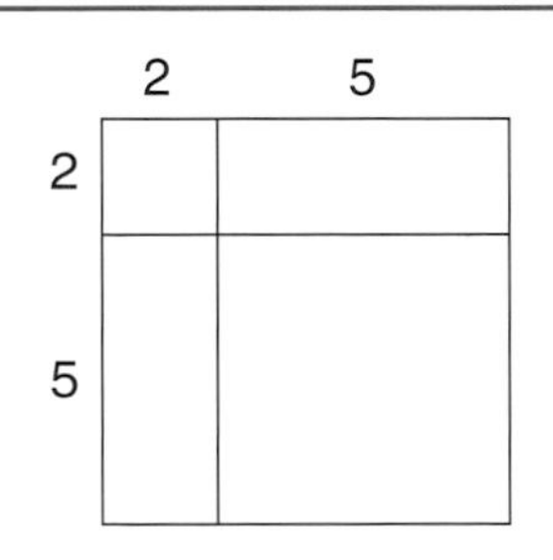

$$(2 + 5)^2 = 2 \cdot 2 + 2 \cdot 5 + 2 \cdot 5 + 5 \cdot 5$$
$$= 4 + 10 + 10 + 25$$
$$= 49$$

(a)

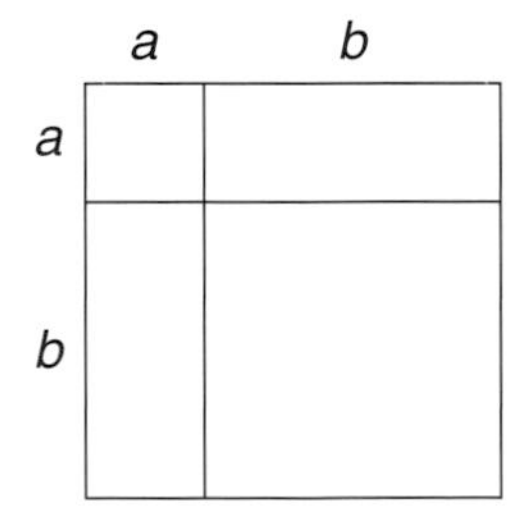

$$(a + b)^2 = a \cdot a + a \cdot b + a \cdot b + b \cdot b$$
$$= a^2 + 2ab + b^2$$

(b)

general case of a square with sides of length $(a + b)$, as shown in figure 6.22b, which demonstrates the identity $(a + b)^2 = a^2 + 2ab + b^2$.

Many investigations in middle-grades geometry can be connected to other school subjects. Nature, art, and the sciences provide opportunities for the observation and the subsequent exploration of geometry concepts and patterns as well as for appreciating and understanding the beauty and utility of geometry. For example, the study in nature or art of golden rectangles (i.e., rectangles in which the ratio of the lengths is the golden ratio, $(1 + \sqrt{5})/2$) or the study of the relationship between the rigidity of triangles and their use in construction helps students see and appreciate the importance of geometry in our world.

Measurement Standard for Grades 6–8

Instructional programs from prekindergarten through grade 12 should enable all students to—

Expectations

In grades 6–8 all students should–

Understand measurable attributes of objects and the units, systems, and processes of measurement

- understand both metric and customary systems of measurement;
- understand relationships among units and convert from one unit to another within the same system;
- understand, select, and use units of appropriate size and type to measure angles, perimeter, area, surface area, and volume.

Apply appropriate techniques, tools, and formulas to determine measurements

- use common benchmarks to select appropriate methods for estimating measurements;
- select and apply techniques and tools to accurately find length, area, volume, and angle measures to appropriate levels of precision;
- develop and use formulas to determine the circumference of circles and the area of triangles, parallelograms, trapezoids, and circles and develop strategies to find the area of more-complex shapes;
- develop strategies to determine the surface area and volume of selected prisms, pyramids, and cylinders;
- solve problems involving scale factors, using ratio and proportion;
- solve simple problems involving rates and derived measurements for such attributes as velocity and density.

Measurement

Students bring to the middle grades many years of diverse experiences with measurement from prior classroom instruction and from using measurement in their everyday lives. In the middle grades, students should build on their formal and informal experiences with measurable attributes like length, area, and volume; with units of measurement; and with systems of measurement.

Important aspects of measurement in the middle grades include choosing and using compatible units for the attributes being measured, estimating measurements, selecting appropriate units and scales on the basis of the precision desired, and solving problems involving the perimeter and area of two-dimensional shapes and the surface area and volume of three-dimensional objects. Students should also become proficient at measuring angles and using ratio and proportion to solve problems involving scaling, similarity, and derived measures.

Measurement concepts and skills can be developed and used throughout the school year rather than treated exclusively as a separate unit of study. Many measurement topics are closely related to what students learn in geometry. In particular, the Measurement and Geometry Standards span several important middle-grades topics, such as similarity, perimeter, area, volume, and classifications of shape that depend on side lengths or angle measures. Measurement is also tied to ideas and skills in number, algebra, and data analysis in such topics as the metric system of measurement, distance-velocity-time relationships, and data collected by direct or indirect measurement. Finally, many measurement concepts and skills can be both learned and applied in students' study of science in the middle grades.

Measurement concepts and skills can be developed and used throughout the school year rather than treated exclusively as a separate unit of study.

Understand measurable attributes of objects and the units, systems, and processes of measurement

From earlier instruction in school and life experience outside school, middle-grades students know that measurement is a process that assigns numerical values to spatial and physical attributes such as length. Students have some familiarity with metric and customary units, especially for length. For example, they should know some common equivalences within these systems, such as 100 centimeters equals 1 meter and 36 inches equals 3 feet, which equals 1 yard. In the middle grades they should become proficient in converting measurements to different units within a system, recognizing new equivalences, such as 1 square yard equals 9 square feet and 1 cubic meter equals 1 000 000 cubic centimeters. Work in the metric system ties nicely to students' emerging understanding of, and proficiency in, decimal computation and the use of scientific notation to express large numbers. When moving between the metric and customary systems, students are likely to find approximate equivalents—a quart is a little less than a liter and a yard is a little less than a meter—both useful and memorable.

Students in grades 6–8 should become proficient in selecting the appropriate size and type of unit for a given measurement situation. They should know that it makes sense to use liters rather than milliliters when determining the amount of refreshments for the school dance but that milliliters may be quite appropriate when measuring a small amount of a liquid for a science experiment.

In the middle grades, students expand their experiences with measurement. Although students may have developed an initial understanding of area and volume, many will need additional experiences in measuring directly to deepen their understanding of the area of two-dimensional shapes and the surface area and volume of three-dimensional objects. Even in the middle grades, some measurement of area and volume by actually covering shapes and filling objects can be worthwhile for many students. Through such experiences, teachers can help students clarify concepts associated with these topics. For example, many students experience some confusion about why square units are always used to measure area and cubic units to measure volume, especially when the shapes or objects being measured are not squares or cubes. If they move rapidly to using formulas without an adequate conceptual foundation in area and volume, many students could have underlying confusions such as these that would interfere with their working meaningfully with measurements.

Frequent experiences in measuring surface area and volume can also help students develop sound understandings of the relationships among attributes and of the units appropriate for measuring them. For example, some students may hold the misconception that if the volume of a three-dimensional shape is known, then its surface area can be determined. This misunderstanding appears to come from an incorrect overgeneralization of the very special relationship that exists for a cube: If the volume of a cube is known, then its surface area can be uniquely determined. For example, if the volume of a cube is 64 cubic units, then its surface area is 96 square units. But this relationship is not true for rectangular prisms or for other three-dimensional objects in general. To address and correct this misunderstanding, a teacher can have students use a fixed number of stacking cubes to build different rectangular prisms and then record the corresponding surface area of each arrangement. Because the number of cubes is the same, the volume is identical for all, but the surface area varies. Although a single counterexample is

sufficient to demonstrate mathematically that volume does not determine surface area, one example may not dispel the misconception for all students. Some students will benefit from repeating this activity with several different fixed volumes. Students can reap an additional benefit from this activity by considering how the shapes of rectangular prisms with a fixed volume are related to their surface areas. By observing patterns in the tables they construct for different fixed volumes, students can note that prisms of a given volume that are cubelike (i.e., whose linear dimensions are nearly equal) tend to have less surface area than those that are less cubelike. Experiences such as this contribute to a general understanding of the relationship between shape and size, extend students' earlier work in patterns of variation in the perimeters and areas of rectangles, and lay a foundation for a further examination of surface area and volume in calculus.

Students who have had experience in determining and using benchmark angles are less likely to misread a protractor.

Apply appropriate techniques, tools, and formulas to determine measurements

When students measure an object, the result should make sense; estimates and benchmarks can help students recognize when a measurement is reasonable. Students can use their sense of the size of common units to estimate measurements; for example, the height of the classroom door is about two meters, it takes about ten minutes to walk from the middle school to the high school, or the textbook weighs about two pounds. They should also be able to use commonly understood benchmarks to estimate large measurements; for instance, the distance between the middle school and the high school is about the length of ten football fields.

Students should become proficient in composing and decomposing two- and three-dimensional shapes in order to find the lengths, areas, and volumes of various complex objects. In addition, they should develop an understanding of different angle relationships and be proficient in measuring angles. Toward this end, they should learn to use a protractor to measure angles directly. Just as lower-grades students need help learning to use a ruler to measure length, so middle-grades students also need help with the mechanics of using a protractor—aligning it properly with the vertex and sides of the angle to be measured and reading the correct size of the angle on the scale. Students who have had experience in determining and using benchmark angles are less likely to misread a protractor. Estimating that an angle is less than 90 degrees should prevent a student from misreading a measurement of 150 degrees for a 30-degree angle. Students can develop a repertoire of benchmark angles, including right angles, straight angles, and 45-degree angles. They should be able to offer reasonable estimates for the measurement of any angle between 0 degrees and 180 degrees. Checking the reasonableness of a measurement should be a part of the process.

In the middle grades, students should also develop an understanding of precision and measurement error. By examining and discussing how objects are measured and how the results are expressed, teachers can help their students understand that a measurement is precise only to one-half of the smallest unit used in the measurement. That is, when students say that the length of a book, to the nearest quarter inch, is 12 1/4 inches, they should be aware that the measurement could be off by 1/8 inch. Thus, the absolute error in the measurement is ±1/8 inch in this instance. Similarly, if they use a protractor to measure angles to the nearest degree, they will be precise within 1/2 degree.

An understanding of the concepts of perimeter, area, and volume is initiated in lower grades and extended and deepened in grades 6–8. Whenever possible, students should develop formulas and procedures meaningfully through investigation rather than memorize them. Even formulas that are difficult to justify rigorously in the middle grades, such as that for the area of a circle, should be treated in ways that help students develop an intuitive sense of their reasonableness.

One particularly accessible and rich domain for such investigation is areas of parallelograms, triangles, and trapezoids. Students can develop formulas for these shapes using what they have previously learned about how to find the area of a rectangle, along with an understanding that decomposing a shape and rearranging its component parts without overlapping does not affect the area of the shape. For example, figure 6.23 illustrates how students could use their knowledge of the area of a rectangle to generate formulas for the area of a parallelogram and a triangle. Once students develop these formulas, they can also generate a formula for the area of a trapezoid. A teacher might have students begin working with isosceles trapezoids and then try to generalize their formula for any trapezoid. As suggested in figure 6.24, students can use what they know about rectangles, parallelograms, and triangles in several different ways. They might decompose an isosceles trapezoid into two triangles and a rectangle and rearrange these shapes to form a rectangle, or they might duplicate the trapezoid and arrange the two shapes to form a parallelogram.

Fig. **6.23.**

Students can use their knowledge of the area of a rectangle to generate a formula for the area of a parallelogram (a) and for the area of a triangle (b).

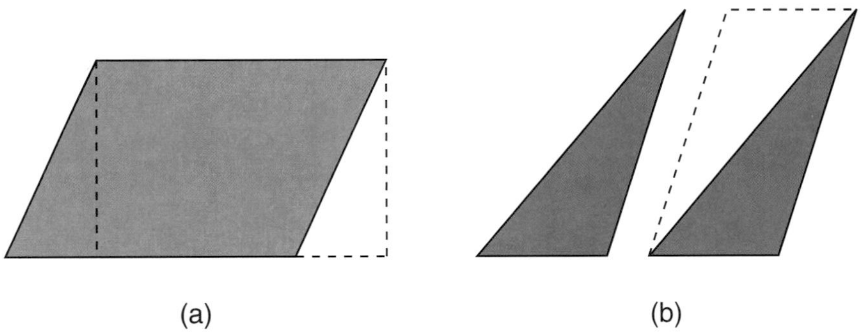

(a)

(b)

Fig. **6.24.**

An isosceles trapezoid can be decomposed and rearranged or duplicated in order to find a formula for its area.

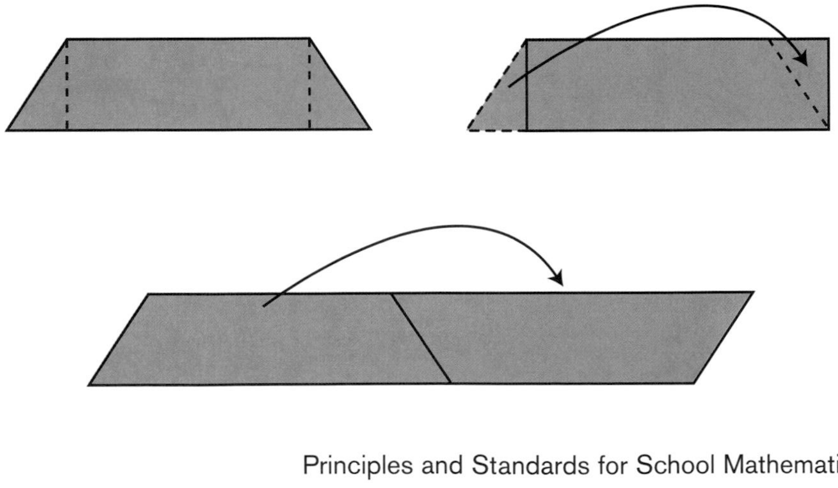

Students in the middle grades should also develop formulas for the surface areas and volumes of three-dimensional objects. Teachers can help students develop formulas for the volumes of prisms, pyramids, and cylinders and for the surface areas of right prisms and cylinders by having them construct models, measure the dimensions, estimate the areas and volumes, and look for patterns and relationships related to lengths, areas, and volumes. In their work with three-dimensional objects, students can make use of what they know about two-dimensional shapes. For example, they can relate the surface area of a three-dimensional object to the area of its two-dimensional net. Students might determine the surface area of a cylinder by determining the area of its net (the two-dimensional figure produced by cutting the cylinder and laying it flat), which consists of a rectangle and two circles (see fig. 6.25). Thus, students can develop a formula for the surface area of a cylinder because they can find the areas of the circles and the rectangle that compose its net.

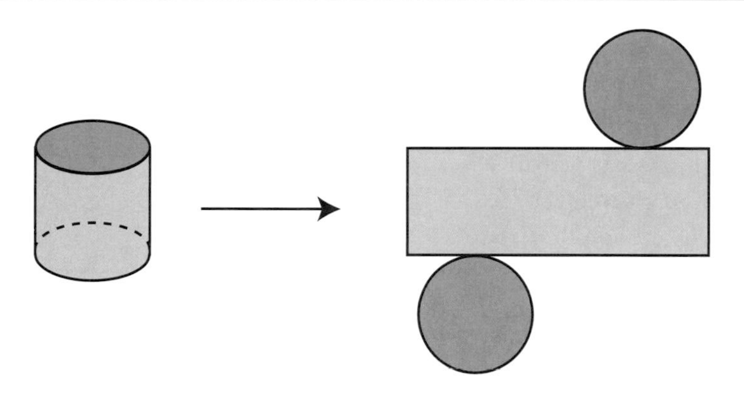

Fig. **6.25.**

Students can determine the surface area of a cylinder by determining the area of its net.

It is important that middle-grades students understand similarity, which is closely related to their more general understanding of proportionality and to the idea of correspondence. Students can use measurement to explore the meaning of similarity and later to apply the concept to solve problems. The important observation that the measurements of the corresponding angles of similar shapes are equal is often a starting point for work with similarity. Measurement is also useful for determining the relationships between the side lengths and the perimeters and areas of similar shapes and the surface areas and volumes of similar objects. Students need to understand that the perimeters of pairs of similar shapes are proportional to their corresponding side lengths but that their areas are proportional to the squares of the corresponding side lengths. Similarly, through investigation they should recognize that the surface areas of similar objects are proportional to the squares of the lengths of their corresponding sides but that their volumes are proportional to the cubes of those lengths.

Problems that involve constructing or interpreting scale drawings offer students opportunities to use and increase their knowledge of similarity, ratio, and proportionality. Such problems can be created from many sources, such as maps, blueprints, science, and even literature. For example, in *Gulliver's Travels*, a novel by Jonathan Swift, many passages suggest problems related to scaling, similarity, and proportionality. Another interesting springboard for such problems is "One Inch Tall," a poem by Shel Silverstein (1974) (see fig. 6.26).

Fig. **6.26**

"One Inch Tall," by Shel Silverstein (1974). Used with permission.

ONE INCH TALL

If you were only one inch tall, you'd ride a worm to school.
The teardrop of a crying ant would be your swimming pool.
A crumb of cake would be a feast
And last you seven days at least,
A flea would be a frightening beast
If you were one inch tall.

If you were only one inch tall, you'd walk beneath the door,
And it would take about a month to get down to the store.
A bit of fluff would be your bed,
You'd swing upon a spider's thread,
And wear a thimble on your head
If you were one inch tall.

You'd surf across the kitchen sink upon a stick of gum.
You couldn't hug your mama, you'd just have to hug her thumb.
You'd run from people's feet in fright,
To move a pen would take all night,
(This poem took fourteen years to write—
'Cause I'm just one inch tall).

—*Shel Silverstein*

In connection with the poem, a teacher could pose a problem like the following:

> Use ratios and proportions to help you decide whether the statements in Shel Silverstein's poem are plausible. Imagine that you are the person described in the poem, and assume that all your body parts changed in proportion to the change in your height. Choose one of the following to investigate and write a complete report of your investigation, including details of any measurements you made or calculations you performed:
>
> - In the poem the author says that you could ride a worm to school. Is this statement plausible? Would it be true that you could ride a worm if you were 1 inch tall? Use the fact that common earthworms are about 5 inches long with diameters of about 1/4 inch.
> - In the poem the author says that you could wear a thimble on your head. Would this be true if you were only 1 inch tall? Use one of the thimbles in the activity box to help you decide.

To solve this problem, students need to use proportionality to imagine a scale model of a student shrunk to a height of 1 inch. They need to consider the resulting circumference of a student's head or the resulting length of a student's legs in relation to the diameter of a cross section of a worm. Because middle-grades students' heights vary greatly, scale factors will vary greatly within a class, which can generate a lively discussion. A student who is 4 feet 8 inches tall would use a 56:1 ratio as a scale factor; in contrast, a student who is 5 feet tall would use a scale factor based on a 60:1 ratio. After discussing this problem and pointing out that an author can legitimately use poetic license to create images that do not conform to reality, a teacher might extend the investigation by asking students to evaluate the plausibility of other statements in the poem that intrigue them. Alternatively, a teacher might select a statement, for example, "And it would take about a month to get down to the store," which refers to a rate given as a distance-time relationship. Students could use a stopwatch and a tape measure to get distance-time readings for a typical student. They could then determine how far away the store would need to be in order for the assertion to be plausible, given a proportional change in the rate of walking for a student shrunk to 1 inch.

Students should also have opportunities to consider other kinds of rates, such as monetary exchange rates, which can afford practice with decimal computation and experience with ratios and rates expressed as single numbers. Experience with cost-per-item rates are also valuable; see the examples in the "Problem Solving" and "Algebra" sections in this chapter.

Teachers can use technological tools such as computer-based laboratories (CBLs) to expand the set of measurement experiences, especially those involving rates and derived measures, and to relate measurement to other topics in the curriculum. For example, using the CBL to measure a student's distance from an object as she walks away from or toward it and plotting the corresponding points on a distance-time graph can be very instructive. Different paths generate different graphs. Different start-end points and variations in speed can also affect the graphs. Students could generate many such graphs with specific kinds of variation and then discuss the graphs to help them relate this experience to their developing understandings of linear relationships, proportionality, and slopes and rates of change. Questions such as the following might be useful:

- For which graphs does the relationship between distance and time appear to be linear? For which is the relationship nonlinear? Why?
- For the graphs that depict a linear relationship, how does the speed at which the person walks appear to affect the graph? Why?
- Which graphs portray a proportional relationship between distance and time? Are there any graphs that depict proportional relationships that are not also linear? Are there any that depict linear relationships that are not also proportional? Why?
- Would it be possible to generate a distance-time graph that depicts a relationship that is linear but not proportional? That is proportional but not linear? Why?

A teacher can use such experiences, whether in the mathematics class or in collaboration with a science teacher, not only to enrich students' understanding of topics in measurement but also to provide a springboard for the study of data representation and analysis.

Data Analysis and Probability STANDARD for Grades 6–8

Instructional programs from prekindergarten through grade 12 should enable all students to—

Expectations

In grades 6–8 all students should–

Standard	Expectations
Formulate questions that can be addressed with data and collect, organize, and display relevant data to answer them	• formulate questions, design studies, and collect data about a characteristic shared by two populations or different characteristics within one population; • select, create, and use appropriate graphical representations of data, including histograms, box plots, and scatterplots.
Select and use appropriate statistical methods to analyze data	• find, use, and interpret measures of center and spread, including mean and interquartile range; • discuss and understand the correspondence between data sets and their graphical representations, especially histograms, stem-and-leaf plots, box plots, and scatterplots.
Develop and evaluate inferences and predictions that are based on data	• use observations about differences between two or more samples to make conjectures about the populations from which the samples were taken; • make conjectures about possible relationships between two characteristics of a sample on the basis of scatterplots of the data and approximate lines of fit; • use conjectures to formulate new questions and plan new studies to answer them.
Understand and apply basic concepts of probability	• understand and use appropriate terminology to describe complementary and mutually exclusive events; • use proportionality and a basic understanding of probability to make and test conjectures about the results of experiments and simulations; • compute probabilities for simple compound events, using such methods as organized lists, tree diagrams, and area models.

Data Analysis and Probability

Prior to the middle grades, students should have had experiences collecting, organizing, and representing sets of data. They should be facile both with representational tools (such as tables, line plots, bar graphs, and line graphs) and with measures of center and spread (such as median, mode, and range). They should have had experience in using some methods of analyzing information and answering questions, typically about a single population.

In grades 6–8, teachers should build on this base of experience to help students answer more-complex questions, such as those concerning relationships among populations or samples and those about relationships between two variables within one population or sample. Toward this end, new representations should be added to the students' repertoire. Box plots, for example, allow students to compare two or more samples, such as the heights of students in two different classes. Scatterplots allow students to study related pairs of characteristics in one sample, such as height versus arm span among students in one class. In addition, students can use and further develop their emerging understanding of proportionality in various aspects of their study of data and statistics.

Middle-grades students should formulate questions and design experiments or surveys to collect relevant data.

Formulate questions that can be addressed with data and collect, organize, and display relevant data to answer them

Middle-grades students should formulate questions and design experiments or surveys to collect relevant data so that they can compare characteristics within a population or between populations. For example, a teacher might ask students to examine how various design characteristics of a paper airplane—such as its length or the number of paper clips attached to its nose—affect the distance it travels and its consistency of flight. Students would then plan experiments in which they collect data that would allow them to compare the effects of particular design features. In addition to helping students design their experiments logically, the teacher should help them consider other factors that might affect the data, such as wind or inconsistencies in launching the planes.

Because laboratory experiments involving data collection are part of the middle-grades science curriculum, mathematics teachers may find it useful to collaborate with science teachers so that they are consistent in their design of experiments. Such collaboration could be extended so that students might collect the data for an experiment in science class and analyze it in mathematics class.

In addition to collecting their own data, students should learn to find relevant data in other resources, such as Web sites or print publications. *Consumer Reports*, for example, regularly compares the characteristics of various products, such as the quality of peanut butter; the longevity of rechargeable batteries; or the cost, size, and fuel efficiency of automobiles. When using data from other sources, students need to determine which data are appropriate for their needs, understand how the data were gathered, and consider limitations that could affect interpretation.

Middle-grades students should learn to use absolute- and relative-frequency bar graphs and histograms to represent the data they collect and to decide which form of representation is appropriate for different

Fig. **6.27.**

A relative-frequency histogram for data for a paper airplane with one paper clip

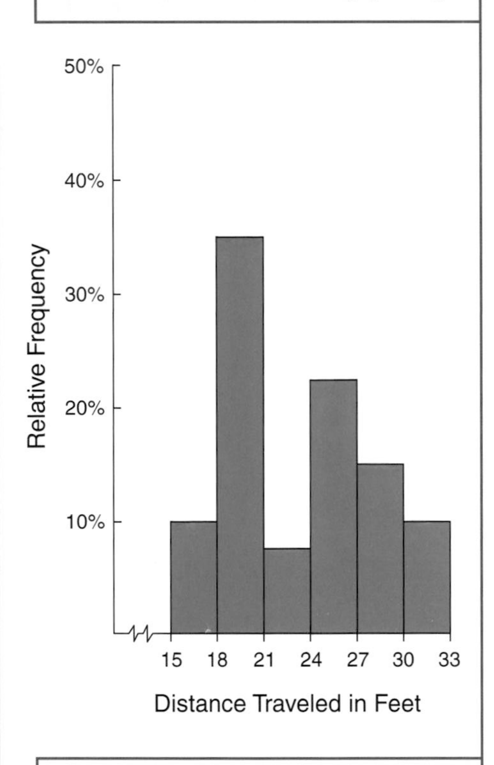

purposes. For example, suppose students were considering the following question:

> Compare the distance traveled by a paper airplane constructed using one paper clip with the distance traveled by a plane that is built with two paper clips. Which one travels farther when thrown indoors?

In an experiment conducted to answer this question, one student might throw one of the airplanes forty times while team members measure and record the distance traveled each time. The group might later do the same for the other paper airplane. The teacher might then have the students use a relative-frequency histogram to represent the data, as shown in figure 6.27. For comparison, the teacher might suggest that students display both sets of data using box plots, as in figure 6.28.

Select and use appropriate statistical methods to analyze data

In the middle grades, students should learn to use the mean, and continue to use the median and the mode, to describe the center of a set of data. Although the mean often quickly becomes the method of choice

Fig. **6.28.**

Box plots for the paper airplane data

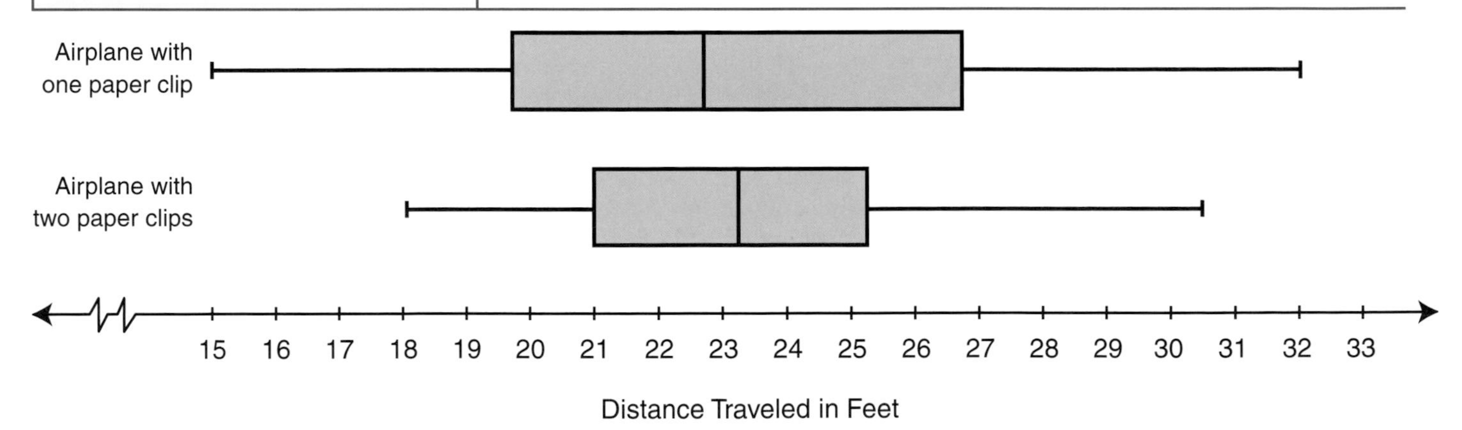

for students when summarizing a data set, their knack for computing the mean does not necessarily correspond to a solid understanding of its meaning or purpose (McClain 1999). Students need to understand that the mean "evens out" or "balances" a set of data and that the median identifies the "middle" of a data set. They should compare the utility of the mean and the median as measures of center for different data sets. As several authors have noted (e.g., Uccellini [1996]; Konold [forthcoming]), students often fail to apprehend many subtle aspects of the mean as a measure of center. Thus, the teacher has an important role in providing experiences that help students construct a solid understanding of the mean and its relation to other measures of center.

Students also need to think about measures of center in relation to the spread of a distribution. In general, the crucial question is, How do changes in data values affect the mean and median of a set of data? To examine this question, teachers could have students use a calculator to create a table of values and compute the mean and median. Then they could change one of the data values in the table and see whether the values of the mean and the median are also changed. These relationships can be effectively demonstrated using software through which students can control a data value and observe how the mean and median are affected. For example, using software that produces line plots for data sets, students could plot a set of data and mark the mean and median on the line. The students could then change one data value and observe how the mean and median change. By repeating this process for various data points, they can notice that changing one data value usually does not affect the median at all, unless the moved value is at the middle of the data set or moves across the middle, but that every change in a value affects the mean. Thus, the mean is more likely to be influenced by extreme values, since it is affected by the actual data values, but the median involves only the relative positions of the values. Other similar problems can be useful in helping students understand the different sensitivities of the mean and median; for example, the mean is very sensitive to the addition or deletion of one or two extreme data points, whereas the median is far less sensitive to such changes.

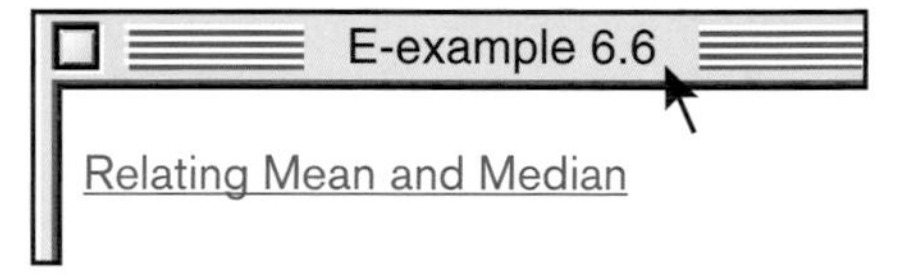

Students should consider how well different graphs represent important characteristics of data sets. For example, they might notice that it is easier to see symmetry or skewness in a graph than in a table of values. Graphs, however, can lose some of the features of the data, as can be demonstrated by generating a family of histograms for a single set of data, using different bin sizes: the different histograms may convey different pictures of the symmetry, skewness, or variability of the data set. Another example is seen when comparing a histogram and a box plot for the same data, such as those for the one-clip plane in figures 6.27 and 6.28. Box plots do not convey as much specific information about the data set, such as where clusters occur, as histograms do. But box plots can provide effective comparisons between two data sets because they make descriptive characteristics such as median and interquartile range readily apparent.

The teacher has an important role in providing experiences that help students construct a solid understanding of the mean and its relation to other measures of center.

Develop and evaluate inferences and predictions that are based on data

In collecting and representing data, students should be driven by a desire to answer questions on the basis of the data. In the process, they should make observations, inferences, and conjectures and develop new questions. They can use their developing facility with rational numbers and proportionality to refine their observations and conjectures. For example, when considering the relative-frequency histogram in figure 6.27, most students would observe that "the paper plane goes between 15 and 21 feet about as often as it goes between 24 and 33 feet," but such an observation is not very precise about the frequency. A teacher could press students to make more-precise statements: "The plane goes between 15 and 21 feet about 45 percent of the time."

Box plots are useful when making comparisons between populations. A teacher might pose the following question about the box plots in figure 6.28:

> From the box plots (in fig. 6.28), which type of plane appears to fly farther? Which type of plane is more consistent in the distance it flies?

From the relative position of the two graphs, students can infer that the two-clip plane generally flies slightly farther than the one-clip plane. Students can answer the second question by using the spreads of the data portrayed in the box plots to argue that the one-clip plane is more variable in the distance it travels than the two-clip plane.

Scatterplots are useful for detecting and examining relationships between two characteristics of a population. For example, a teacher might ask students to consider if a relationship exists between the length and the width of warblers' eggs (activity adapted from Encyclopaedia Britannica Educational Corporation [1998, pp. 104–19]). She might provide the students with data and ask the students to make a scatterplot in which each point displays the length and the width of an egg, as shown in figure 6.29. Most students will note that the relationship between the length and the width of the eggs seems to be direct (or positive); that is, longer eggs also tend to be wider. Many students will also note that the points on this scatterplot approximate a straight line, thus suggesting a nearly linear relationship between length and width. To make this relationship even more apparent, the teacher could have students draw an approximate line of fit for the data, as has been done in figure 6.29. Students could apply their developing understanding of the slope of a line to determine that

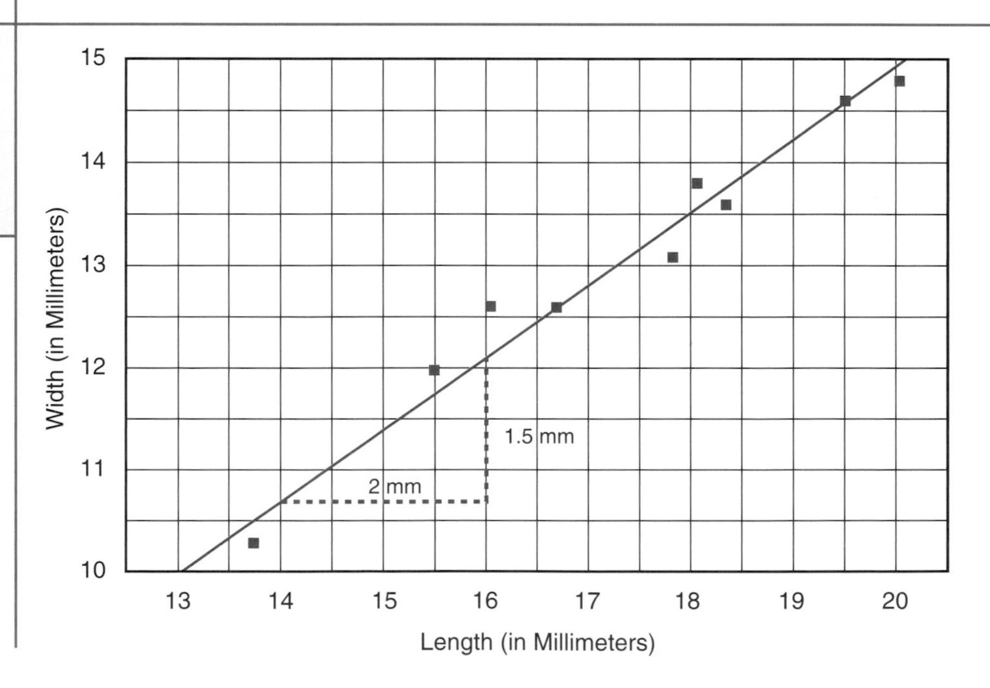

Fig. **6.29.**

A scatterplot showing the relationship between the length and the width of warblers' eggs (Encyclopaedia Britannica Educational Corporation 1998, p. 109) the slope is approximately three-fourths and that therefore the ratio of the width to the length of warblers' eggs is approximately 3:4.

Teachers can also help students learn to use scatterplots to consider the relationship between two characteristics in different populations. For example, students could measure the height and arm span for groups of middle school and high school students and then make a scatterplot in which the points for middle school students are plotted with one color and the points for high school students are plotted with a second color. Students can make observations about the differences between the two samples, such as that students in the high school sample are generally taller than those in the middle school sample. They can also use the plots to examine possible similarities. In particular, if students draw an approximate line of fit for each set of points, they can determine whether the slopes are approximately equal (i.e., the lines are approximately parallel), which would indicate that the relationship between height and arm span is about the same for both middle school and high school students.

Scatterplots are useful for detecting and examining relationships between two characteristics of a population.

Because linearity is an important idea in the middle grades, students should encounter many scatterplots that have a nearly linear shape. But teachers should also have students explore plots that represent nonlinear relationships. For example, in connection with their study of geometry and measurement, students could measure the lengths of the bases of several similar triangles and use formulas to find their areas or graph paper to estimate their areas. Creating a scatterplot of the lengths of the bases and the areas will make evident the quadratic relationship between length and area in similar figures.

Teachers should encourage students to plot many data sets and look for relationships in the plots; computer graphing software and graphing calculators can be very helpful in this work. Students should see a range of examples in which plotting data sets suggests linear relationships, nonlinear relationships, and no apparent relationship at all. When a scatterplot suggests that a relationship exists, teachers should help students determine the nature of the relationship from the shape and direction of the plot. For example, for an apparently linear relationship, students could use their understanding of slope to decide whether the relationship is direct or inverse. Students should discuss what the relationships they have observed might reveal about the sample, and they should also discuss whether their conjectures about the sample might apply to larger populations containing the sample. For example, if a sample consists of students from one sixth-grade class in a school, how valid might the inferences made from the sample be for all sixth graders in the school? For all middle-grades students in the school? For all sixth graders in the city? For all sixth graders in the country? Such discussions can suggest further studies students might undertake to test the generality of their conjectures.

Understand and apply basic concepts of probability

Teachers should give middle-grades students numerous opportunities to engage in probabilistic thinking about simple situations from which students can develop notions of chance. They should use appropriate terminology in their discussions of chance and use probability to

Computer simulations may help students avoid or overcome erroneous probabilistic thinking.

make predictions and test conjectures. For example, a teacher might give students the following problem:

> Suppose you have a box containing 100 slips of paper numbered from 1 through 100. If you select one slip of paper at random, what is the probability that the number is a multiple of 5? A multiple of 8? Is not a multiple of 5? Is a multiple of both 5 and 8?

Students should be able to use basic notions of chance and some basic knowledge of number theory to determine the likelihood of selecting a number that is a multiple of 5 and the likelihood of not selecting a multiple of 5. In order to facilitate classroom discussion, the teacher should help students learn commonly accepted terminology. For example, students should know that "selecting a multiple of 5" and "selecting a number that is not a multiple of 5" are *complementary* events and that because 40 is in the set of possible outcomes for both "selecting a multiple of 5" and "selecting a multiple of 8," they are not *mutually exclusive* events.

Teachers can help students relate probability to their work with data analysis and to proportionality as they reason from relative-frequency histograms. For example, referring to the data displayed in figure 6.27, a teacher might pose questions like, How likely is it that the next time you throw a one-clip paper airplane, it goes at least 27 feet? No more than 21 feet?

Although the computation of probabilities can appear to be simple work with fractions, students must grapple with many conceptual challenges in order to understand probability. Misconceptions about probability have been held not only by many students but also by many adults (Konold 1989). To correct misconceptions, it is useful for students to make predictions and then compare the predictions with actual outcomes.

Computer simulations may help students avoid or overcome erroneous probabilistic thinking. Simulations afford students access to relatively large samples that can be generated quickly and modified easily. Technology can thus facilitate students' learning of probability in at least two ways: With large samples, the sample distribution is more likely to be "close" to the actual population distribution, thus reducing the likelihood of incorrect inferences based on empirical samples. With easily generated samples, students can focus on the analysis of the data rather than be distracted by the demands of data collection. If simulations are used, teachers need to help students understand what the simulation data represent and how they relate to the problem situation, such as flipping coins.

Although simulations can be useful, students also need to develop their probabilistic thinking by frequent experience with actual experiments. Many can be quite simple. For example, students could be asked to predict the probability of various outcomes of flipping two coins sixty times. Some students will incorrectly expect that there are three equally likely outcomes of flipping two coins once: two heads, two tails, and one of each. If so, they may predict that each of these will occur about twenty times. If groups of students conducted this experiment, they could construct a relative-frequency bar graph from the pooled data for the entire class. Then they could discuss whether the results of the experiment are consistent with their predictions. If students are accustomed to reasoning from and about data, they will understand that discrepancies between predictions and outcomes from a large and representative sample must be taken seriously. The detection of discrepancies can lead to learning when students turn to classmates and their teacher for alternative ways to think about the possible results of flipping two coins (or other similar compound events). Teachers can then introduce students to various methods—organized lists, tree diagrams, and area models—to help them understand and compute the probabilities of compound events.

Using a problem like the following, a teacher might assess students' understanding of probability in a manner that includes data analysis and reveals possible misconceptions:

> For the one-clip paper airplane, which was flight-tested with the results shown in the relative-frequency histogram (in fig. 6.27), what is the probability that exactly one of the next two throws will be a dud (i.e., it will travel less that 21 feet) and the other will be a success (i.e., it will travel 21 feet or more)?

To solve this problem, students would need to understand the data representation in figure 6.27 and use ratios to estimate that there is about a 45 percent chance that a throw will be a dud and about a 55 percent chance that it will be a success. Then they would need to use some method for handling the compound event and deal with the fact that there are two ways it might occur. Students who understand all that is required might produce a tree diagram like the one in figure 6.30 to show that the total probability is 198/400, or .495, since each of the two possibilities—"dud first, then success" and "success first, then dud"—has a probability of 99/400.

Fig. **6.30.**

A tree diagram for determining the probability of a compound event, given sample data

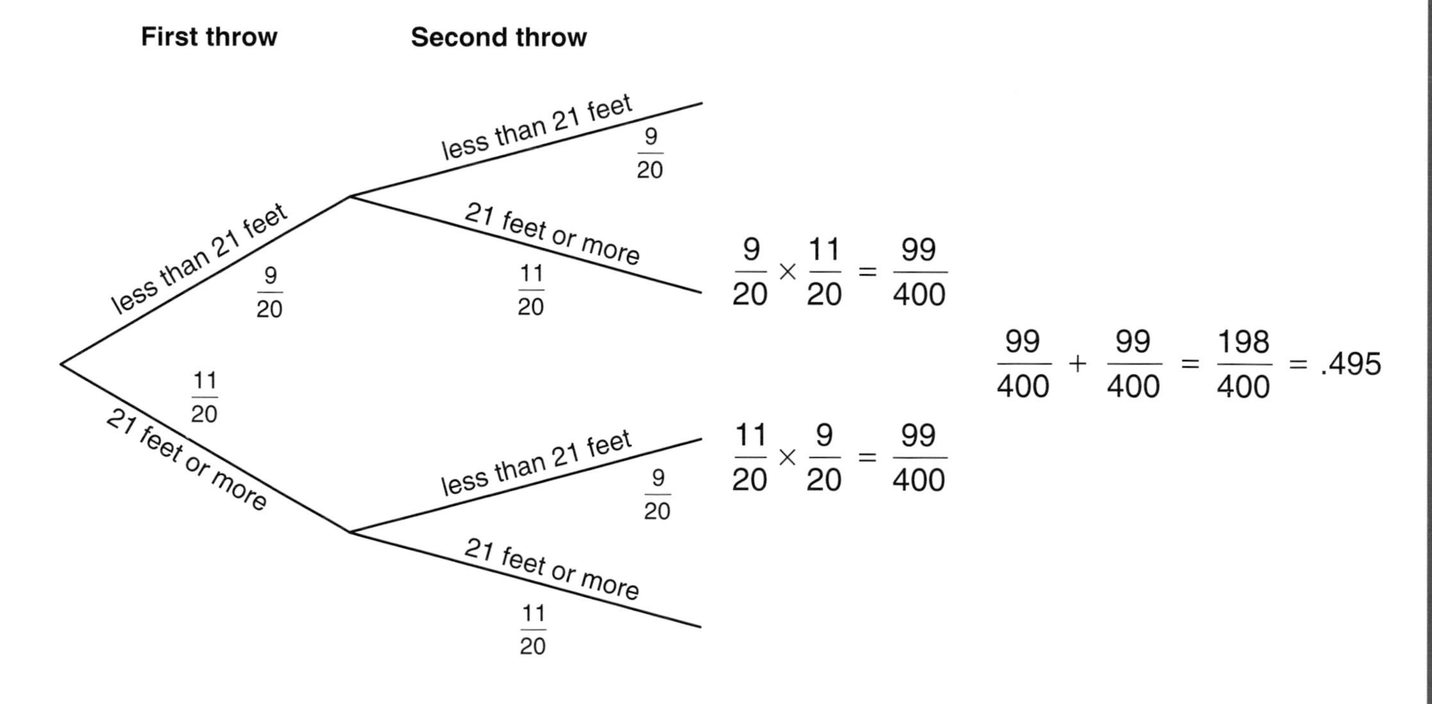

Problem Solving Standard for Grades 6–8

Instructional programs from prekindergarten through grade 12 should enable all students to—

Build new mathematical knowledge through problem solving

Solve problems that arise in mathematics and in other contexts

Apply and adapt a variety of appropriate strategies to solve problems

Monitor and reflect on the process of mathematical problem solving

Through problem solving, students can experience the power and utility of mathematics. Problem solving is central to inquiry and application and should be interwoven throughout the mathematics curriculum to provide a context for learning and applying mathematical ideas. Middle-grades students whose curriculum is based on the Standards in this document will benefit from frequent opportunities for both independent and collaborative problem-solving experiences. They will engage profitably in complex investigations, perhaps occasionally working for several days on a single problem and its extensions.

What should problem solving look like in grades 6 through 8?

Problem solving in grades 6–8 should promote mathematical learning. Students can learn about, and deepen their understanding of, mathematical concepts by working through carefully selected problems that allow applications of mathematics to other contexts. Many interesting problems can be suggested by everyday experiences, such as reading literature or using cellular telephones, in-line skates, kites, and paper airplanes.

Instruction in grades 6–8 should take advantage of the expanding mathematical capabilities of students to include more-complex problems that integrate such topics as probability, statistics, geometry, and rational numbers. Situations and approaches should build on and extend the mathematical understanding, skills, and language that students have acquired.

Well-chosen problems can be particularly valuable in developing or deepening students' understanding of important mathematical ideas. Consider the following problem that might be used by a teacher who wants her students to think about various ways to use ratios and proportions:

> A baseball team won 48 of its first 80 games. How many of its next 50 games must the team win in order to maintain the ratio of wins to losses?

Students can solve this problem in many ways. One student might express the ratio of wins in the first 80 games as 48/80 and note that the ratio is a little more than one-half; that is, the team wins a little more than half the time. She might then estimate that in the next 50 games the team should win about 28 games. She could compare the resulting ratio of 28/50 to the given ratio of 48/80 and adjust her estimate until the two ratios are equivalent. Another student might look at the ratio of wins to losses, 48:32, and simplify it to 3:2. Restating this result as "3 wins in every 5 games" and noting that there are 10 sets of 5 games in the 50 games to be played, he could conclude that 30 games is the solution. Yet another student might use a proportion, $48/80 = x/50$, to find the solution. A fourth student might use percents or decimals (as newspapers do when reporting the "standings" of baseball teams). This student might divide 48 by 80 and represent the ratio as 60 percent and then find 60 percent of 50 games, or represent it as 0.600 and multiply by 50, to determine that 30 games must be won to maintain the success rate. Such problems help students develop and use a variety of problem-solving strategies and approaches, and sharing these methods within the classroom affords students opportunities to assess the strengths and limitations of alternative approaches to considering them.

Well-chosen problems can be particularly valuable in developing or deepening students' understanding of important mathematical ideas.

After students have had similar experiences with ratios, rates, and proportions in grades 6 and 7, a teacher who wanted to extend and deepen her eighth-grade students' understanding of the topics might use a problem like the following:

> Over the past few weeks, the American Movie Corporation has introduced two new kinds of candy at the concession stands in movie theaters in town. For three weeks, two theaters have offered Apple Banana Chews. For two weeks, five other theaters sold Mango Orange Nips. Only one of the two types of candy was sold at each theater, and all the theaters showed the same movies and had roughly the same attendance each week during the introductory period. During that period, 660 boxes of Apple Banana Chews and 800 boxes of Mango Orange Nips were sold. Suppose you have been hired by the company to help them determine which candy sold better. Use the information to decide which type of candy was more popular, and carefully and completely explain the basis for your answer.

This problem can help students see the need to go beyond superficial approaches and to dig deeply into their understanding of ratios and rates. For some students, an initial response will be that Mango Orange Nips (MONs) were more popular because more were sold. In an early class discussion, other students might point out to them that such a direct comparison is misleading because the two types of candy were sold at different numbers of theaters and for different amounts of time. From this discussion, a need to probe more deeply into the relationships in the problem will be apparent. Students are likely next to consider ratios or rates, which they have learned to use to express quantitative relationships. By inviting individuals or groups of students to present possible solutions, the teacher can initiate a lively discussion of competing approaches and arguments. Some students will consider the average number of candies sold for each theater—330 boxes of ABCs (Apple Banana Chews) per theater versus 160 boxes of MONs per theater—and conclude that ABCs were more popular. Other students will consider a different rate, namely, the average number of candies sold each week—220 boxes of ABCs per week versus 400 boxes of MONs per week—and conclude that MONs were more popular. Because each of these answers is different and seems to be based on a sensible approach, neither answer can be argued to be "better" than the other. So students can see that they must go beyond these simple rates to answer the question. The teacher can then help them develop a more complex rate—the average number of boxes of candy per theater per week—that incorporates all the information in the problem and yields a defensible solution: the rates are 110 for ABCs and 80 for MONs, so ABCs are the better sellers.

Teachers should regularly ask students to formulate interesting problems based on a wide variety of situations, both within and outside mathematics. Teachers should also give students frequent opportunities to explain their problem-solving strategies and solutions and to seek general methods that apply to many problem settings. These experiences should engender in students important problem-solving dispositions—an orientation toward problem finding and problem posing; an interest in, and capacity for, explaining and generalizing; and a propensity for reflecting on their work and monitoring their solutions. They should be expected to explain their ideas and solutions in words first, and then teachers can help them learn to use conventional mathematical symbols or their own forms of representations, as appropriate, to convey their thinking.

Problems in the middle grades can and should respond to students' questions and engage their interests.

The availability of technology—in the form of computers and scientific or graphing calculators—allows middle-grades students to deal with "messy," complex problems. The technology can alleviate much of the drudgery that until recently often constrained middle-school mathematics to using only problems with "nice numbers." Computers, calculators, and electronic data-gathering devices, such as calculator-based laboratories (CBLs) or rangers (CBRs), offer means of gathering or analyzing data that in years past might have been considered too troublesome to deal with. Similarly, classroom Internet connections make it possible for students to find information for use in posing and solving a wide variety of problems. For example, students might be interested in investigating whether it is cost-effective to recycle aluminum cans at their school, or they might explore weather patterns in different regions. Graphing calculators and easy-to-use computer software enable students to move between different representations of data and to compute with large quantities of data and with messy numbers, both large and small, with relative ease. As a result, problems in the middle grades can and should respond to students' questions and engage their interests.

What should be the teacher's role in developing problem solving in grades 6 through 8?

Students' learning about and through problem solving and their dispositions toward mathematics are shaped by teachers' instructional decisions and actions. Teachers can make problem solving an integral part of the class's mathematical activity by choosing interesting problems that incorporate important mathematical ideas from the curriculum. To help students develop a problem-solving orientation, teachers can allow them to choose or create some of the problems to be solved. Teachers can help build students' problem-analysis skills by including tasks that have extraneous information or insufficient information. And they can challenge students with problems that have more than one answer, such as the following (adapted from Gelfand and Shen [1993], p. 3):

> Make a sum of 1000, using some eights (8s) with some plus signs (+s) inserted.

Because this problem can be solved in more than one way, students could find several solutions (888 + 88 + 8 + 8 + 8 = 1000 is one solution, and 888 + 8 + 8 + 8 + 8 + 8 + 8 + 8 + 8 + 8 + 8 + 8 + 8 + 8 + 8 = 1000 is another). They could then analyze these solutions and discuss whether others exist.

Teachers motivate students by encouraging communication and collaboration and by urging students to seek complete solutions to challenging problems. Recognizing students' contributions can add to their motivation. Some teachers, for example, find it effective to name a problem, conjecture, or solution method after the student who proposed it (e.g., Tamela's problem).

Research suggests that an important difference between successful and unsuccessful problem solvers lies in their beliefs about problem solving, about themselves as problem solvers, and about ways to approach solving problems (Kroll and Miller 1993). For example, many students have developed the faulty belief that all mathematics problems could be solved quickly and directly. If they do not immediately know how to solve a problem, they will give up, which supports a view of themselves as incompetent problem solvers. Furthermore, many students believe there is just one "right" way to solve any mathematics problem. Not only do these students become dependent on the teacher or an answer key for a verification of their solution, but they also fail to appreciate the excitement and insight that can come from recognizing and connecting very different ways to solve a problem. To counteract negative dispositions, teachers can help students develop a tendency to contemplate and analyze problems before attempting a solution and then persevere in finding a solution.

Some teachers find it effective to name a problem, conjecture, or solution method after the student who proposed it.

The essence of problem solving is knowing what to do when confronted with unfamiliar problems. Teachers can help students become reflective problem solvers by frequently and openly discussing with them the critical aspects of the problem-solving process, such as understanding the problem and "looking back" to reflect on the solution and the process (Pólya 1957). Through modeling, observing, and questioning, the teacher can help students become aware of their activity as they solve problems. For example, consider how the following problem might be used to develop students' skill in problem solving (Schroeder and Lester 1989, p. 40):

> On centimeter graph paper outline all the shapes that have an area of 14 square cm and a perimeter of 24 cm. For each shape you draw, at least one side of each square must share a side with another square.

Students may initially assume that the shapes referred to are rectangles. Under that assumption, students can discover that the problem has no solution. A teacher might allow this line of thinking to surface early so that it can be addressed. Once students understand that shapes other than rectangles are possible, they might approach the problem by experimenting with a few shapes, using graph paper or 14 square cutouts. If students do not begin to recognize that haphazard experimentation is not likely to produce a complete solution, questioning by the teacher might help them. The teacher can help students develop a systematic way to keep track of the shapes that have been tried. The teacher can also ask provocative questions that encourage students to find all possibilities: What makes two shapes "different"? Are shapes different if one is a flip of the other? What strategies can be used to create new shapes from old ones in a way that preserves both the area and the perimeter? In this problem, students draw on their knowledge of various of geometric ideas, such as area, perimeter, and congruence and transformations that preserve area and perimeter. Moreover, they engage in a process that is applicable to a wide variety of problems: gradually understanding a problem more deeply and then working systematically to determine all possible solutions. As research has shown, effective problem solvers move flexibly among aspects of the problem-solving process as they work through a problem (Kroll and Miller 1993).

Students should reflect on their problem solving and consider how it might be modified, elaborated, streamlined, or clarified.

Although it is not the main focus of problem solving in the middle grades, learning about problem solving helps students become familiar with a number of problem-solving heuristics, such as looking for patterns, solving a simpler problem, making a table, and working backward. These general strategies are useful when no known approach to a problem is readily apparent. These processes may have been used in the elementary grades, but middle-grades students need additional experience and instruction in which they consider how to use these strategies appropriately and effectively.

Students also should be encouraged to monitor and assess themselves. Good problem solvers realize what they know and don't know, what they are good at and not so good at; as a result they can use their time and energy wisely. They plan more carefully and more effectively and take time to check their progress periodically. These habits of mind are important not only in making students better problem solvers but also in helping students become better learners of mathematics.

For several reasons, students should reflect on their problem solving and consider how it might be modified, elaborated, streamlined, or clarified: Through guided reflection, students can focus on the mathematics involved in solving a problem, thus solidifying their understanding of the concepts involved. They can learn how to generalize and extend problems, leading to an understanding of some of the structure underlying mathematics. Students should understand that the problem-solving process is not finished until they have looked back at their solution and reviewed their process.

An important aspect of a problem-solving orientation toward mathematics is making and examining conjectures raised by solving a problem and posing follow-up questions. For example, according to the Pythagorean relationship, if squares are built on the legs and the hypotenuse of any right triangle, then the areas of the squares on the legs will together sum to the area of the square on the hypotenuse. This well-known relationship, summarized with the formula $a^2 + b^2 = c^2$, where a and b are the lengths of the triangle's legs and c is the length of its hypotenuse, is used frequently to solve numerical and algebraic problems. It can be the source of much interesting problem posing and generalization for middle-grades students. A teacher might orchestrate a discussion in which students pose a variety of "what if" questions about variants and extensions of the Pythagorean relation (Brown and Walter 1983), for example, Would the area relationship hold if we built something other than squares on the sides of right triangles, say for equilateral triangles? Or regular hexagons? Or semicircles? Will the areas still sum in the same way? Such conjectures can easily be examined by using interactive geometry software, which can also facilitate students' search for a counterexample to disprove a conjecture. Although formal proof of a generalization of the Pythagorean relationship may be beyond the reach of most students in the middle grades, some students might be able to use their developing understanding of proportionality and similarity to argue that the generalization holds because the areas of similar figures are proportional to the square of the lengths of their corresponding sides.

By reflecting on their solutions, such as in this extension of the Pythagorean relationship, students use a variety of mathematical skills, develop a deeper insight into the structure of mathematics, and gain a disposition toward generalizing. The teacher can ensure that classroom discussion continues until several solution paths have been considered, discussed, understood, and evaluated. It should become second nature for students to talk about connections among problems; to propose, critique, and value alternative approaches to solving problems; and to be adept in explaining their approaches.

Reasoning and Proof STANDARD for Grades 6–8

Instructional programs from prekindergarten through grade 12 should enable all students to—

Recognize reasoning and proof as fundamental aspects of mathematics

Make and investigate mathematical conjectures

Develop and evaluate mathematical arguments and proofs

Select and use various types of reasoning and methods of proof

Reasoning is an integral part of doing mathematics. Students should enter the middle grades with the view that mathematics involves examining patterns and noting regularities, making conjectures about possible generalizations, and evaluating the conjectures. In grades 6–8 students should sharpen and extend their reasoning skills by deepening their evaluations of their assertions and conjectures and using inductive and deductive reasoning to formulate mathematical arguments. They should expand the audience for their mathematical arguments beyond their teacher and their classmates. They need to develop compelling arguments with enough evidence to convince someone who is not part of their own learning community.

What should reasoning and proof look like in grades 6 through 8?

In the middle grades, students should have frequent and diverse experiences with mathematics reasoning as they—

- examine patterns and structures to detect regularities;
- formulate generalizations and conjectures about observed regularities;
- evaluate conjectures;
- construct and evaluate mathematical arguments.

Students should discuss their reasoning on a regular basis with the teacher and with one another, explaining the basis for their conjectures and the rationale for their mathematical assertions. Through these experiences, students should become more proficient in using inductive and deductive reasoning appropriately.

Students can use inductive reasoning to search for mathematical relationships through the study of patterns. Consider an example from a classroom in which rising seventh-grade students were studying figurate numbers (drawn from classroom observation and partially described in Malloy [1997]).

> The teacher began by explaining triangular numbers and then asked the students to generate representations for the first five triangular numbers. The students visualized the structure of the numbers to

draw successive dot triangles, each time adding at the bottom a row containing one more dot than the bottom row in the previous triangle (see fig. 6.31). Next the teacher asked the students to predict (without drawing) how many dots would be needed for the next triangular number. Reflecting on what they had done to generate the sequence thus far, they quickly concluded that the sixth triangular number would have six more dots than the fifth triangular number. These students were engaged in recursive reasoning about the structure of this sequence of numbers, using the just-formed number to generate the next number. This approach was repeated for several more "next" numbers in the sequence, and it worked well.

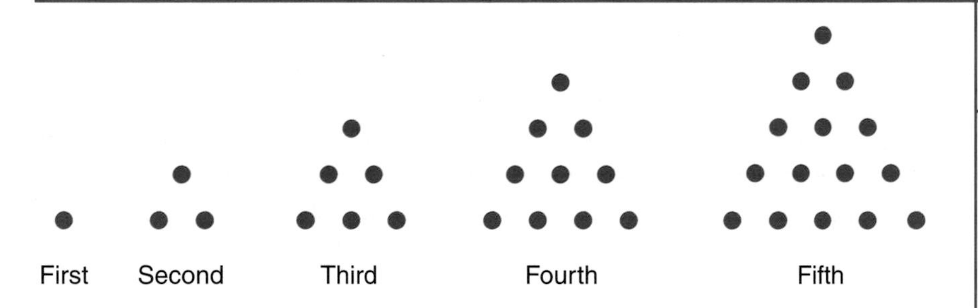

Fig. **6.31.**
First five triangular numbers

The teacher then asked the students to find the 100th term in the sequence. Most students knew that the value of the 100th term is 100 more than the value of the 99th term, but because they did not already know the value of the 99th term, they were not able to find the answer quickly. The teacher suggested that they make a chart to record their observations about triangular numbers and to look for a pattern or a relationship to help them find the 100th triangular number. The students began with a display that reflected what they had already observed (see fig. 6.32). They examined the display for additional patterns. Tamika commented that she thought there was a pattern relating the differences and the numbers. She explained that if the consecutive differences are multiplied, the product is twice the number that is "between" them in the display; for example, the product of 4 and 5 is twice as large as 10.

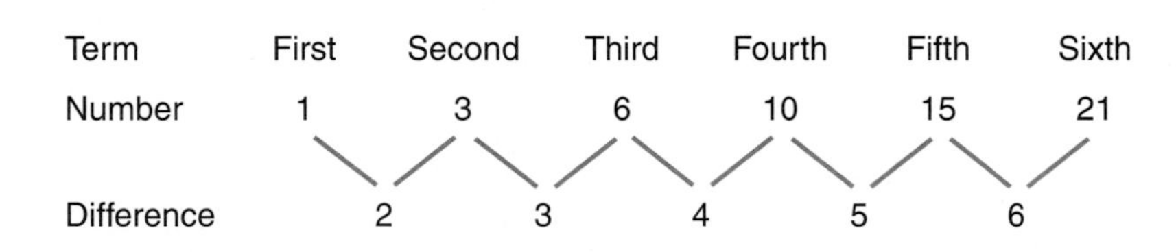

Fig. **6.32.**
Triangular numbers

The teacher asked the students to check to see if Tamika's observation was true for other numbers in the display. After they verified the observation, the teacher asked them to use this method to find the next triangular number. Some students were unable to see how it could be done, but Curtis used Tamika's observation as follows: "Using Tamika's method, the seventh number is (7)(8)/2, which is 28." Several students checked this answer by using the recursive method of adding 7 to the sixth triangular number to find the seventh triangular number (21 + 7 = 28). The teacher then asked the students to check Tamika's method for the next few triangular numbers to verify that it worked in those instances. She next asked if Tamika's method could be used to find the 100th triangular number. Darnell said, "If Tamika is right, the hundredth triangular number should be (100)(101)/2."

In general, the students agreed that the method of multiplying and dividing by 2 was useful because it seemed to work and because it did not require knowing the nth term in order to find the $(n + 1)$th term. However, some students were not convinced that the method was correct. It lacked the intuitive appeal of the recursive method they used first, and it did not appear to have a mathematical basis. The teacher decided that it was worth additional class time to develop a mathematical argument to support Tamika's method. She began by asking students to notice that each triangular number is the sum of consecutive whole numbers, which they readily saw from the dot triangles. Then the teacher demonstrated Gauss's method for finding the sum of consecutive whole numbers, applying it to the first seven whole numbers. She asked the students to add the numbers from 1 to 7 to those in the reversed sequence, 7 to 1, as shown in figure 6.33, to see that the seventh triangular number—1 + 2 + 3 + 4 + 5 + 6 + 7— could also be expressed as (7)(8)/2. After the students completed this exercise, the teacher asked them to express the general relationship in words. They struggled, but they came up with this general rule: If you want to find a particular triangular number, you multiply your number by the next number and divide by 2. The students wrote the rule this way: (number)(number + 1)/2.

Fig. **6.33.**

Gauss's method—the sum of the first seven counting numbers

$$\begin{array}{r} 1+2+3+4+5+6+7 \\ 7+6+5+4+3+2+1 \\ \hline 8+8+8+8+8+8+8 \end{array}$$

Students can see that the sums of the pairs of addends can be represented as 7 × 8, or 56.

Because each number is listed twice, they divide 56 by 2, resulting in (7)(8)/2 = 56/2 = 28.

The example illustrates what reasoning and proof can look like in the middle grades. Although mathematical argument at this level lacks the formalism and rigor often associated with mathematical proof, it shares many of its important features, including formulating a plausible conjecture, testing the conjecture, and displaying the associated reasoning for evaluation by others. The teacher and students used inductive reasoning to reach a generalization. They noted regularities in a pattern (growth of triangular numbers), formulated a conjecture about the regularities (Tamika's rule), and developed and discussed a convincing argument about the truth of the conjecture.

Middle-grades students can develop arguments to support their conclusions in varied topics, such as number theory, properties of geometric shapes, and probability. For example, students who encounter the rules of divisibility by 2 and by 3 in number theory know that even numbers are divisible by 2 and numbers whose digits add to a number divisible by 3 are divisible by 3. A teacher might ask students to formulate a rule for divisibility by 6 and develop arguments to support their rule.

Some students might begin by listing some multiples of 6: 12, 18, 24, and 30. They could examine the numbers and try to detect patterns resembling those in other rules they have learned. Students might observe that all the numbers are even, which allows them to infer divisibility by 2. They could also look at the sums of the digits of the multiples and notice that the sums of the digits are all divisible by 3, just as in the test for divisibility by 3. Noting that $2 \cdot 3 = 6$, they might conclude that if the number is divisible both by 2 and by 3, then it must be divisible by 6, which might lead them to form the following conjecture for determining whether a number is divisible by 6: Check to see if the number is even and if the sum of its digits is divisible by 3.

The teacher should also challenge students to consider possible limitations of their reasoning. For example, she could ask them to use 12 as an example to consider whether it is always true that the product of two factors of a number is itself a factor of that number. The students should note that although 6 and 4 are both factors of 12, $6 \cdot 4$ is not. In this way, the teacher can help students become appropriately cautious in making inferences about divisibility on the basis of factors. Such an exploration should lead to the correct generalization that combining criteria for divisibility, which worked with divisibility by 6, works only when the two factors are relatively prime.

What should be the teacher's role in developing reasoning and proof in grades 6 through 8?

Teachers in the middle grades can help students appreciate and use the power of mathematical reasoning by regularly engaging students in thinking and reasoning in the classroom. Fostering a mathematically thoughtful environment is vital to supporting the development of students' facility with mathematical reasoning.

The teacher plays an important role by creating or selecting tasks that are appropriate to the ages and interests of middle-grades students and that call for reasoning to investigate mathematical relationships. Tasks that require the generation and organization of data to make, validate, or refute a conjecture are often appropriate. For example, the examination of patterns associated with figurate numbers discussed above shows how a teacher can use the task both to stimulate student investigation and to develop facility with mathematical reasoning and argumentation. Suitable tasks can arise in everyday life, although many will arise within mathematics itself.

Students need to know the limitations of inductive reasoning as well as its possibilities.

Teachers also serve as monitors of students' developing facility with reasoning. In order to use inductive reasoning appropriately, students need to know its limitations as well as its possibilities. Because many elementary and middle-grades tasks rely on inductive reasoning, teachers should be aware that students might develop an incorrect expectation that patterns always generalize in ways that would be expected on the basis of the regularities found in the first few terms. The following hypothetical example shows how a teacher could help students develop a healthy appreciation for the power and limits of inductive reasoning.

A teacher asks students to determine how many segments of different lengths can be made by connecting pegs on a square geoboard that is 5 units on each side (a 5×5 square geoboard). Because the number of segments is large and some students will have difficulty being systematic in representing the segments on their geoboards, the teacher encourages the students to examine simpler cases to develop a systematic way to generate the different segments. The students approach this task by examining the number of segments on various subsquares on a 5×5 geoboard, looking at the growth from a 1×1 square to a 4×4 square, as shown in figure 6.34.

Fig. **6.34.**

Segments of different lengths on a geoboard

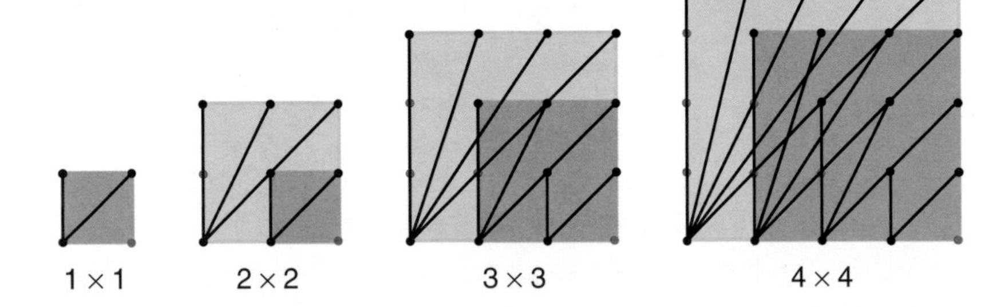

The teacher helps the students see that each successive square contains the previous square within it. Thus, the number of segments on a 3×3 square can be found by adding the number of segments found on a 2×2 square to the number of new segments that can be created using the "new" pegs within the 3×3 square. The teacher has the students verify—by direct measurement or by treating the diagonal lengths as hypotenuses of right triangles—that all segments are really of different lengths. The students then record the number of segments of different lengths in each square and note the pattern of growth, as shown in figure 6.35.

Fig. **6.35.**

Students can record in a table like this one data about the number of segments of different lengths on a geoboard.

Size of Square	Number of Segments of Different Lengths: Old + New	Total Number of Different Lengths
1 × 1	2	2
2 × 2	2 + 3	5
3 × 3	(2 + 3) + 4	9
4 × 4	(2 + 3 + 4) + 5	14
5 × 5	?	?

The teacher orchestrates a class discussion about the numbers in the table. Most students quickly detect a pattern of growth and are prepared to predict the answer for a 5×5 geoboard—20 different segments—because $(2 + 3 + 4 + 5) + 6 = 14 + 6 = 20$. In fact, many students are prepared to state a more general conjecture: The number of segments for an $N \times N$ square geoboard is the sum $2 + 3 + \cdots + (N + 1)$.

After the students have made their prediction, the teacher asks them to check its accuracy by actually making all the possible segments of different lengths on a 5 × 5 geoboard, as in figure 6.36. Because of their prior experience in systematically generating all possible segments, most of the students are able to find all the possibilities. In fact, most recognize that they need only to count the "new" segments and check to be sure that the segments are of different lengths from one another and from the segments already counted in the previous cases. The students note that there are twenty segments, as predicted, and most are content with the observation that all the new segments are of different lengths. But some students discover that two segments—*AB* and *CD* in figure 6.36—are both five units long. Thus, there are only nineteen different lengths, rather than twenty as predicted. Most of the students are surprised at this result, although they recognize that it is correct.

Fig. **6.36.**

Line segments on a 5 × 5 geoboard

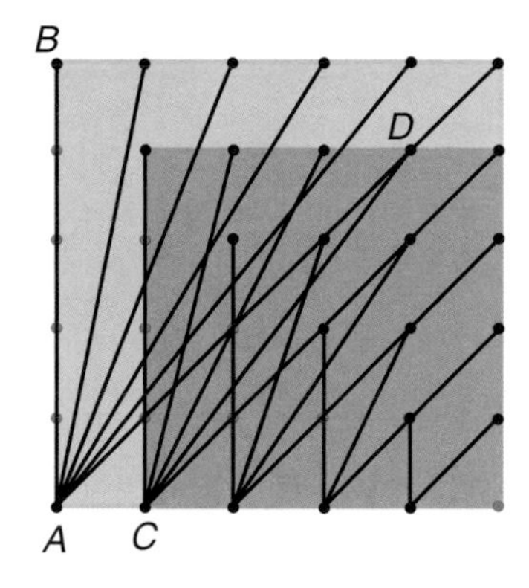

A teacher can use an example such as this as a powerful reminder that students should be cautious when generalizing inductively from a small number of cases, because not all patterns generalize in ways that we might wish or expect from early observations. This important lesson allows students to develop a healthy skepticism in their work with patterns and generalization.

Teachers need to monitor students' developing facility not only with inductive reasoning but also with deductive reasoning. In the middle grades, students begin to consider assertions such as the following: The diagonals of any given rectangle are equal in length. (See the "Geometry" section of this chapter for more discussion of how this assertion might be generated and verified by students.) An assertion such as this is tricky, at least in part because it is an implicitly conditional statement: If a shape is a rectangle, then its diagonals are equal in length. Thus, it is probably not surprising that some students will misapply this idea by inferring that any quadrilateral with diagonals of equal length must be a rectangle. Doing so reflects the erroneous view that if a statement is true then its converse is true. In this instance, the converse is not true because nonrectangular isosceles trapezoids also have diagonals of equal length, as do many other quadrilaterals.

Teachers in the middle grades need to be mindful of complexities in logical thinking and be alert in order to help students reason correctly. In this example, a teacher might have students use dynamic geometry software to investigate which types of quadrilaterals have diagonals of equal length. The software could allow students to see changes in the lengths of the diagonals instantly as they change the shape of the quadrilateral. A teacher might have students investigate quadrilaterals in general and particular types of quadrilaterals, including rectangles, squares, parallelograms, rhombuses, and trapezoids. The teacher might ask students to note which shapes have diagonals of equal length. If no one found such a shape, the teacher could ask them to construct an isosceles trapezoid with a given set of vertices, and the students would then see that this trapezoid has diagonals of equal length. This type of investigation can lead students to understand that even when a statement is true, its converse may be false.

Communication Standard for Grades 6–8

Instructional programs from prekindergarten through grade 12 should enable all students to—

Organize and consolidate their mathematical thinking through communication

Communicate their mathematical thinking coherently and clearly to peers, teachers, and others

Analyze and evaluate the mathematical thinking and strategies of others

Use the language of mathematics to express mathematical ideas precisely

In classrooms where students are challenged to think and reason about mathematics, communication is an essential feature as students express the results of their thinking orally and in writing. This type of environment is desirable at all grade levels, but there are a few distinctive features of such classrooms in the middle grades. For example, the mathematics under discussion in grades 6–8 is generally more complex and perhaps more abstract than the mathematics in the lower grades.

A second distinctive feature relates to the norms for evaluating the thinking of members of a classroom learning community. When students in grades 6–8 explain their thinking, they can be held to standards that are more stringent than would likely be applied to younger students, though not as demanding as might be applied in high school. Each student should be expected not only to present and explain the strategy he or she used to solve a problem but also to analyze, compare, and contrast the meaningfulness, efficiency, and elegance of a variety of strategies. Explanations should include mathematical arguments and rationales, not just procedural descriptions or summaries (Yackel and Cobb 1996).

A third distinguishing feature pertains to the social norms in a middle-grades classroom rather than to the content of the students' discussions. During adolescence, students are often reluctant to do anything that causes them to stand out from the group, and many middle-grades students are self-conscious and hesitant to expose their thinking to others. Peer pressure is powerful, and a desire to fit in is paramount. Teachers should build a sense of community in middle-grades classrooms so students feel free to express their ideas honestly and openly, without fear of ridicule.

What should communication look like in grades 6 through 8?

Consider an extended example (adapted from Silver and Smith [1997]) of mathematical communication in a middle-grades classroom.

> The students began by working collaboratively in pairs to solve the following problem, adapted from Bennett, Maier, and Nelson (1998):
>
> > A certain rectangle has length and width that are whole numbers of inches, and the ratio of its length to its width is 4 to 3. Its area is 300 square inches. What are its length and width?

As the students worked on the problem, the teacher circulated around the room, monitoring the work of the pairs and responding to their questions. She also noted different approaches that were used by the students and made decisions about which students she would ask to present solutions.

After most students had a chance to solve the problem, the teacher asked Lee and Randy to present their method. They proceeded to the overhead projector to explain their work. After briefly restating the problem, Lee indicated that 3 times 4 is equal to 12 and that they needed "a number that both 3 and 4 would go into." The teacher asked why they had multiplied 3 by 4. Randy replied that the ratio of the length to the width was given as "4 to 3" in the problem. Lee went on to say that they had determined that "3 goes into 15 five times and that 4 goes into 20 five times." Since 15 times 20 is equal to 300, the area of the given rectangle, they concluded that 15 inches and 20 inches were the width and length of the rectangle.

The teacher asked if there were questions for Lee or Randy. Echoing the teacher's query during the presentation of the solution, Tyronne said that he did not understand their solution, particularly where the 12 had come from and how they knew it would help solve the problem. Neither Lee nor Randy was able to explain why they had multiplied 3 by 4 or how the result was connected to their solution. The teacher then indicated that she also wondered how they had obtained the 15 and the 20. The boys reiterated that they had been looking for a number "that both 3 and 4 went into." In reply, Darryl asked how the boys had obtained the number 5. Lee and Randy responded that 5 was what "3 and 4 go into." At this point, Keisha said "Did you guys just guess and check?" Lee and Randy responded in unison, "Yeah!" Although Lee and Randy's final answer was correct and although it contained a kernel of good mathematical insight, their explanation of their solution method left other students confused.

To address the confusion generated by Lee and Randy, the teacher decided to solicit another solution. Because the teacher had seen Rachel and Keisha use a different method, she asked them to explain their approach. Keisha made a sketch of a rectangle, labeling the length 4 and the width 3. She explained that the 4 and 3 were not really the length and width of the rectangle but that the numbers helped remind her about the ratio. Then Rachel explained that she could imagine 12 squares inside the rectangle because 3 times 4 is equal to 12, and she drew lines to subdivide the rectangle accordingly. Next she explained that the area of the rectangle must be equally distributed in the 12 "inside" squares. Therefore, they divided 300 by 12 to determine that each square contains 25 square inches. At the teacher's suggestion, Rachel wrote a 25 in each square in the diagram to make this point clear. Keisha then explained that in order to find the length and width of the rectangle, they had to determine the length of the side of each small square. She argued that since the area of each square was 25 square inches, the side of each square was 5 inches. Then, referring to the diagram in figure

Teachers should build a sense of community in middle-grades classrooms so students feel free to express their ideas honestly and openly, without fear of ridicule.

Fig. **6.37.**

Rachel's and Keisha's method

15	5	25	25	25	25
	5	25	25	25	25
	5	25	25	25	25
		5	5	5	5
		20			

6.37, she explained that the length of the rectangle was 20 inches, since it consisted of the sides of four squares. Similarly, the width was found to be 15 inches. To clarify their understanding of the solution, a few students asked questions, which were answered well by Keisha and Rachel.

At this point the teacher might ask the girls if they think their approach would work for similar problems: What if the ratio were not 4 to 3? What if the area were not 300? Other students might be invited to ask questions: What would happen if the product of the length-width ratio numbers does not divide evenly into the area of the rectangle? Such questions could generate lively exchange that would include several students and could invite comparison to methods used by other students. The teacher could encourage students to consider generalizations and work to engage the entire class in this kind of thinking. For homework, students might be asked to come up with some possible generalizations. In the final few minutes of the class, the students could record in their journals their observations about what they had learned during the lesson along with any lingering questions they might have.

What should be the teacher's role in developing communication in grades 6–8?

The previous example illustrates several important facets of the teacher's role in supporting communication, particularly a whole-class discussion, which was portrayed in the example. One is establishing norms within a classroom learning community that support the learning of all students. Another is selecting and using worthwhile mathematical tasks that allow significant communication to occur. And a third is guiding classroom discussion on the basis of what is learned by monitoring students' learning.

The middle-grades mathematics teacher should strive to establish a communication-rich classroom in which students are encouraged to share their ideas and to seek clarification until they understand. In such a classroom community, communication is central to teaching and learning mathematics and to assessing students' knowledge. The focus in such classrooms is trying to make sense of mathematics together. Explaining, questioning, debating, and sense making are thus natural and expected activities. To achieve this kind of classroom, teachers need to establish an atmosphere of mutual trust and respect, which can be gained by supporting students as they assume substantial responsibility for their own mathematics learning and that of their peers. When teachers build such an environment, students understand that it is acceptable to struggle with ideas, to make mistakes, and to be unsure. This attitude encourages them to participate actively in trying to understand what they are asked to learn because they know that they will not be criticized personally, even if their mathematical thinking is critiqued.

Communication should be focused on worthwhile mathematical tasks. Teachers should identify and tasks that—

- relate to important mathematical ideas;
- are accessible to multiple methods of solution;
- allow multiple representations;
- afford students opportunities to interpret, justify, and conjecture.

Although the task in the example was in many ways quite simple, it provided students with an opportunity to use their understanding of area and ratio—important ideas in the middle grades. The task was simple enough that all students could do it, difficult enough to challenge students to think and reason, and rich enough to allow students to engage at different levels.

Teachers also need to monitor students' learning in order to direct classroom discourse appropriately. Facilitating students' mathematics learning through classroom discussion requires skill and good judgment. In the example, to be sure her objectives were met, the teacher skillfully steered the "mathematical direction" of the conversation by calling on particular students to present a different solution. Teachers must consider numerous issues in orchestrating a classroom conversation. Who speaks? When? Why? For how long? Who doesn't? Why not? It is important that everyone have opportunities to contribute, although it is not necessary to give equal speaking time to all students.

Clearly, the students in the example were accustomed to being asked regularly, not only by the teacher but also by other students, to explain their mathematical thinking and reasoning. The teacher and several students pressed for justification and explanation as each solution was presented. Because not all students regularly participate in whole-class discussions, teachers need to monitor their participation to ensure that some are not left entirely out of the discussion for long periods. But encouraging all students to speak can sometimes conflict with advancing the mathematical goals of a lesson because students' contributions may occasionally be irrelevant or lack mathematical substance. But even when this happens, the teacher and students can derive some benefit. It can be productive for the teacher to pick up on and probe incorrect or

Facilitating students' mathematics learning through classroom discussion requires skill and good judgment.

incomplete responses. Only by examining misconceptions and errors can students deal with them appropriately.

Teachers can use class discussions as opportunities for ongoing assessment of their teaching and of students' learning. Making mental notes about missed teaching opportunities and about students' difficulties or confusions may help in making decisions about follow-up lessons.

Classroom communication can contribute to multiple goals. The lesson in the example generated at least four instructional directions. In the next few lessons, the teacher and her class explored the use of algebraic representations and solution methods for related problems as they sought a method that would work for a larger range of values for the areas and the ratios of the sides. The teacher later examined with the class how the area of a rectangle is affected when its length and width are each multiplied by the same factor. The teacher's informal assessment of the students' understanding during work on the task led her to take Lee and Randy aside later in the week to help them clarify their thinking. Finally, to acquaint her students with the scoring guidelines that would be used on the state proficiency test, she had them prepare written solutions for this problem and then score one another's solutions using the guidelines. In this way, they learned valuable lessons about the need for accuracy, precision, and completeness in their written communication.

Working in pairs is often a very effective approach with students in the middle grades.

Teachers can use oral and written communication in mathematics to give students opportunities to—

- think through problems;
- formulate explanations;
- try out new vocabulary or notation;
- experiment with forms of argumentation;
- justify conjectures;
- critique justifications;
- reflect on their own understanding and on the ideas of others.

To help students reflect on their learning, teachers can ask them to write commentaries on what they learned in a lesson or a series of lessons and on what remains unclear to them. To strive for clarity in explaining their ideas, students can write a letter to a younger student explaining a difficult concept (e.g., "Here's what it means for two figures to be similar. Let me start with rectangles. . . ."). In the example, journal writing and individual homework offered all the students opportunities for individual reflection and communication. Working in pairs also afforded opportunities for communication. This approach is often very effective with students in the middle grades because they can try out their ideas in the relative privacy of a small group before opening themselves up to the entire class.

Even when students are working in small groups, the teacher has an important role to play in ensuring that the discourse contributes to the mathematics learning of the group members and helps to further the teacher's mathematical goals. For example, when students ask questions about the task requirements or about the correctness of their work, the teacher should respond in ways that keep their focus on thinking and reasoning rather than only on "getting the right answer." Teachers should resist students' attempts to have the teacher "do the thinking for them." In the incident related in the example when the students were working in pairs, the teacher generally responded to questions with suggestions (e.g., "Try to think of some way to use a diagram") or her own questions (e.g., "What do you know about the relationship between the area of a rectangle and its length and width? How can you use what you know?"). Teachers also must be sure that all members are participating in the group and understanding its work.

Connections Standard for Grades 6–8

Instructional programs from prekindergarten through grade 12 should enable all students to—

Recognize and use connections among mathematical ideas

Understand how mathematical ideas interconnect and build on one another to produce a coherent whole

Recognize and apply mathematics in contexts outside of mathematics

Thinking mathematically involves looking for connections, and making connections builds mathematical understanding. Without connections, students must learn and remember too many isolated concepts and skills. With connections, they can build new understandings on previous knowledge. The important mathematical foci in the middle grades—rational numbers, proportionality, and linear relationships—are all intimately connected, so as middle-grades students encounter diverse new mathematical content, they have many opportunities to use and make connections.

This chapter on grades 6–8 mathematics contains numerous illustrations of mathematical connections. Many of the formulas students develop and use in the "Measurement" section draw on their knowledge of algebra, geometry, and measurement. The kite example in the "Geometry" section engages students in examining the perimeter and area of similar figures to investigate proportional relationships. Several examples in the "Data Analysis" section illustrate how gathering, representing, and analyzing data can help students develop insights into other mathematical ideas, including variation and change, probability, and ratio and proportion. The "cellular telephone" problem in the "Algebra" section demonstrates how connections among various forms of representation provide insights into patterns and regularities in problem situations. Clearly, rich problem contexts involve connections to other disciplines (e.g., science, social studies, art) as well as to the real world and to the daily life experiences of middle-grades students.

What should connections look like in grades 6 through 8?

Mathematics classes in the middle grades should continually provide opportunities for students to experience mathematics as a coherent whole through the curriculum used and the questions teachers and classmates ask. Students reveal the ways they are connecting ideas when they answer questions such as, What made you think of that? Why does that make sense? Where have we seen a problem like this before? How are these ideas related? Did anyone think about this in a different way? How does today's work relate to what we have done in earlier units of study? From these discussions, students can develop new connections and enhance their own understanding of mathematics by listening to their classmates' thinking.

If curriculum and instruction focus on mathematics as a discipline of connected ideas, students learn to expect mathematical ideas to be related. Rich mathematical tasks prompt students to use and develop mathematical understandings and connections. Challenging problems encourage students to think about how familiar concepts and procedures can be applied in new situations. In classrooms where students are expected to reason mathematically and to communicate clearly about significant mathematical tasks, new ideas surface quite naturally as extensions of previously learned mathematics. With prompting from their teacher, students routinely ask themselves, "How is this problem like what I have done before? How is it different?"

Without connections, students must learn and remember too many isolated concepts and skills.

Consider an expanded version of a summary (adapted from NCTM, Algebra Working Group [1998, p. 155]) of a lesson on ratio and proportion. The intent of this lesson was to begin developing students' understanding of methods for comparing ratios. The students had not previously been taught such methods, so the teacher wanted to uncover whether and how students could apply what they had already learned about number and ratio. The lesson was centered on the following task, which was adapted from Lappan et al. (1998, p. 27):

> Southwestern Middle School Band is hosting a concert. The seventh-grade class is in charge of refreshments. One of the items to be served is punch. The school cook has given the students four different recipes calling for sparkling water and cranberry juice.

RECIPE A	RECIPE B
2 cups cranberry juice 3 cups sparkling water	4 cups cranberry juice 8 cups sparkling water

RECIPE C	RECIPE D
3 cups cranberry juice 5 cups sparkling water	1 cup cranberry juice 4 cups sparkling water

1. Which recipe will make punch that has the strongest cranberry flavor? Explain your answer.
2. Which recipe will make punch that has the weakest cranberry flavor? Explain your answer.
3. The band director says that 120 cups of punch are needed. For each recipe, how many cups of cranberry juice and how many cups of sparkling water are needed? Explain your answer.

The students worked on the first two questions in groups of two or three. When the groups had finished, they came together as a whole class to share and explain their answers.

The groups had attempted to figure out which recipe has the strongest cranberry flavor in different ways. Some examined the part-whole relationships of the number of cups of juice to the total number of cups in the recipe (these ratios are 2/5, 4/12, 3/8, 1/5 for recipes A–D, respectively). Others looked at the part-part ratios of juice to water (2/3, 4/8, 3/5, 1/4). Still others, failing to consider that the recipes, as given, make different amounts of punch, incorrectly considered only the number of cups of juice in each recipe (2, 4, 3, 1). After questioning and challenging one another's solutions and comparing methods, the class decided to move on to the last question to see if they could resolve the differences in their answers. Each group was assigned to determine the amounts of juice and water needed for just one of the recipes. Below are four of the strategies the groups used to work through this part of the problem.

Group with recipe A

We figured out that each recipe would make 5 cups: 2 of juice and 3 of water. So to make 120 cups, it would take 120 divided by 5, and that is 24, the number of recipes needed. Since we need 2 cups of juice and 3 cups of water for one recipe, we need $2 \times 24 = 48$ cups of juice and $3 \times 24 = 72$ cups of water for the 24 recipes. And since 48 cups of juice + 72 cups of water makes 120 cups of punch that must be right.

Group with recipe B

We thought that 4 cups of juice and 8 cups of water is the same ratio as 1 cup of juice and 2 cups of water. We then thought about the 120 cups of punch as divided into three groups of 40 cups each: $40 + 40 + 40 = 120$. We need 1 part juice, so that is 40 cups, and 2 parts water, so that is 80 cups. This makes 120 cups of punch, and you still have a ratio of 1 part juice to 2 parts water.

Group with recipe C

We tried to double the recipe, but that was not enough. So we added another batch and that still was not enough. So we just kept adding recipes and seeing how many total cups of punch we had. We kept up this pattern until we got 120 cups. So we had [a table like that shown in fig. 6.38]. That means we had 45 cups of juice and 75 cups of water.

Fig. **6.38.**

Chart for punch recipes

Cups of juice	3	6	9	12	15	18	21	24	27	30	33	36	39	42	45
Cups of water	5	10	15	20	25	30	35	40	45	50	55	60	65	70	75
Cups of punch	8	16	24	32	40	48	56	64	72	80	88	96	104	112	120

Later in the class discussion, this group noticed that they could have gone directly from 3/5 to 45/75 by multiplying the numerator and denominator by 15 because they needed 15 recipes.

Group with recipe D

We tried various numbers. First we tried 20 cups of juice. This means we needed 4 times as much water or 80 cups of water. But this was too small because $20 + 80$ is only 100. So we tried 30 cups of juice, so that meant $30 \times 4 = 120$ cups of water. This time we had too much punch, $30 + 120 = 150$. Next we tried 25 cups of juice. And $25 \times 4 = 100$, so we had 100 cups of water. But this made 125 cups of punch, which was close but too much. So we tried 24 cups of juice, which needed $24 \times 4 = 96$ cups of water. This worked because $24 + 96 = 120$ cups of punch.

After the groups had shared their approaches to the third question, the teacher continued the conversation by encouraging the class to talk about the similarities and differences among the strategies.

The "making punch" problem had brought numerous mathematical ideas to the forefront: fractions, ratios, proportions, operations, magnitude, scaling, number sense, patterns, and so on. By bringing previously understood mathematical ideas or processes to bear on this problem, the students were developing understandings that laid a foundation for the later study of such topics as rates of change and linear relationships.

Since the task required the students to explain their strategies, all the students had an opportunity to enhance their understanding of ratios by listening to the others' different ideas. For example, the group with recipe D used a "guess and check" approach to solve the problem. The group with recipe C made a table and used the ideas of scaling ratios and adding iteratively in the same way that students find equivalent fractions. The groups with recipes A and B thought about comparing quantities and using ratios.

None of the students mentioned that the answers to the first two questions would have been more obvious if they had solved the third problem first. For each recipe, we can add the number of cups of cranberry juice to the number of cups of water to determine how much punch one recipe makes. We divide this number into 120 to determine the multiples—24, 10, 15, and 24, respectively—of the ingredients that are needed. Because recipes A–D use 2, 4, 3, and 1 cup of cranberry juice initially, they will use 48, 40, 45, and 24 cups of cranberry juice, respectively, when multiplied to serve 120 people. Clearly, recipe D has the weakest cranberry flavor and recipe A has the strongest. This finding confirms the students' previous answers and approaches.

What should be the teacher's role in developing connections in grades 6 through 8?

The teacher's role includes selecting problems that connect mathematical ideas within topics and across the curriculum; it also includes helping students build on their current mathematical ideas to develop new ideas. The teacher's orchestration of the "making punch" lesson allowed the students to make the connections explicit and to focus on the relationships and commonalities among their strategies. The teacher took advantage of an opportunity to foster the students' disposition to look for connections as well as to use connections. In situations like this, it is essential that the teacher recognize and understand the mathematical concepts being developed, not just to teach the abstract manipulation but also to orchestrate the conversation. The teacher needs to be able to make quick decisions about next steps. It is also important to encourage students to use words and notation appropriately to support their understanding of new concepts, such as proportionality and algebra.

It is sometimes quite effective to revisit a problem to help students connect familiar ideas to new concepts or skills. Indeed, the "making punch" problem has potential for connections to proportionality and linearity. For instance, students could make a graph, plotting values

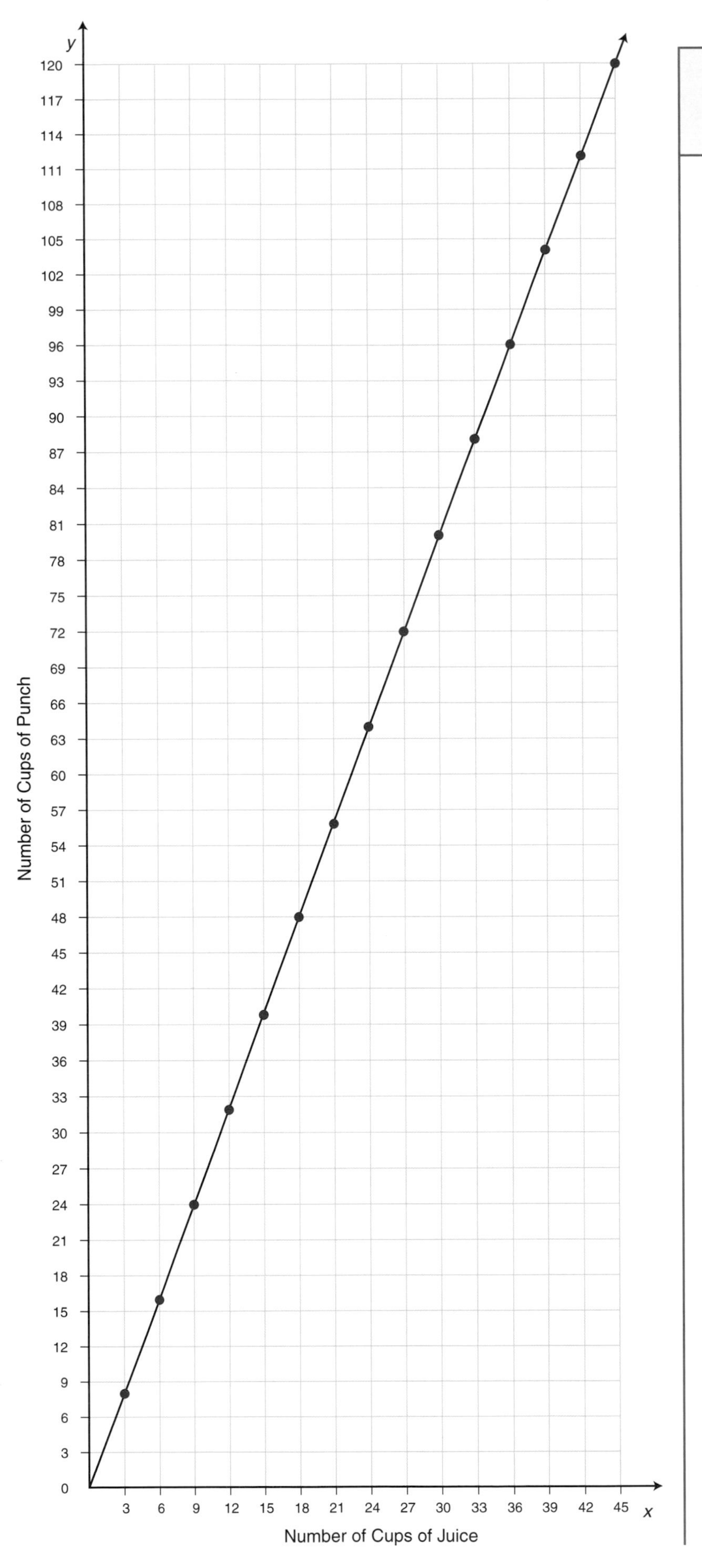

Fig. **6.39.**

A graph of the numbers of cups of juice and punch reveals a linear relationship.

from the first and third rows of the table in figure 6.38. These points lie on a line. If y represents the total number of cups of punch and x the number of cups of juice, then the line has equation $y = (8/3)x$ (see fig. 6.39). To answer question 3—which asks how much juice and how much water are needed for 120 cups of punch—students can substitute 120 for y in this equation and compute x to find the number of cups of juice. They would subtract this number from 120 to find the amount of sparkling water needed. This equation works equally well for finding how much juice is needed for any other quantity of punch, so they see the power of expressing a relationship in general terms. Here the slope of the line is the ratio of the amount of punch to the amount of juice. Of course, having used this method to answer question 3 for all recipes, students can easily answer the other two questions. A teacher might also revisit the "making punch" problem to assess students' understanding of tabular, graphical, and symbolic representations for linear relationships.

Teachers can also enhance students' understanding of mathematics by using other disciplines as sources of problems. Science and social studies are rich in opportunities to learn about measurement, data, and algebra; art and computer graphics can be used to make sense of shape, symmetry, similarity, and transformations of geometric figures. Environmental studies offers a context for the study of large numbers (analyzing population growth), measurement (finding the percent of different types of trash in landfills by volume and considering recycling alternatives), or data and statistics (studying the effects of pollution on animal and plant populations).

In many schools, teachers are interested in fostering interdisciplinary studies. The mathematics teacher may work with teachers of other subjects to develop integrated units of study. For example, middle-grades science classes might study populations of wildlife such as deer, fish, eagles, or sharks (see Curcio and Bezuk [1994]). If students will be expected to use sampling techniques in science class to determine the population of a species, it is important that the mathematics and science teachers discuss students' understanding of different sampling techniques and of the idea of randomness. It is also important that science teachers understand that students are likely to use scaling and equivalent ratios to estimate the total population rather than cross multiplication and formal algebraic symbol manipulation to find their solutions.

In the same spirit, mathematics teachers can build on and connect to disciplines other than science and social studies. For example, language arts teachers can describe the strategies they teach for writing convincing arguments. The mathematics teachers may then be able to help students use the strategies when appropriate in formulating mathematical arguments. They may also be better able to help students recognize and analyze forms of argumentation and justification that are peculiar to mathematics. Again, students benefit from teachers' efforts to understand how other subjects are taught and to make connections between the subjects explicit.

Conversations about students' experiences, understandings, and familiarity with procedures give teachers of other subjects an opportunity to learn about elements of the mathematics curriculum, such as algorithms and the level of abstract symbol manipulation that students might use. Without such conversations, those who are not mathematics teachers may expect students to understand and use procedures that are not part of their repertoire or teachers may fail to build on ideas with which students are already conversant. Students may miss an opportunity to apply and extend their reasoning skills or to see that mathematical ideas can be used in other disciplines. This is not to imply that merely applying mathematics in science, social studies, or any other discipline constitutes a sufficient middle-grades mathematics curriculum. The point is that interdisciplinary experiences serve as ways to revisit mathematical ideas and they help students see the usefulness of mathematics both in school and at home. If all the middle-grades teachers in a school do their best to connect content areas, mathematics and other disciplines will be seen as permeating life and not as just existing in isolation.

Merely applying mathematics in other disciplines does not constitute a sufficient mathematics curriculum.

Representation Standard for Grades 6–8

Instructional programs from prekindergarten through grade 12 should enable all students to—

Create and use representations to organize, record, and communicate mathematical ideas

Select, apply, and translate among mathematical representations to solve problems

Use representations to model and interpret physical, social, and mathematical phenomena

Representation is central to the study of mathematics. Students can develop and deepen their understanding of mathematical concepts and relationships as they create, compare, and use various representations. Representations—such as physical objects, drawings, charts, graphs, and symbols—also help students communicate their thinking.

Representations are ubiquitous in the middle-grades mathematics curriculum proposed here. The study of proportionality and linear relationships is intertwined both with students' learning to use variables flexibly in order to represent unknowns and with their learning to employ tables, graphs, and equations as tools for representation and analysis. Middle-grades students who are taught with this Standard in mind will learn to recognize, compare, and use an array of representational forms for fractions, decimals, percents, and integers. They also will learn to use representational forms such as exponential and scientific notation when working with large and small numbers and to use a variety of graphical tools to represent and analyze data sets.

What should representation look like in grades 6 through 8?

Students in the middle grades solve many problems in which they create and use representations to organize and record their thinking about mathematical ideas. For example, they use representations to develop or apply their understanding of proportionality when they make or interpret scale drawings of figures or scale models of objects, when they connect the geometric notion of similarity with numerical ratios, and when they draw relative-frequency histograms for data sets. While solving challenging problems, students might use standard representations, but they can also develop and use nonstandard representations that work well for a particular problem.

When solving problems involving proportionality, students can create representations that blend visual and numerical information to depict relationships among quantities. Consider the following problem:

> The Copy Cat printing shop has a printer that uses only black, red, and blue cartridges. All the cartridges print the same number of pages. The black cartridges are replaced 4 times as often as the red ones. And during the time in which 3 red cartridges need to be replaced, 5 blue cartridges will also need to be replaced.

1. What fraction of Copy Cat's printing is in black?
2. What percent of the printing is in blue?
3. In a month, 60 black cartridges are used. What is the total number of red and blue cartridges used in that month?

Students can use a variety of approaches to represent and solve this problem, including both standard and nonstandard forms of representation. Some students will find it natural to develop and use a discrete model like the one shown in figure 6.40. With such a representation, students can merge the information from the two ratios in the problem statement. They can see that in every set of 20 cartridges used at Copy Cat, 12 are black, 3 are red, and 5 are blue. They can conclude that 12/20 (or 6/10, 3/5, or 0.6) of the printing is in black, which answers the first question. To answer the second question, a student might imagine replicating this set of 20 five times to see that Copy Cat uses 25 blue cartridges in every 100 used. Thus, 25 percent of the printing is in blue. To answer the third question, students could note that a set of 60 black cartridges comprises 5 sets of 12 cartridges and that a total of 8 red and blue cartridges are used in the time that 12 black cartridges are used. Thus, it follows that 40 red and blue cartridges—15 red and 25 blue—are used in the time that it takes to use 60 black cartridges.

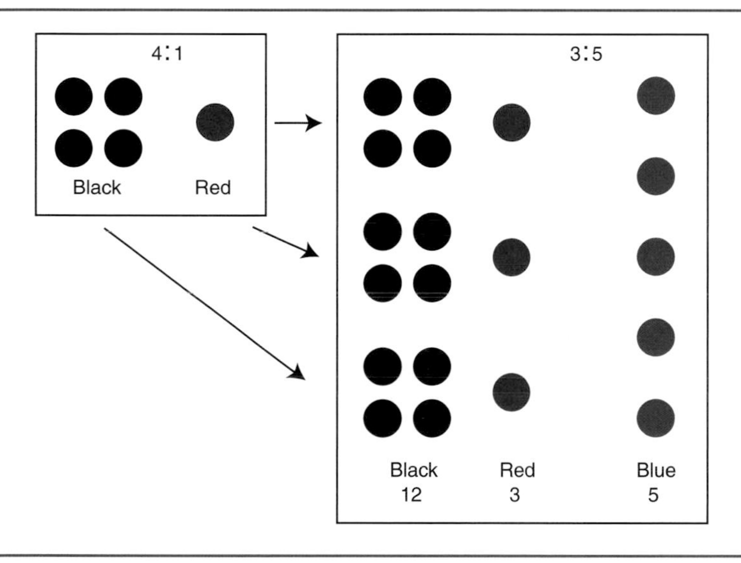

Fig. **6.40.**

This representation for the "printing cartridges" problem combines visual and numerical information.

The power of representations that blend visual and numerical information can be appreciated in solving many problems involving ratios, proportions, and percents. Consider the following problem:

> A group of students has $60 to spend on dinner. They know that the total cost, after adding tax and tip, will be 25 percent more than the food prices shown on the menu. How much can they spend on the food so that the total cost will be $60?

Of the various ways this problem might be represented, many students would find the representation in figure 6.41 useful. In this figure, a rectangular bar represents the total of $60. This total must include the price of the food plus 25 percent more for tax and tip. To show this relationship, the bar is segmented into five equal parts, of which four represent the price of the food and one the tax and tip. Because there are five equal parts and the total is $60, each part must be $12. Therefore, the total price allowed for food is \$48. This type of visual representation for numerical quantities is quite adaptable and can be used to solve many problems involving fractions, percents, ratios, and proportions (see, e.g., some problems in Curriculum Development Institute of Singapore 1997, or Bennett, Maier, and Nelson 1988). For example, the representation in figure 6.41 could also help students see and understand that when one quantity is 125 percent of a second quantity, then the second is 80 percent of the first.

Fig. **6.41.**

One possible representation for the "food, tax, and tip" problem

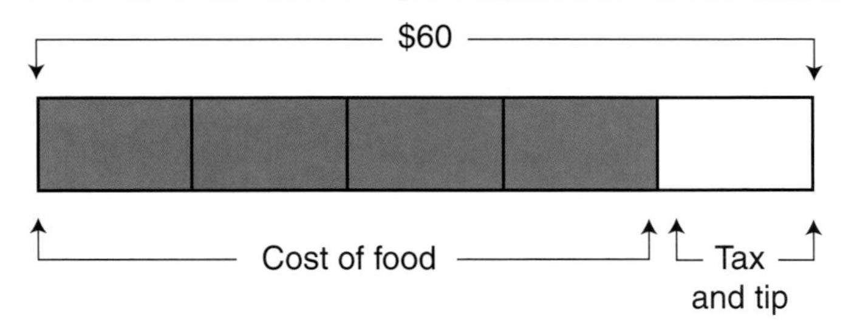

The study of linear functions, with the associated patterns and relationships, is another major focus in the middle grades. By considering problems in a variety of contexts, students should become familiar with a range of representations for linear relationships, including tables, graphs, and equations. Students need to learn to use these representations flexibly and appropriately. In the "Algebra" section of this chapter, several examples are discussed, notably the "cellular telephone" problem, through which students could develop a repertoire of representations.

Students also need to examine relationships among representations for linear functions. The use of graphing calculators or appropriate computer software can greatly facilitate such an examination and can allow students to see such important relationships as the one between the value of k in the equation $y = kx$ and the slope of the corresponding line.

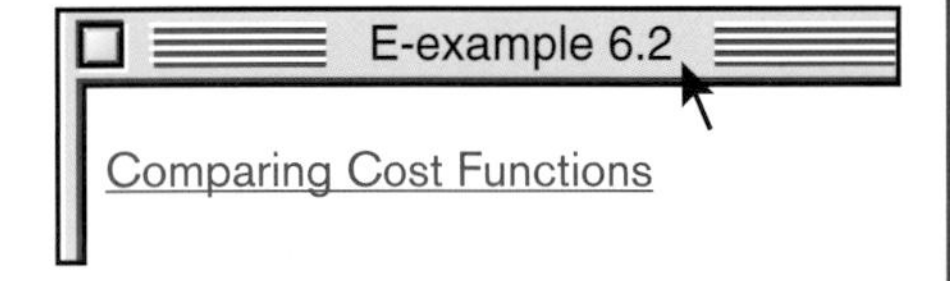

Students will be better able to solve a range of algebra problems if they can move easily from one type of representation to another. In the middle grades, students often begin with tables of numerical data to examine a pattern underlying a linear function, but they should also learn to represent those data in the form of a graph or equation when they wish to characterize the generalized linear relationship. Students should also become flexible in recognizing equivalent forms of linear equations and expressions. This flexibility can emerge as students gain experience with multiple ways of representing a contextualized problem. For example, consider the following problem, which is adapted from Ferrini-Mundy, Lappan, and Phillips (1997):

> A rectangular pool is to be surrounded by a ceramic-tile border. The border will be one tile wide all around. Explain in words, with numbers or tables, visually, and with symbols the number of tiles that will be needed for pools of various lengths and widths.

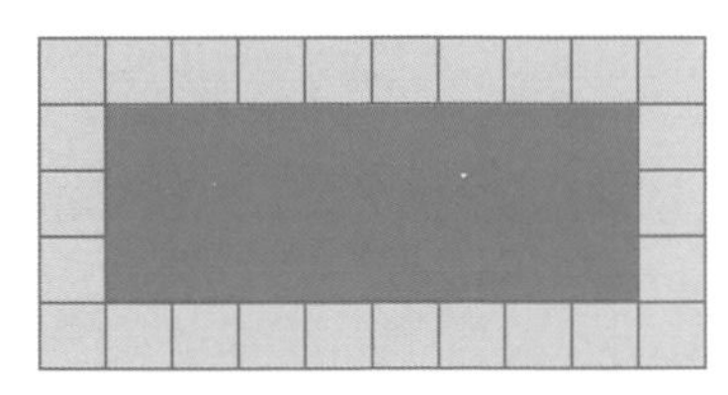

Some students would solve this problem by using a table to record the values for various lengths and widths of rectangular pools and for the corresponding number of tiles on the border. From the table, they could discern a generalization and then express it as an equation, as is suggested in the response and the accompanying work shown in figure 6.42.

The formula for the number of tiles is $T = 2(L + W) + 4$. I made a table with columns for L, W, and T. I drew some pictures. Then I counted the tiles for those pictures, filled in the numbers in the table, and looked for a pattern. It's easy! You always just add the length to the width, double that answer, and then add 4.

Pool Length	Pool Width	Number of Tiles
1	1	8
2	1	10
3	1	12
3	2	14
3	3	16
3	4	18

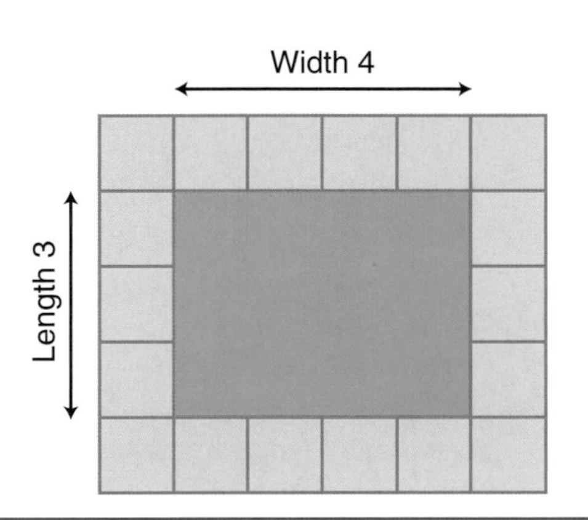

Fig. **6.42.**
Student work for the "tiled pool" problem

Other students might reason about the situation geometrically (or visually) rather than numerically. Here are three other possible student responses:

> I drew several pictures and saw this pattern. You need $L + 2$ tiles across the top and the same number across the bottom. And you also need W tiles on the left and W tiles on the right. So all together, the number of tiles needed is $T = 2(L + 2) + 2W$.
>
> I pictured it in my head. First, place one tile at each of the corners of the pool. Then you just need L tiles across the top and the bottom, and W tiles along each of the sides. So all together, the number of tiles needed is $4 + 2L + 2W$.
>
> You can find the number of tiles needed by finding the area of the whole rectangle (pool plus tile border) and then subtracting off the area of the pool. The area of the pool and deck together is $(L + 2)(W + 2)$. The area of the pool alone is LW. So all together, the number of tiles needed is $(L + 2)(W + 2) - LW$.

These three responses differ in the way in which particular geometric (visual) features are considered. For example, in the first two solutions, the tile border is related to the perimeter of the large rectangle comprising the tiles and the pool but with different decompositions of the perimeter. In contrast, the third response considers the area of the large rectangle and derives the number of tiles as the area of the border, which is legitimate because the tiles are unit squares.

By working on problems like the "tiled pool" problem, students gain experience in relating symbolic representations of situations and relationships to other representations, such as tables and graphs. They also see that several apparently different symbolic expressions often can be used to represent the same relationship between quantities or variables in a situation. The latter observation sets the stage for students to understand equivalent

Middle-grades students have opportunities to solve relatively large-scale, motivating, and significant problems that involve modeling.

symbolic expressions as different symbolic forms that represent the same relationship. In the "tiled pool" problem, for example, a class could discuss why the four expressions obtained for the total number of tiles should be equivalent. They could then examine ways to demonstrate the equivalence symbolically. For example, they might observe from their sketches that adding two lengths to two widths ($2L + 2W$) is actually the same as adding the length and width and then doubling: $2(L + W)$. They should recognize this pictorial representation for the distributive property of multiplication over addition—a useful tool in rewriting variable expressions and solving equations. In this way, teachers may be able to develop approaches to algebraic symbol manipulation that are meaningful to students.

Finally, it is important that middle-grades students have opportunities to use their repertoire of mathematical representations to solve relatively large-scale, motivating, and significant problems that involve modeling physical, social, or mathematical phenomena. The goal of this sort of mathematical modeling is for students to gain experience in using the mathematics they know and an appreciation of its utility for understanding and solving applied problems. For example, students might decide to investigate problems associated with trash disposal and recycling by collecting data on the volume of paper discarded in their classroom or home over a period of weeks or months. After organizing their data using graphs, tables, or charts, the students could think about which representations are most useful for illuminating regularities in the data. From their observations, the students might be able to offer thoughtfully justified estimates of the volume and types of paper discarded in their entire school, school district, or city in a week, month, or year. Drawing on what they have learned in science and social studies, they might then make recommendations for reducing the flow of paper into landfills or incinerators.

What should be the teacher's role in developing representation in grades 6 through 8?

Mathematics teachers help students learn to use representations flexibly and appropriately by encouraging them as they create and use representations to support their thinking and communication. Teachers help students develop facility with representations by listening, questioning, and making a sincere effort to understand what they are trying to communicate with their drawings or writings, especially when idiosyncratic, unconventional representations are involved. Teachers need to use sound professional judgment when deciding when and how to help students move toward conventional representations. Although using conventional representational forms has many advantages, introducing representations before students are able to use them meaningfully can be counterproductive.

Teachers play a significant role in helping students develop meaning for important forms of representation. For example, middle-grades students need many experiences to develop a robust understanding of the very complex notion of variable. Teachers can help students move from a limited understanding of variable as a placeholder for a single number to the idea of variable as a representation for a range of possible values by providing experiences that use variable expressions to describe numerical data (Demana and Leitzel 1988).

Teachers need to give students experiences in using a wide range of visual representations and introduce them to new forms of representations that are useful for solving certain types of problems. Vertex-edge graphs, for instance, can be used to represent abstract relationships among people or objects in many different kinds of situations. Take a situation in which several students might be working in different groups (for math review, history research, and a science project) or involved in different activities (basketball team and band). Each group wants to arrange a different meeting time to accommodate the students who are involved in more than one group. To help solve this scheduling problem, a teacher might suggest that students make a graph in which the vertices represent the groups and an edge between two groups indicates that there is some student who is a member of both groups. Figure 6.43 shows a possible vertex-edge graph involving the five groups, where an edge represents a relationship—common membership. This graph illustrates that no student participates in both the math-review group and history-research group (e.g., there is no edge joining those vertices) and that at least one student participates in both the math-review group and band practice (e.g., there is an edge between those vertices). The information in this graph can identify the potential scheduling conflicts so they can be avoided by scheduling all connected activities at different times.

Another new type of representation that teachers might wish to introduce their students to is a NOW-NEXT equation, which can be used to define relationships among variables iteratively. The equation NEXT = NOW + 10 would mean that each term in a pattern is found by adding 10 to the previous term. This notational form can be used as an alternative to the equation form of the general term when a recursive relationship is being highlighted. The data in figure 6.44 can be represented in a summarized form both as $y = 10x$ (where x must be a whole number) and as NEXT = NOW + 10.

The teacher thus has an important role in helping middle-grades students develop confidence and competence both in creating their own representations when they are needed to solve a challenging problem and in selecting flexibly and appropriately from an extensive repertoire of conventional representations. Whether helping students to use their own invented representations or introducing them to conventional forms, teachers should help students use representations meaningfully. By encouraging students to discuss the graphs, pictures, or symbols they are using in their work, teachers can monitor their developing fluency with representations. When students see how others interpret what they have written and how others have represented the same ideas, they can evaluate representations thoughtfully and recognize characteristics that make a representation flexible, appropriate, and useful. Through such a process, most students will come to appreciate the simplicity and effectiveness of conventional forms of representation and the role of representations in enabling communication with others.

Fig. **6.43.**

A vertex-edge graph used to identify potential scheduling conflicts

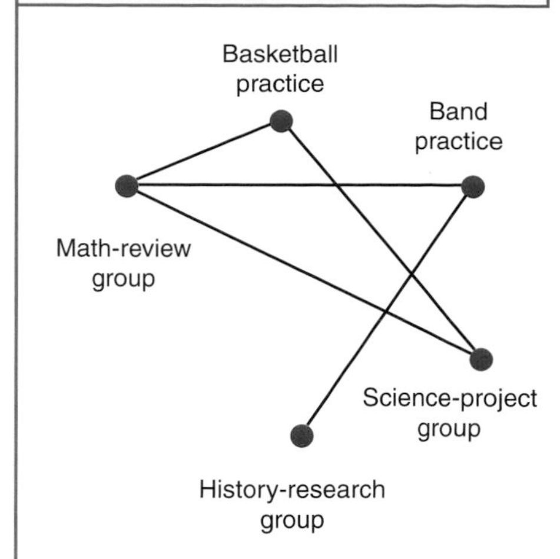

Fig. **6.44.**

Values for the terms in a pattern represented by a NOW = NEXT equation

x	y
1	10
2	20
3	30
4	40
⋮	⋮

$y = 10x$

NEXT = NOW +10

To ensure that students will have a wide range of career and educational choices, the secondary school mathematics program must be both broad and deep.

Chapter 7

Standards for Grades 9–12

Students in secondary school face choices and decisions that will determine the course of their lives. As they approach the end of required schooling, they must have the opportunity to explore their career interests—which may change during high school and later—and their options for postsecondary education. To ensure that students will have a wide range of career and educational choices, the secondary school mathematics program must be both broad and deep.

The high school years are a time of major transition. Students enter high school as young teenagers, grappling with issues of identity and with their own mental and physical capacities. In grades 9–12, they develop in multiple ways—becoming more autonomous and yet more able to work with others, becoming more reflective, and developing the kinds of personal and intellectual competencies that they will take into the workplace or into postsecondary education.

These Standards describe an ambitious foundation of mathematical ideas and applications intended for all students. Through its emphasis on fundamental mathematical concepts and essential skills, this foundation would give all students solid preparation for work and citizenship, positive mathematical dispositions, and the conceptual basis for further study. In grades 9–12, students should encounter new classes of functions, new geometric perspectives, and new ways of analyzing data. They should begin to understand aspects of mathematical form and structure, such as that all quadratic functions share certain properties, as do all functions of other classes—linear, periodic, or exponential. Students should see the interplay of algebra, geometry, statistics, probability, and discrete mathematics and various ways that mathematical phenomena can be represented. Through their high school experiences, they stand to develop deeper understandings of the fundamental mathematical concepts of function and relation, invariance, and transformation.

All students are expected to study mathematics each of the four years that they are enrolled in high school.

In high school, students should build on their prior knowledge, learning more-varied and more-sophisticated problem-solving techniques. They should increase their abilities to visualize, describe, and analyze situations in mathematical terms. They need to learn to use a wide range of explicitly and recursively defined functions to model the world around them. Moreover, their understanding of the properties of those functions will give them insights into the phenomena being modeled. Their understanding of statistics and probability could provide them with ways to think about a wide range of issues that have important social implications, such as the advisability of publicizing anecdotal evidence that can cause health scares or whether DNA "fingerprinting" should be considered strong or weak evidence.

Secondary school students need to develop increased abilities in justifying claims, proving conjectures, and using symbols in reasoning. They can be expected to learn to provide carefully reasoned arguments in support of their claims. They can practice making and interpreting oral and written claims so that they can communicate effectively while working with others and can convey the results of their work with clarity and power. They should continue to develop facility with such technological tools as spreadsheets, data-gathering devices, computer algebra systems, and graphing utilities that enable them to solve problems that would require large amounts of computational time if done by hand. Massive amounts of information—the federal budget, school-board budgets, mutual-fund values, and local used-car prices—are now available to anyone with access to a networked computer (Steen 1997). Facility with technological tools helps students analyze these data. A great deal is demanded of students in the program proposed here, but no more than is necessary for full quantitative literacy.

All students are expected to study mathematics each of the four years that they are enrolled in high school, whether they plan to pursue the further study of mathematics, to enter the workforce, or to pursue other postsecondary education. The focus on conceptual understanding provides the underpinnings for a wide range of careers as well as for further study, as Hoachlander (1997, p. 135) observes:

> Most advanced high school mathematics has rigorous, interesting applications in the work world. For example, graphic designers routinely use geometry. Carpenters apply the principles of trigonometry in their work, as do surveyors, navigators, and architects.... Algebra pervades computing and business modeling, from everyday spreadsheets to sophisticated scheduling systems and financial planning strategies. Statistics is a mainstay for economists, marketing experts, pharmaceutical companies, and political advisers.

With the experience proposed here in making connections and solving problems from a wide range of contexts, students will learn to adapt flexibly to the changing needs of the workplace. The emphasis on facility with technology will result in students' ability to adapt to the increasingly technological work environments they will face in the years to come. By learning to think and communicate effectively in mathematics, students will be better prepared for changes in the workplace that increasingly demand teamwork, collaboration, and communication (U.S. Department of Labor 1991; Society for Industrial and Applied Mathematics 1996). Note that these skills are also needed increasingly by people who will pursue careers in mathematics or science. With its emphasis on fundamental concepts, thinking and reasoning, modeling, and communicating, the core is a foundation for the study of more-advanced mathematics. Consider, for example, the recommendations for precalculus courses generated at the Preparing for a New Calculus conference (Gordon et al. 1994, p. 56):

> Courses designed to prepare students for the new calculus should:
>
> - cover fewer topics ... with more emphasis on fundamental concepts.
> - place less emphasis on complex manipulative skills.
> - teach students to think and reason mathematically, not just to perform routine operations....
> - emphasize modeling the real world and develop problem-solving skills.
> - make use of all appropriate calculator and computer technologies....
> - promote experimentation and conjecturing.
> - provide a solid foundation in mathematics that prepares students to read and learn mathematical material at a comparable level on their own.

A central theme of *Principles and Standards for School Mathematics* is connections. Students develop a much richer understanding of mathematics and its applications when they can view the same phenomena from multiple mathematical perspectives. One way to have students see mathematics in this way is to use instructional materials that are intentionally designed to weave together different content strands. Another means of achieving content integration is to make sure that courses oriented toward any particular content area (such as algebra or geometry) contain many integrative problems—problems that draw on a variety of aspects of mathematics, that are solvable using a variety of methods, and that students can access in different ways.

High school students with particular interests could study mathematics that extends beyond what is recommended here in various ways. One approach is to include in the program material that extends these ideas in depth or sophistication. Students who encounter these kinds of enriched curricula in heterogeneous classes will tend to seek different levels of understanding. They will, over time, learn new ways of thinking from their peers. Other approaches make use of supplementary courses. For instance, students could enroll in additional courses concurrent with the program. Or the material proposed in these Standards could be included in a three-year program that allows students to take supplementary courses in the fourth year. In any of these approaches, the curriculum can be designed so that students can complete the foundation proposed here and choose from additional courses such as computer science, technical mathematics, statistics, and calculus. Whatever the approach taken, all students learn the same core material while some, if they wish, can study additional mathematics consistent with their interests and career directions.

These Standards are demanding. It will take time, patience, and skill to implement the vision they represent. The content and pedagogical demands of curricula aligned with these Standards will require extended and sustained professional development for teachers and a large degree of administrative support. Such efforts are essential. We owe our children no less than a high degree of quantitative literacy and mathematical knowledge that prepares them for citizenship, work, and further study.

Number and Operations

STANDARD for Grades 9–12

Instructional programs from prekindergarten through grade 12 should enable all students to—

Expectations

In grades 9–12 all students should–

Instructional programs from prekindergarten through grade 12 should enable all students to—	In grades 9–12 all students should–
Understand numbers, ways of representing numbers, relationships among numbers, and number systems	• develop a deeper understanding of very large and very small numbers and of various representations of them; • compare and contrast the properties of numbers and number systems, including the rational and real numbers, and understand complex numbers as solutions to quadratic equations that do not have real solutions; • understand vectors and matrices as systems that have some of the properties of the real-number system; • use number-theory arguments to justify relationships involving whole numbers.
Understand meanings of operations and how they relate to one another	• judge the effects of such operations as multiplication, division, and computing powers and roots on the magnitudes of quantities; • develop an understanding of properties of, and representations for, the addition and multiplication of vectors and matrices; • develop an understanding of permutations and combinations as counting techniques.
Compute fluently and make reasonable estimates	• develop fluency in operations with real numbers, vectors, and matrices, using mental computation or paper-and-pencil calculations for simple cases and technology for more-complicated cases. • judge the reasonableness of numerical computations and their results.

Number and Operations

In high school, students' understanding of number is the foundation for their understanding of algebra, and their fluency with number operations is the basis for learning to operate fluently with symbols. Students should enter high school with an understanding of the basic operations and fluency in using them on integers, fractions, and decimals. In grades 9–12, they will develop an increased ability to estimate the results of arithmetic computations and to understand and judge the reasonableness of numerical results displayed by calculators and computers. They should use the real numbers and learn enough about complex numbers to interpret them as solutions of quadratic equations.

High school students should understand more fully the concept of a number system, how different number systems are related, and whether the properties of one system hold in another. Their increased ability to use algebraic symbolism will enable them to make generalizations about properties of numbers that they might discover. They can study and use vectors and matrices. They need to develop deeper understandings of counting techniques, which further develops the conceptual underpinnings for the study of probability.

Understand numbers, ways of representing numbers, relationships among numbers, and number systems

High school students should become increasingly facile in dealing with very large and very small numbers as part of their deepening understanding of number. Such numbers occur frequently in the sciences; examples are Avogadro's number (6.02×10^{23}) in chemistry or the very small numbers used in describing the size of the nucleus of a cell in biology. As citizens, students will need to grasp the difference between $1 billion, the cost of a moderate-sized government project, and $1 trillion, a significant part of the national budget.

They need to become familiar with different ways of representing numbers. As part of their developing technological facility, students should become adept at interpreting numerical answers on calculator or computer displays. They should recognize 1.05168475E-12 as a very small number given in scientific notation, 6.66666667 as the *approximate* result of dividing 20 by 3, and ERROR as a response for either an invalid operation or a number that overflows the capacity of the device.

Students' understanding of the mathematical development of number systems—from whole numbers to integers to rational numbers and then on to real and complex numbers—should be a basis for their work in finding solutions for certain types of equations. Students should understand the progression and the kinds of equations that can and cannot be solved in each system. For example, the equation $3x = 1$ does not have an integer solution but does have a rational-number solution; the equation $x^3 = 2$ does not have a rational-number solution but does have a real-number solution; and the equation $x^2 + 4 = 0$ does not have a real-number solution but does have a complex-number solution.

High school students should understand fully the concept of a number system.

Whereas middle-grades students should have been introduced to irrational numbers, high school students should develop an understanding of the system of real numbers. They should understand that given an origin and a unit of measure, every point on a line corresponds to a real number and vice versa. They should understand that irrational numbers can only be approximated by fractions or by terminating or repeating decimals. They should understand the difference between rational and irrational numbers. Their understanding of irrational numbers needs to extend beyond π and $\sqrt{2}$.

High school students can use their understanding of numbers to explore new systems, such as vectors and matrices. By working with examples that include forces or velocities, students can learn to appreciate vectors as a means of simultaneously representing magnitude and direction. Using matrices, students can also see connections among major strands of mathematics: they can use matrices to solve systems of linear equations, to represent geometric transformations (some of which can involve creating computer graphics), and to represent and analyze vertex-edge graphs.

Properties that hold in some systems may not hold in others. So teachers and students should explicitly discuss the associative, commutative, and distributive properties, and students should learn to examine whether those properties hold in the systems they study. The exploration of the properties of matrices may be particularly interesting, since the system of matrices is often the first that students encounter in which multiplication is not commutative.

In grades 9–12, students can use algebraic arguments in many areas, including in their study of number. Consider, for example, a simple number-theory problem such as the following:

> What can you say about the number that results when you subtract 1 from the square of an odd integer?

It is easy to verify that the number that results is even, and that it is divisible by 4. But if students try a few examples, such as starting with 3 or 5, they may note that the results they obtain are divisible by 8. They might wonder if this property will hold in general. Proving that it does involves finding useful representations. If students decide to express an arbitrary odd integer as $2n + 1$ and the resulting number as A, some quick computations show that $A = (2n + 1)^2 - 1 = 4n^2 + 4n = 4(n)(n + 1)$. The observation that either n or $(n + 1)$ must be even gives an additional factor of 2, showing that A must be divisible by 8. Working such problems deepens students' understanding of number while providing practice in symbolic representation, reasoning, and proof.

Understand meanings of operations and how they relate to one another

As high school students' understanding of numbers grows, they should learn to consider operations in general ways, rather than only in particular computations. The questions in figure 7.1 call for reasoning about the properties of the numbers involved rather than for following procedures to arrive at exact answers. Such reasoning is important in judging the reasonableness of results. Although the questions can be approached by substituting approximate values for the numbers represented by a through h, teachers should encourage students to arrive at and justify their conclusions by thinking about properties of numbers.

For example, to determine the point whose coordinate is closest to *ab*, a teacher might suggest considering the sign of *ab* and whether the magnitude of *ab* is greater or less than that of *b*. Likewise, students should be able to explain why, if *e* is positioned as given in figure 7.1, the magnitude of $\sqrt{e}$ is greater than that of *e*. Listening to students explain their reasoning gives teachers insights into the sophistication of their arguments as well as their conceptual understanding.

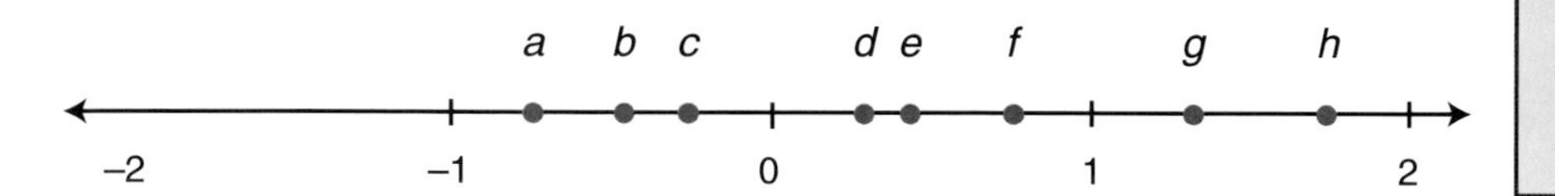

Given the points with coordinates *a, b, c, d, e, f, g,* and *h* as shown,
Which point is closest to *ab*? To $|c|$? To $1/f$? To $\sqrt{e}$? To $\sqrt{h}$?
Explain your reasoning.

Fig. **7.1.**

These questions call for reasoning without exact values.

Developing understandings of the properties of numbers can also help students solve problems like the following:

1. The graphs of the functions $f(x) = x$, $g(x) = \sqrt{x}$, $h(x) = x^2$, and $j(x) = x^3$ are shown [in fig. 7.2]. Identify which function corresponds to which graph and explain why.
2. Given $f(x) = 30/x^2$ and $a > 0$, which is larger: $f(a)$ or $f(a + 2)$? Explain why.

Fig. **7.2.**

Graphs of four functions

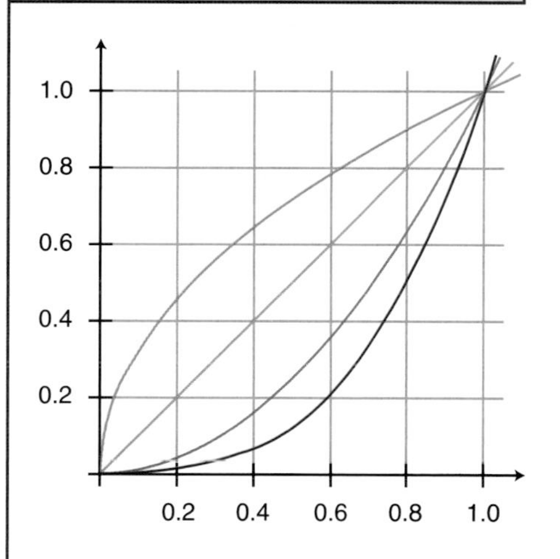

The first question is answered easily if one notices that if x is between 0 and 1, $x^3 < x^2 < x < \sqrt{x}$. To answer the second, a student would have to recognize that x^2 is an increasing function for positive x and then be able to reason that if x is positive, $30/x^2$ decreases as x increases. Hence, $f(a) > f(a + 2)$ for $a > 0$.

Students should also extend their understanding of operations to number systems that are new to them. They should learn to represent two-dimensional vectors in the coordinate plane and determine vector sums (see fig. 7.3). Dynamic geometry software can be used to illustrate the properties of vector addition. As students learn to represent systems of equations using matrices, they should recognize how operations on the matrices correspond to manipulations of such systems.

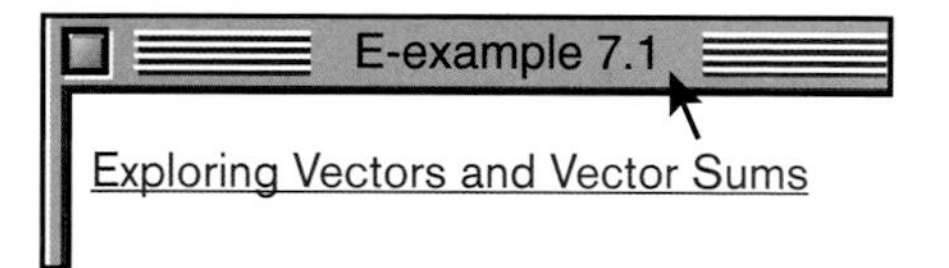

The organized lists and tree diagrams that students will have used in the elementary and middle grades to count outcomes or compute probabilities can be used in high school to work on permutations and combinations. Consider, for example, the task of determining how many

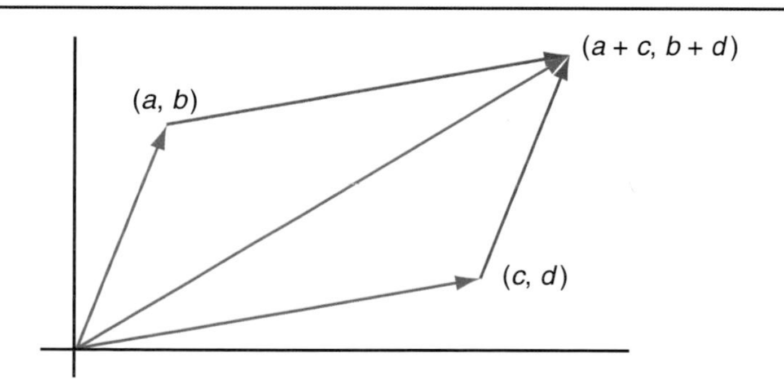

Fig. **7.3.**

A simple vector sum

two-person committees can be chosen from a group of seven people. Students should learn that the tree diagram they draw to represent the number of possibilities has a multiplicative structure: there are seven main branches, representing the first choice of committee member, and six branches off each of those, representing the choice of the second member. They also need to understand that this method of enumerating committees results in "double counting": each committee of the form (person 1, person 2) is also represented as (person 2, person 1). Hence the number of two-person subcommittees is

$$\binom{7}{2} = \frac{7 \times 6}{2}.$$

The students should also understand and be able to explain why the number of two-person committees is the same as the number of five-person committees that can be chosen from a group of seven people. This kind of reasoning provides the conceptual underpinnings for work in probability.

Students should be able to decide whether a problem calls for a rough estimate, an approximation, or an exact answer.

Compute fluently and make reasonable estimates

Students should be able to decide whether a problem calls for a rough estimate, an approximation to an appropriate degree of precision, or an exact answer. They should select a suitable method of computing from among mental mathematics, paper-and-pencil computations, and the use of calculators and computers and be proficient with each method. Electronic computation technologies provide opportunities for students to work on realistic problems and to perform difficult computations, for example, computing roots and powers of numbers or performing operations with vectors and matrices. However, students must be able to perform relatively simple mental computations as the basis for making reasonable estimates and sensible predictions and to spot potential sources of error. Suppose, for example, that a student wants to know the cube root of 49, enters the expression 49^1/3 into a calculator, and the number 16.3333333 is displayed in response. The student should note immediately that there has been an error, because the cube root of 49 should be between 3 and 4, and 16 • 16 • 16 is much larger than 49. To have the calculator compute the cube root, the student would need to have entered 49^(1/3).

Algebra Standard for Grades 9–12

Instructional programs from prekindergarten through grade 12 should enable all students to—

Expectations

In grades 9–12 all students should–

Understand patterns, relations, and functions	• generalize patterns using explicitly defined and recursively defined functions; • understand relations and functions and select, convert flexibly among, and use various representations for them; • analyze functions of one variable by investigating rates of change, intercepts, zeros, asymptotes, and local and global behavior; • understand and perform transformations such as arithmetically combining, composing, and inverting commonly used functions, using technology to perform such operations on more-complicated symbolic expressions; • understand and compare the properties of classes of functions, including exponential, polynomial, rational, logarithmic, and periodic functions; • interpret representations of functions of two variables.
Represent and analyze mathematical situations and structures using algebraic symbols	• understand the meaning of equivalent forms of expressions, equations, inequalities, and relations; • write equivalent forms of equations, inequalities, and systems of equations and solve them with fluency–mentally or with paper and pencil in simple cases and using technology in all cases; • use symbolic algebra to represent and explain mathematical relationships; • use a variety of symbolic representations, including recursive and parametric equations, for functions and relations; • judge the meaning, utility, and reasonableness of the results of symbol manipulations, including those carried out by technology.
Use mathematical models to represent and understand quantitative relationships	• identify essential quantitative relationships in a situation and determine the class or classes of functions that might model the relationships; • use symbolic expressions, including iterative and recursive forms, to represent relationships arising from various contexts; • draw reasonable conclusions about a situation being modeled.
Analyze change in various contexts	• approximate and interpret rates of change from graphical and numerical data.

Algebra

In the vision of school mathematics in these Standards, middle-grades students will learn that patterns can be represented and analyzed mathematically. By the ninth grade, they will have represented linear functions with tables, graphs, verbal rules, and symbolic rules and worked with and interpreted these representations. They will have explored some nonlinear relationships as well.

In high school, students should have opportunities to build on these earlier experiences, both deepening their understanding of relations and functions and expanding their repertoire of familiar functions. Students should use technological tools to represent and study the behavior of polynomial, exponential, rational, and periodic functions, among others. They will learn to combine functions, express them in equivalent forms, compose them, and find inverses where possible. As they do so, they will come to understand the concept of a class of functions and learn to recognize the characteristics of various classes.

High school algebra also should provide students with insights into mathematical abstraction and structure. In grades 9–12, students should develop an understanding of the algebraic properties that govern the manipulation of symbols in expressions, equations, and inequalities. They should become fluent in performing such manipulations by appropriate means—mentally, by hand, or by machine—to solve equations and inequalities, to generate equivalent forms of expressions or functions, or to prove general results.

The expanded class of functions available to high school students for mathematical modeling should provide them with a versatile and powerful means for analyzing and describing their world. With utilities for symbol manipulation, graphing, and curve fitting and with programmable software and spreadsheets to represent iterative processes, students can model and analyze a wide range of phenomena. These mathematical tools can help students develop a deeper understanding of real-world phenomena. At the same time, working in real-world contexts may help students make sense of the underlying mathematical concepts and may foster an appreciation of those concepts.

High school algebra should provide students with insights into mathematical abstraction and structure.

Understand patterns, relations, and functions

High school students' algebra experience should enable them to create and use tabular, symbolic, graphical, and verbal representations and to analyze and understand patterns, relations, and functions with more sophistication than in the middle grades. In helping high school students learn about the characteristics of particular classes of functions, teachers may find it helpful to compare and contrast situations that are modeled by functions from various classes. For example, the functions that model the essential features of the situations in figure 7.4 are quite different from one another. Students should be able to express them using tables, graphs, and symbols.

For the first situation, students might begin by generating a table of values. If C is the cost in cents of mailing a letter and P is the weight of the letter in ounces, then the function $C = 33 + (P - 1)(22)$ describes C as a function of P for positive integer values of P up through 13. Students

Fig. **7.4.**

Three situations that can be modeled by functions of different classes

Situation 1: In February 2000 the cost of sending a letter by first-class mail was 33¢ for the first ounce and an additional 22¢ for each additional ounce or portion thereof through 13 ounces.

Number of ounces	1	2	3	4	5	...	P
Cost in cents	33	33 + 22	33 + 2(22)	33 + 3(22)	33 + 4(22)	...	33 + (P – 1)(22)

Situation 2: During 1999 the population of the world hit 6 billion. The expected average rate of growth is predicted to be 2 percent a year.

Situation 3: A table of data gives the number of minutes of daylight in Chicago, Illinois, every other day from 1 January 2000 through 30 December 2000.

551, 553, 555, 557, 559, 562, 565, 568, 571, 575, 579, 582, 586, 591, 595, 599, 604, 609, 614, 619, 624, 629, 634, 639, 644, 650, 655, 661, 666, 672, 677, 683, 689, 694, 700, 706, 711, 717, 723, 728, 734, 740, 745, 751, 757, 762, 768, 773, 779, 785, 790, 796, 801, 806, 812, 817, 822, 827, 832, 837, 842, 847, 852, 856, 861, 865, 870, 874, 878, 881, 885, 889, 892, 895, 898, 901, 903, 905, 907, 909, 911, 912, 913, 914, 914, 914, 914, 914, 914, 913, 912, 911 909, 907, 905, 903, 901, 898, 895, 892, 889, 885, 882, 878, 874, 870, 866, 861, 857, 852, 848, 843, 838, 833, 828, 823, 818, 813, 807, 802, 797, 791, 786, 781, 775, 770, 764, 758, 753, 747, 742, 736, 731, 725, 719, 714, 708, 703, 697, 691, 686, 680, 675, 669, 664, 658, 653, 648, 642, 637, 632, 627, 622, 617, 612, 607, 603, 598, 594, 590, 585, 581, 578, 574, 571, 567, 564, 561, 559, 557, 554, 553, 551, 550, 549, 548, 547, 547, 547, 548, 548, 549, 550

should understand that this situation has some linear qualities. For real-number values of P, the points on the graph of $C = 33 + (P - 1)(22)$ lie on a line, and the rate of change is constant at 22 cents per ounce. However, the actual cost of postage and the linear function agree only at positive integer values of P. Students must realize that the graph of postal cost as a function of weight is a step function, as seen in figure 7.5.

For the second situation described in figure 7.4, teachers could encourage students to find a general expression for the function and note how its form differs from the step function that describes the postal cost. Some students might generate an iterative or recursive definition for the function, using the population of a given year (NOW) to determine the population of the next year (NEXT):

$$\text{NEXT} = (1.02) \bullet \text{NOW, start at 6 billion}$$

(See the discussion of NOW-NEXT equations in the "Representation" section of chapter 6.) Moreover, students should be able to recognize that this situation can be represented explicitly by the exponential function $f(n) = 6(1.02)^n$, where $f(n)$ is the population in billions and n is the number of years since 1999. A discussion of whether this formula is likely to be a good model forever would help students see the limitations of mathematical models.

For the third situation, students could begin by graphing the given data. It will help them to know that everywhere on earth except at the equator, the period of sunlight during the day increases for six months of the year and decreases for the other six. From the graph, they should be able to see that the daily increase in daylight is nonconstant over the first half of the year and that the decrease in the second half of the year also is nonconstant. Students could be asked to find a function that models the data well. The teacher could tell them that the length of

Fig. **7.5.**

A comparison of step and linear functions

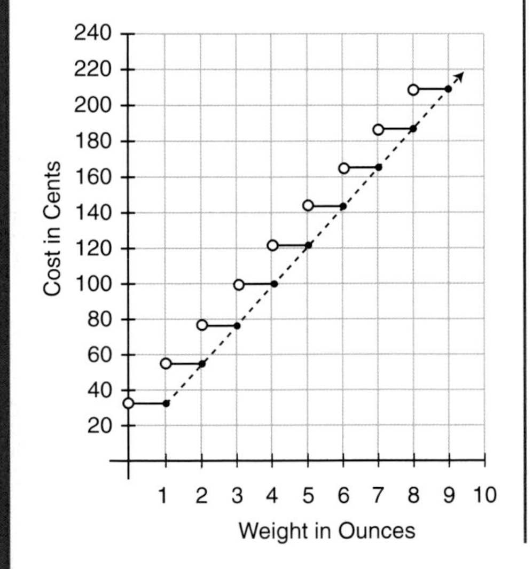

daylight can indeed be modeled by a function of the form $T(t) = T_{ave} + T_A(\theta)\sin(\omega t + \varphi)$, where t is measured in months, T_{ave} = average daylight time = 12 hours; $T_A(\theta)$ = amplitude, depending on latitude θ (changes sign at the equator); ω = frequency = $2\pi/12$, and φ = phase (depending on choice of the initial time, t_0). Students will see such formulas in their physics courses and need to understand that formulas express models of physical phenomena. It is also important to note that the parameters in physical equations have units.

After exploring and modeling each of the three situations individually, students could be asked to compare the situations. For example, they might be asked to find characteristics that are common to two or more of the functions. Some students might note that over the intervals given, the first function is nondecreasing, the second is strictly increasing, and the third both increases and decreases. Students need to be sensitive to the facts that functions that are increasing over some intervals don't necessarily stay increasing and that increasing functions may have very different rates of increase, as these three examples illustrate.

Students could also be asked to consider the advantages and disadvantages of the different ways the three functions were represented. The teacher should help students realize that depending on what one wants to know, different representations of these functions can be more or less useful. For instance, a table may be the most convenient way to initially represent the postage function in the first example. The same may be so for the third example if the goal is to determine quickly how much sunlight there will be on a given day. Despite the convenience of being able to "read" a value directly, however, the table may obscure the periodicity of the phenomenon. The periodicity becomes apparent when the function is represented graphically or symbolically. Similarly, although students may first create tables when presented with the second situation, graphical and symbolic representations of the exponential function may help students develop a better understanding of the nature of exponential growth.

High school students should have substantial experience in exploring the properties of different classes of functions. For instance, they should learn that the function $f(x) = x^2 - 2x - 3$ is quadratic, that its graph is a parabola, and that the graph opens "up" because the leading coefficient is positive. They should also learn that some quadratic equations do not have real roots and that this characteristic corresponds to the fact that their graphs do not cross the x-axis. And they should be able to identify the complex roots of such quadratics.

In addition, students should learn to recognize how the values of parameters shape the graphs of functions in a class. With access to computer algebra systems (CAS)—software on either a computer or calculator that carries out manipulations of symbolic expressions or equations, can compute or approximate values of functions or solutions to equations, and can graph functions and relations—students can easily explore the effects of changes in parameter as a means of better understanding classes of functions. For example, explorations with functions of the form $y = ax^2 + bx + c$ lead to some interesting results. The consequences of changes in the parameters a and c on the graphs of functions are relatively easy to observe. Changes in b are not as obvious: changing b results in a translation of the parabola along a nonvertical line. Moreover, a trace of the vertices of the parabolas formed as b is varied forms a parabola itself. Exploring functions of the form $f(x) = a(x - h)^2 + b(x - h) + c$ and seeing how their graphs change as the value of h is changed also provides a basis for understanding transformations and coordinate changes.

As high school students study several classes of functions and become familiar with the properties of each, they should begin to see that classifying functions as linear, quadratic, or exponential makes sense because the functions in each of these classes share important attributes. Many of these attributes are global characteristics of the functions. Consider, for example, the graphs of the three exponential functions of the form $f(x) = a \cdot b^x + c$, with $a > 0$ and $b > 1$, given in figure 7.6.

Fig. **7.6.**

Graphs of exponential functions of the form $f(x) = a \cdot b^x + c$

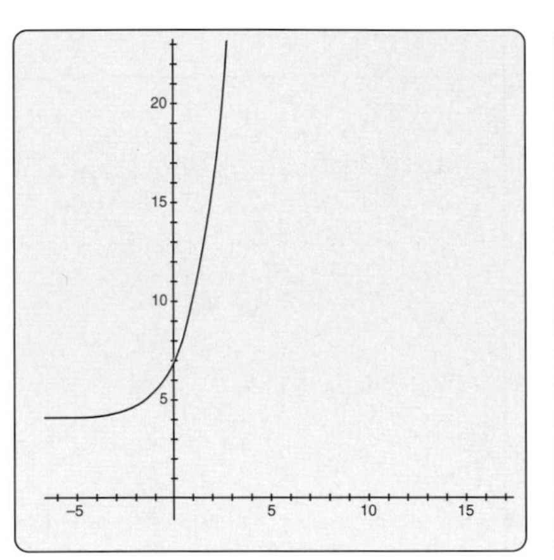

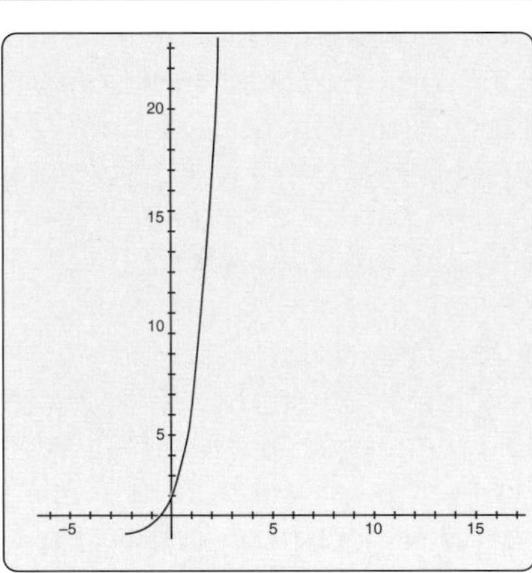

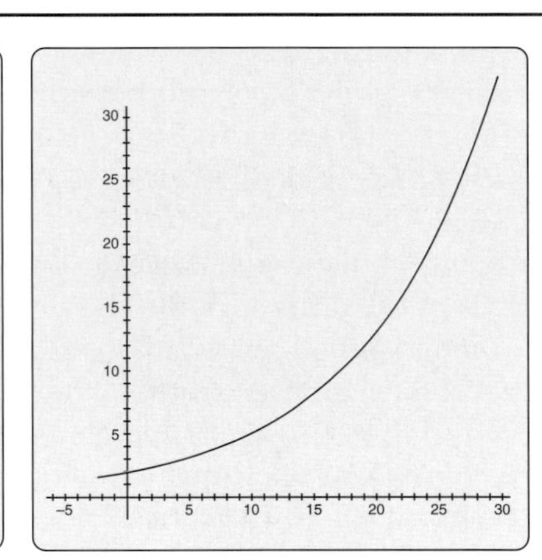

$g(x) = 3 \cdot 2^x + 4$ $h(x) = 2 \cdot 3^x - 1$ $k(x) = 2 \cdot 1.1^x$

To help students notice and describe characteristics of these three functions, teachers might ask, "What happens to each of these functions for large positive values of x? For large negative values of x? Where do they cross the y-axis?" One student might note that the values of each function increase rapidly for large positive values of x. Another student could point out that the y-intercept of each graph appears to be $a + c$. Teachers should then encourage students to explore what happens in cases where $a < 0$ or $0 < b < 1$. Students should find that changing the sign of a will reflect the graph over a horizontal line, whereas changing b to $1/b$ will reflect the graph over the y-axis. The graphs will retain the same shape. This type of exploration should help students see that all functions of the form $f(x) = a \cdot b^x + c$ share certain properties. Through analytic and exploratory work, students can learn the properties of this and other classes of functions.

Represent and analyze mathematical situations and structures using algebraic symbols

Fluency with algebraic symbolism helps students represent and solve problems in many areas of the curriculum. For example, proving that the square of any odd integer is 1 more than a multiple of 8 (see the related discussion in the "Number" section of this chapter) can involve representing odd numbers and operating on that representation algebraically. Likewise, the equations in figure 7.7 suggest an algebraic justification of

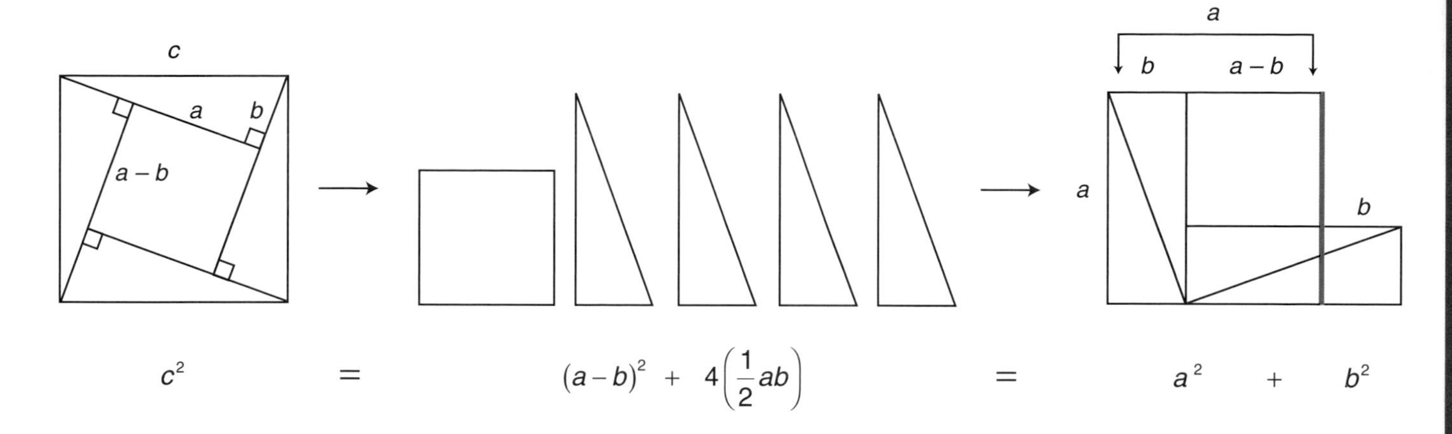

Fig. **7.7.**

An algebraic explanation of a visual proof of the Pythagorean theorem

a visual argument for the Pythagorean theorem. And many geometric conjectures—for example, that the medians of a triangle intersect at a point—can be proved by representing the situation using coordinates and manipulating the resulting symbolic forms (see the "Geometry" section of this chapter). Straightforward algebraic arguments can be used to show how the mean and standard deviation of a data set change if sample measurements are converted from square meters to square feet (see the "Reasoning and Proof" section of this chapter).

Students should be able to operate fluently on algebraic expressions, combining them and reexpressing them in alternative forms. These skills underlie the ability to find exact solutions for equations, a goal that has always been at the heart of the algebra curriculum. Even solving equations such as

$$(x+1)^2 + (x-2) + 7 = 3(x-3)^2 + 4(x+5) + 1$$

requires some degree of fluency. Finding and understanding the meaning of the solution of an equation such as

$$e^{4x} = 4e^{2x} + 3$$

calls for seeing that the equation can be written as a quadratic equation by making the substitution $u = e^{2x}$. (Such an equation deserves careful attention because one of the roots of the quadratic is negative.) Whether they solve equations mentally, by hand, or using CAS, students should develop an ease with symbols that enables them to represent situations symbolically, to select appropriate methods of solution, and to judge whether the results are plausible.

Being able to operate with algebraic symbols is also important because the ability to rewrite algebraic expressions enables students to reexpress functions in ways that reveal different types of information about them. For example, given the quadratic function $f(x) = x^2 - 2x - 3$, some of whose graphical properties were discussed earlier, students should be able to reexpress it as $f(x) = (x - 1)^2 - 4$, a form from which they can easily identify the vertex of the parabola. And they should also be able to express the function in the form $f(x) = (x - 3)(x + 1)$ and thus identify its roots as $x = 3$ and $x = -1$.

The following example of how symbol-manipulation skills and the ability to interpret graphs could work in concert is a hypothetical composite of exploratory classroom activities, inspired by Waits and Demana (1998):

A teacher asks her students to analyze the function

$$f(x) = \frac{2x^2 + 11x + 6}{x - 2}$$

and make as many observations about it as they can. Some students begin by trying to graph the function, plotting points by hand. Some students use a CAS and others perform long division by hand, producing the equivalent form

$$f(x) = 2x + 15 + \frac{36}{(x-2)}.$$

Some graph the original function or the equivalent form on a computer or on graphing calculators; the zoom feature enables them to see various views of the graph, as seen in figure 7.8.

Fig. **7.8.**
Different views of the function
$$f(x) = \frac{2x^2 + 11x + 6}{x - 2}$$

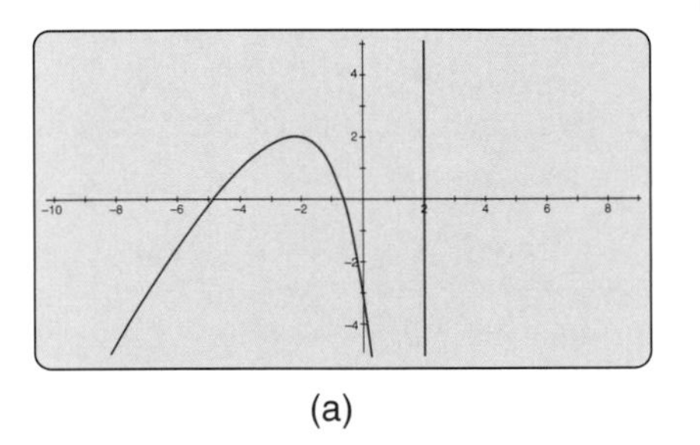

(a)

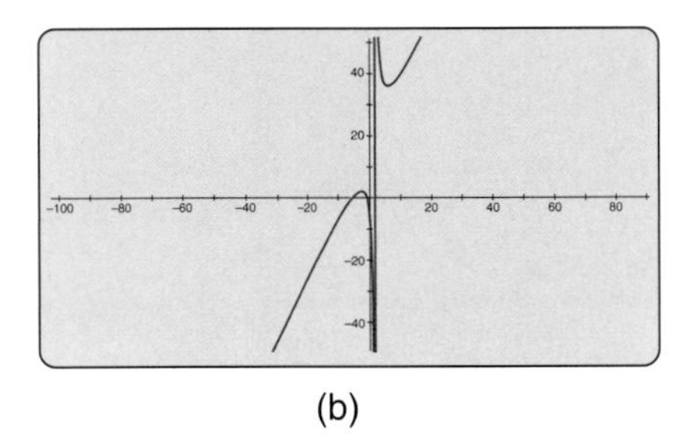

(b)

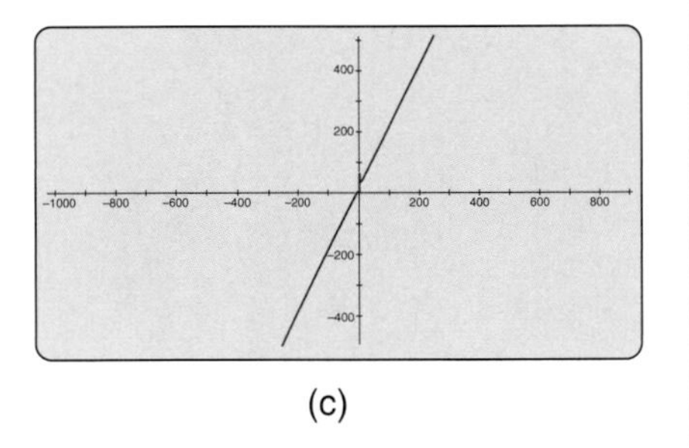

(c)

It is hard to interpret some of the graphs near $x = 2$, a matter the class returns to later. Focusing on a graph where the zoom-out feature has been used a number of times (see fig. 7.8c), some students observe, "The graph looks like a straight line." The teacher asks the class to decide whether it is a line and, if so, what the equation of the line might be. To investigate the question, the teacher suggests that they find several values of $f(x)$ for large positive and negative values of x and use curve-fitting software to find the equation of the line passing through those points. Different groups choose different x-values and, as a result, obtain slightly different values for the slope and the y-intercept. However, when the class discusses their findings, they discover that the lines that fit those points all seemed "close" to the line $y = 2x + 15$. Some students point out that this function is part of the result they obtained after performing the long division.

The class concludes that the line $y = 2x + 15$ is a good *approximation* to $f(x)$ for large x-values but that it is not a perfect fit. This conclusion leads to the question of how the students might combine the graphs of $g(x) = 2x + 15$ and $h(x) = 36/(x - 2)$ to deduce the shape of the graph of $f(x)$. Hand-drawn and computer plots help students explore how the graph of each function "contributes to" the graph of the sum. Examining the behavior of

$$h(x) = \frac{36}{x-2}$$

leads to a discussion of what "really" happens near $x = 2$, of why the function appears to be linear for large values of x, and of the need to develop a sense of how algebraic and graphical representations of functions are related, even when graphing programs or calculators are available.

Students in grades 9–12 should develop understandings of algebraic concepts and skill in manipulating symbols that will serve them in situations that require both. Success in the example shown in figure 7.9, for example, requires more than symbol manipulation. There are several ways to approach this problem, each of which requires understanding algebraic concepts *and* facility with algebraic symbols. For example, to complete the first row of the table, students need only know how to evaluate $f(x)$ and $g(x)$ for a given value of x. However, to complete the second row, students must know what it means to compose functions, including the role of the "inner" and "outer" function and the numbers

If $f(x) = x^2 - 1$ and $g(x) = (x + 1)^2$, complete the table below.

x	$f(x)$	$g(x)$	$f(g(x))$	$g(f(x))$
2			80	16
		4		81

Fig. **7.9.**

A composition-of-functions problem (Adapted from Tucker [1995])

on which they act in a composition. They also must understand how to read the symbols $f(g(x))$ and $g(f(x))$. Students might reason, using an intuitive understanding of the inverse of a function, that because $g(x) = 4$, x must be either 1 or –3. They can then determine that x cannot be 1, because $g(f(1))$ is not 81.

Use mathematical models to represent and understand quantitative relationships

Modeling involves identifying and selecting relevant features of a real-world situation, representing those features symbolically, analyzing and reasoning about the model and the characteristics of the situation, and considering the accuracy and limitations of the model. In the program proposed here, middle-grades students will have used linear functions to model a range of phenomena and explored some nonlinear phenomena. High school students should study modeling in greater depth, generating or using data and exploring which kinds of functions best fit or model those data.

Teachers may find that having students generate data helps generate interest in creating mathematical models. For example, students could conduct an experiment to study the relationship between the time it takes a skateboard to roll down a ramp of fixed length and the height of the ramp (Zbiek and Heid 1990). Teams of students might set ramps at different heights and repeatedly roll skateboards down the ramps and measure the time. Once students have gathered and plotted the data, they can analyze the physical features of the situation to create appropriate mathematical models. Their knowledge of the characteristics of various classes of functions should help them select potential models. In this situation, as the height of the ramp is increased, less time is needed, suggesting that the function is decreasing. Students can discuss the suitability of linear, quadratic, exponential, and rational functions by arguing from their data or from the physics of the situation. Curve-fitting software allows students to generate possible models, which they can examine for suitability on the basis of the data and the situation.

In making choices about what kinds of situations students will model, teachers should include examples in which models can be expressed in iterative, or recursive, form. Consider the following example, adapted from National Research Council (1998, p. 80), of the elimination of a medicine from the circulatory system.

> A student strained her knee in an intramural volleyball game, and her doctor prescribed an anti-inflammatory drug to reduce the swelling. She is to

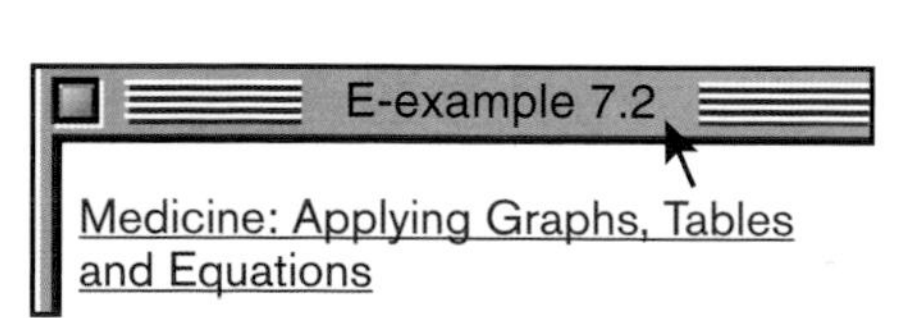

	A	B
1	440	
2	616	
3	686.4	
4	714.56	
5	725.824	
6	730.3296	
7	732.13184	
8	732.852736	
9	733.1410944	
10	733.2564378	
11	733.3025751	
12	733.32103	
13	733.328412	
14	733.3313648	
15	733.3325459	
16	733.3330184	
17	733.3332073	
18	733.3332829	
19	733.3333132	
20	733.3333253	
21	733.3333301	
22	733.333332	
23	733.3333328	
24	733.3333331	
25		

Fig. **7.10.**

A spreadsheet computation of the "drug dosage" problem

take two 220-milligram tablets every 8 hours for 10 days. If her kidneys filtered 60% of this drug from her body every 8 hours, how much of the drug was in her system after 10 days? How much of the drug would have been in her system if she had continued to take the drug for a year?

Teachers might ask students to conjecture about how much of the drug would be in the volleyball player's system after 10 days. They might also ask about whether the drug keeps accumulating noticeably in the athlete's system. Students will tend to predict that it does, and they can be asked to examine the accumulation in their analysis.

Students might begin by calculating a few values of the amount of the drug in the player's system and looking for a pattern. They can proceed to model the situation directly, representing it informally as

NEXT = 0.4(NOW) + 440, start at 440

or more formally as

$$a_1 = 440 \text{ and } a_{n+1} = 0.4a_n + 440 \text{ for } 1 \le n \le 31,$$

where n represents the dose number (dose 31 would be taken at 240 hours, or 10 days) and a_n represents the amount of the drug in the system just *after* the nth dose. By looking at calculator or spreadsheet computations like those in figure 7.10, students should be able to see that the amount of the drug in the bloodstream reaches an after-dosage "equilibrium" value of about 733 1/3 milligrams. Students should learn to express the relationship in one of the iterative forms given above. Then the mathematics in this example can be pursued in various ways. At the most elementary level, the students can simply verify the equilibrium value by showing that 0.4(733 1/3) + 440 = 733 1/3 milligrams. They can be asked to predict what would happen if the initial dose of the anti-inflammatory drug were different, to run the simulation, and to explain the result they obtain.

This investigation opens the door to explorations of finite sequences and series and to the informal consideration of limits. (For example, spreadsheet printouts for "large n" for various dosages strongly suggest that the sequence $\{a_n\}$ of after-dosage levels converges.) Expanding the first few terms reveals that this is a finite geometric series:

$$a_1 = 440 = 440(1)$$

$$a_2 = 440 + 0.4(440) = 440(1 + 0.4)$$

$$a_3 = 440 + 0.4(440) + (0.4)^2(440) = 440\left(1 + 0.4 + (0.4)^2\right)$$

$$a_4 = 440 + 0.4(440) + (0.4)^2(440) + (0.4)^3(440)$$
$$= 440\left(1 + 0.4 + (0.4)^2 + (0.4)^3\right)$$

Students might find it interesting to pursue the behavior of this series.

To investigate other aspects of the modeling situation, students could also be asked to address questions like the following:

- If the athlete stops taking the drug after 10 days, how long does it take for her system to eliminate most of the drug?
- How could you determine a dosage that would result in a targeted after-dosage equilibrium level, such as 500 milligrams?

Students should also be made aware that problems such as this describe only one part of a treatment regimen and that doctors would be alert to the possibility and implications of various complicating factors.

In grades 9–12, students should encounter a wide variety of situations that can be modeled recursively, such as interest-rate problems or situations involving the logistic equation for growth. The study of recursive patterns should build during the years from ninth through twelfth grade. Students often see trends in data by noticing change in the form of differences or ratios (How much more or less? How many times more or less?). Recursively defined functions offer students a natural way to express these relationships and to see how some functions can be defined recursively as well as explicitly.

Analyze change in various contexts

Increasingly, discussions of change are found in the popular press and news reports. Students should be able to interpret statements such as "the rate of inflation is decreasing." The study of change in grades 9–12 is intended to give students a deeper understanding of the ways in which changes in quantities can be represented mathematically and of the concept of rate of change.

The "Algebra" section of this chapter began with examples of three different real-world contexts in which very different kinds of change occurred. One situation was modeled by a step function, one by an exponential function, and one by a periodic function. Each of these functions changes in different ways over the interval given. As discussed earlier, students should recognize that the step function is nonlinear but that it has some linear qualities. To many students, the kind of change described in the second situation sounds linear: "Each year the population changes by 2 percent." However, the change is 2 percent of the previous year's population; as the population grows, the increase grows as well. Students should come to realize that functions of this type grow *very* rapidly. In the third example, students can see that not only is the function periodic but because it is, its rate of change is periodic as well.

Increasingly, discussions of change are found in the popular press and news reports.

Chapter 6 gives an example in which middle-grades students are asked to compare the costs of two different pricing schemes for telephone calls: a flat rate of $0.45 a minute versus a rate of $0.50 a minute for the first 60 minutes and $0.10 a minute for each minute thereafter. In examples of this type, the dependent variable typically changes (over some interval) a fixed amount for each unit change in the independent variable. In high school, students should analyze situations in which quantities change in much more complex ways and in which the relationships between quantities and their rates of change are more subtle. Consider, for example, the situation (adapted from Carlson [1998, p. 147]) in figure 7.11.

Working problems of this type builds on the understandings of change developed in the middle grades and lays groundwork for the study of calculus. Because students tend to confuse velocity with position, teachers should help them think carefully about which variables are represented in the diagram and about how they change. First, for example, students must realize that the variable on the vertical axis is velocity, rather than position. To answer part *a* of the question, they need to reason that because the velocity of car A is greater than that of car B at every point in the interval $0 < t < 1$, car A has necessarily traveled a

The given graph represents velocity vs. time for two cars. Assume that the cars start from the same position and are traveling in the same direction.

(*a*) State the relationship between the position of car A and that of car B at $t = 1$ hr. Explain.

(*b*) State the relationship between the velocity of car A and that of car B at $t = 1$ hr. Explain.

(*c*) State the relationship between the acceleration of car A and that of car B at $t = 1$ hr. Explain.

(*d*) How are the positions of the two cars related during the time interval between $t = 0.75$ hr. and $t = 1$ hr.? (That is, is one car pulling away from the other?) Explain.

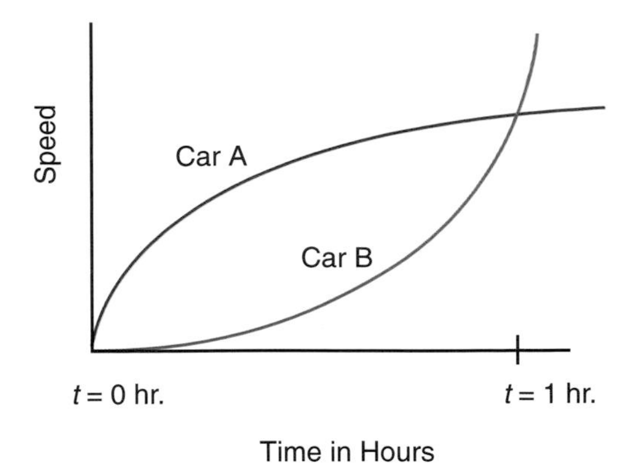

Fig. **7.11.**

A problem requiring a sophisticated understanding of change

greater distance than car B. They can read the answer to part *b* directly off the graph: at $t = 1$ hour, both cars are traveling at the same velocity. Answering part *c* calls for at least an intuitive understanding of instantaneous rate of change. Acceleration is the rate of change of velocity. At $t = 1$ hour, the velocity of car B is increasing more rapidly than that of car A, so car B is accelerating more rapidly than car A at $t = 1$ hour. Part *d* is particularly counterintuitive for students (Carlson 1998). Since car B is accelerating more rapidly than car A near $t = 1$ hour, students tend to think that car B is "catching up" with car A, and it is, although it is still far behind. Some will interpret the intersection of the graphs to mean that the cars meet. Teachers need to help students focus on the relative velocities of the two cars. Questions such as "Which car is moving faster over the interval from $t = 0.75$ hour to $t = 1$ hour?" can help students realize that car A is not only ahead of car B but moving faster and hence pulling away from car B. Car B starts catching up with car A only after $t = 1$ hour.

Geometry Standard for Grades 9–12

Instructional programs from prekindergarten through grade 12 should enable all students to—

Expectations

In grades 9–12 all students should–

Standard	Expectations
Analyze characteristics and properties of two- and three-dimensional geometric shapes and develop mathematical arguments about geometric relationships	• analyze properties and determine attributes of two- and three-dimensional objects; • explore relationships (including congruence and similarity) among classes of two- and three-dimensional geometric objects, make and test conjectures about them, and solve problems involving them; • establish the validity of geometric conjectures using deduction, prove theorems, and critique arguments made by others; • use trigonometric relationships to determine lengths and angle measures.
Specify locations and describe spatial relationships using coordinate geometry and other representational systems;	• use Cartesian coordinates and other coordinate systems, such as navigational, polar, or spherical systems, to analyze geometric situations; • investigate conjectures and solve problems involving two- and three-dimensional objects represented with Cartesian coordinates.
Apply transformations and use symmetry to analyze mathematical situations	• understand and represent translations, reflections, rotations, and dilations of objects in the plane by using sketches, coordinates, vectors, function notation, and matrices; • use various representations to help understand the effects of simple transformations and their compositions.
Use visualization, spatial reasoning, and geometric modeling to solve problems	• draw and construct representations of two- and three-dimensional geometric objects using a variety of tools; • visualize three-dimensional objects from different perspectives and analyze their cross sections; • use vertex-edge graphs to model and solve problems; • use geometric models to gain insights into, and answer questions in, other areas of mathematics; • use geometric ideas to solve problems in, and gain insights into, other disciplines and other areas of interest such as art and architecture.

Geometry

In programs that adopt the recommendations in *Principles and Standards*, middle-grades students will have explored and discovered relationships among geometric shapes, often using dynamic geometry software. Using features of polygons and polyhedra, they will have had experience in comparing and classifying shapes. High school students should conduct increasingly independent explorations, which will allow them to develop a deeper understanding of important geometric ideas such as transformation and symmetry. These understandings will help students address questions that have always been central to the study of Euclidean geometry: Are two geometric figures congruent, and if so, why? Are they similar, and if so, why? Given that a geometric object has certain properties, what other properties can be inferred?

Geometry offers a means of describing, analyzing, and understanding the world and seeing beauty in its structures. Geometric ideas can be useful both in other areas of mathematics and in applied settings. For example, symmetry can be useful in looking at functions; it also figures heavily in the arts, in design, and in the sciences. Properties of geometric objects, trigonometric relationships, and other geometric theorems give students additional resources to solve mathematical problems.

Geometry offers a means of describing, analyzing, and understanding the world and seeing beauty in its structures.

High school students should develop facility with a broad range of ways of representing geometric ideas—including coordinates, networks, transformations, vectors, and matrices—that allow multiple approaches to geometric problems and that connect geometric interpretations to other contexts. Students should recognize connections among different representations, thus enabling them to use these representations flexibly. For example, in one set of circumstances it might be most useful to think about an object's properties from the perspective of Euclidean geometry, whereas in other circumstances, a coordinate or transformational approach might be more useful. This ability to use different representations advantageously is part of students' developing geometric sophistication.

Geometry has always been a rich arena in which students can discover patterns and formulate conjectures. The use of dynamic geometry software enables students to examine many cases, thus extending

Representation | Connections | Communication | Reasoning & Proof | Problem Solving | Data Analysis & Probability | Measurement | **Geometry** | Algebra | Number & Operations

their ability to formulate and explore conjectures. Judging, constructing, and communicating mathematically appropriate arguments, however, remain central to the study of geometry. Students should see the power of deductive proof in establishing the validity of general results from given conditions. The focus should be on producing logical arguments and presenting them effectively with careful explanation of the reasoning, rather than on the form of proof used (e.g., paragraph proof or two-column proof). A particular challenge for high school teachers is to integrate technology in their teaching as a way of encouraging students to explore ideas and develop conjectures while continuing to help them understand the need for proofs or counterexamples of conjectures.

Analyze characteristics and properties of two- and three-dimensional geometric shapes and develop mathematical arguments about geometric relationships

Students should enter high school understanding the properties of, and relationships among, basic geometric objects. In high school, this knowledge can be extended and applied in various ways. Students should become increasingly able to use deductive reasoning to establish or refute conjectures and should be able to use established knowledge to deduce information about other situations. For example, a teacher might ask students to solve problems like that in figure 7.12.

Fig. **7.12.**

A geometric problem requiring deduction and proof

In this figure, $\overline{AB} \parallel \overline{DE}$ and $\overline{DF} \perp \overline{CE}$. Determine the perimeter of $\triangle ABC$ and the perimeter of $\triangle CDE$. Explain completely how you found your answers and how you know they are correct.

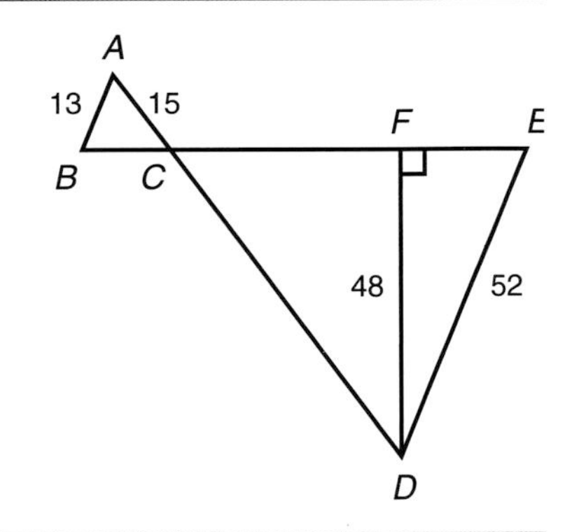

Students need to put together a number of logical deductions in order to solve this problem. The following proof demonstrates an ability to select and focus on important elements in the diagram, and it shows a solid understanding of the concepts involved and how they can be assembled to solve the problem. Note particularly how the fictional student finds different connections to be sure her reasoning is sound.

> First, I noticed that since $\overline{AB}$ and $\overline{DE}$ are parallel, angles B and E must be congruent. Also, angles ACB and DCE are congruent, since they are vertical. So now I know that the two triangles (ABC corresponds to DEC) are similar by angle-angle similarity. But that tells me that their corresponding sides are proportional. Since $DE = 4(AB)$, I know that all the sides of triangle DEC are 4 times as large as the corresponding sides of triangle ABC, so $CD = 4(15) = 60$.
>
> Now I just need to find the other side of triangle DEC to find its perimeter. But $\overline{DF}$ makes it into 2 right triangles, so I can use the

Pythagorean theorem on each of those. $FE^2 + 48^2 = 52^2$, so FE is 20. (Actually, I just noticed that this is just 4 times a 5-12-13 triangle, but I saw that too late.) Then looking at CDF, this is 12 times a 3-4-5 triangle, so CF must be 36. (I checked using the Pythagorean theorem and got the same answer.) So the perimeter is $52 + 60 + 56 = 168$.

Once I find the perimeter of ABC, I'm done. But that's easy, since the scale factor from DEC to ABC is 25%. I can just divide 168 by 4 and get 42. The reason that works is that each of the sides of ABC is 25% of its corresponding side in DEC, so the whole perimeter of ABC will be 25% of DEC. We already proved that in class anyway.

High school students should begin to organize their knowledge about classes of objects more formally. Finding precise descriptions of conditions that characterize a class of objects is an important first step. For example, students might define a trapezoid as a quadrilateral with at least one pair of parallel sides. They should realize that such a definition includes parallelograms, rectangles, and squares as special classes of trapezoids. Students might also ask, "How much information do I need to be sure a quadrilateral is a trapezoid? Do I need also to know something about its diagonals and angles? Can I get by with just some of this information?" As their ability to make logical deductions grows, students should be able to develop characterizations that follow directly from the properties of parallel lines and similar triangles. Alternatively, the class of trapezoids could be characterized in terms of its diagonals: If the diagonals of a quadrilateral cut each other so that the ratios of the corresponding segments of the diagonal are equal, then the quadrilateral is a trapezoid.

A teacher might ask a class to consider, on the basis of this characterization, how trapezoids are related to other classes of quadrilaterals. In considering parallelograms, students may note that the diagonals bisect each other, so each is cut in a 1:1 ratio and therefore the parts are proportional. The obvious conclusion that parallelograms (and many other classes of quadrilaterals) may be considered a special kind of trapezoid may seem unusual to those who think of a trapezoid as a quadrilateral with *exactly* one pair of parallel sides. However, it is important for students to see that the definition chosen will determine the conclusions that can be drawn.

One of the most important challenges in mathematics teaching has to do with the roles of evidence and justification, especially in increasingly technological environments. Using dynamic geometry software, students can quickly generate and explore a range of geometric examples. If they have not learned the appropriate uses of proof and mathematical argumentation, they might argue that a conjecture must be valid simply because it worked in all the examples they tried. Despite the possibility of students' developing such a misconception, in a classroom in which students understand the roles of experimentation, conjecture, and proof, being able to generate and explore many examples can result in deeper and more-extended mathematical investigations than might otherwise be possible. The following hypothetical example illustrates how students might investigate relationships in a dynamic geometry environment and justify or refute conclusions.

Using dynamic geometry software, students can quickly generate and explore a range of geometric examples.

The students are asked to draw a triangle, construct a new triangle by joining the midpoints of its three sides, and calculate the ratio of

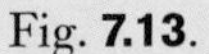

Fig. **7.13.**

Exploring and extending the results of connecting the midpoints of the adjacent sides of polygons

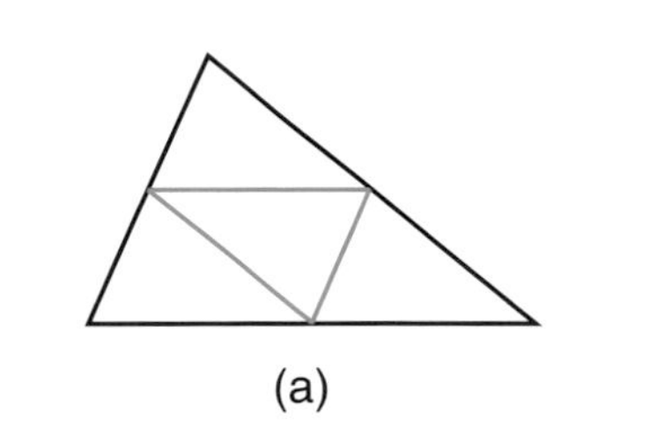

(a)

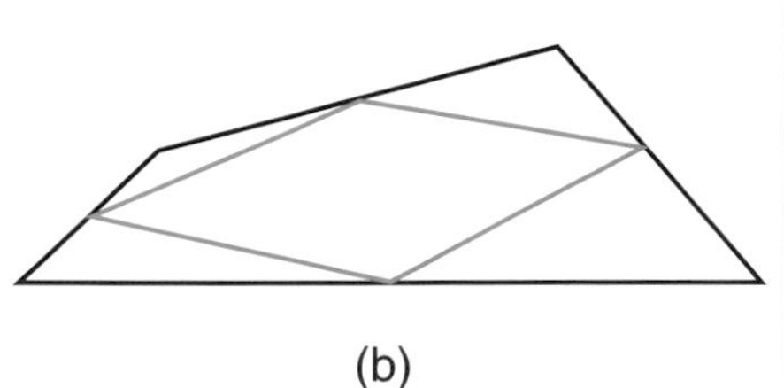

(b)

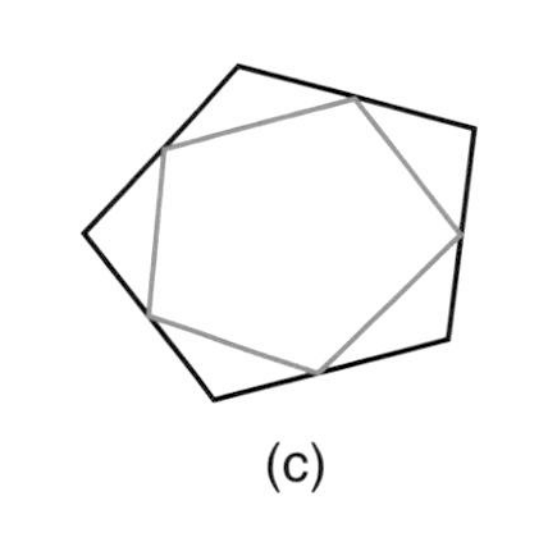

(c)

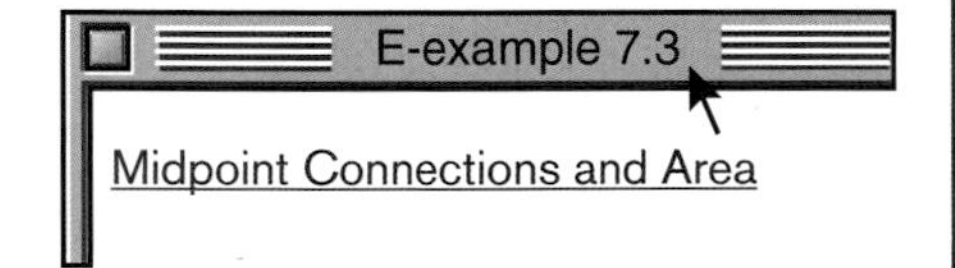

the area of the midpoint triangle to the area of the original triangle (see fig. 7.13a). As they drag one vertex to create many different triangles, the students notice that the ratio of the two areas appears to remain constant at 0.25.

Jake says he thinks that this relationship will always hold. He says that since the base of each of the four small triangles is a midline, each side of the midpoint triangle should be half as long as the parallel side of the large triangle. Each midline cuts the altitude in half, so the height of each small triangle is half that of the large triangle. Dividing each of these lengths by 2 divides the area by 4, so the area of the small triangle is one-fourth of the area of the large one.

Berta agrees with Jake's answer and thinks she can show that it must be true. She explains how she has extended the midlines and the sides of the triangles to form three pairs of parallel lines and is now able to determine many pairs of congruent angles. She reasons, using parallelism, that the corresponding sides are congruent and determines that the three small triangles formed at the vertices of the original triangle are congruent by angle-side-angle. She is confident that the midpoint triangle should be congruent to the other three, but when the teacher asks her how she can be sure, she is unable to give an explanation. The teacher asks one question: "Do you know anything about the sides of that triangle?" Her friend Dawn quickly notes that all its sides are shared with the sides of the other three, which indicates that it would have to be the same size.

Hope has a somewhat different way of looking at the situation. She notices that the lengths of the corresponding sides of the midpoint triangle and the original triangle are in a ratio of 1:2, so they must be similar. Thus, the area of the midpoint triangle must be one-fourth the area of the original on the basis of the class's earlier observation that the areas of similar triangles are related by the square of their scale factor. The teacher asks the class to think about the relationship between Hope's method and Jake's method.

The students decide to test whether a constant ratio exists for the area of a "midpoint" quadrilateral to the area of a convex quadrilateral (see figure 7.13b). It appears that the area ratio in this case is 0.5. They are able to prove this relationship by dividing a quadrilateral into two triangles and employing the methods they used in the previous investigation. The students begin to wonder whether they have discovered a big idea. Does a constant ratio hold for other polygons? For the first few convex pentagons they try (see fig. 7.13c), the area ratio appears to be constant at 0.7 (see Zbiek [1996] for a discussion about how this problem can be solved using technology). When they are unable to generate a proof, they decide to check more examples. When they do so, they begin to see some variation in the ratios. This development is disappointing to the students, who were hoping that the result they proved for triangles and quadrilaterals would be general. Their teacher points out that they really should be encouraged by their results. They have made a series of conjectures, produced proofs of some conjectures, and produced counterexamples to show when other conjectures do not hold. That kind of careful thinking, he says, is truly mathematical.

Applied problems can furnish both rich contexts for using geometric ideas and practice in modeling and problem solving. For example, right-triangle trigonometry is useful in solving a range of practical problems. Teachers can introduce students to problems such as the following, which is adapted from Hamilton and Hamilton (1993):

> People working in the building trades sometimes need to divert their construction around obstacles. Although they often use simple offsets of 90°, 45°, or 30°, other angles are sometimes needed in areas where space is limited.
>
> A construction worker needs to reroute an underground pipe in order to avoid the root systems of two trees. She needs to raise the path of her pipe 23 inches over a distance of over 86 inches [see fig. 7.14], and then continue on a course parallel to that of the original pipe. What angles should she cut in order to accomplish this?

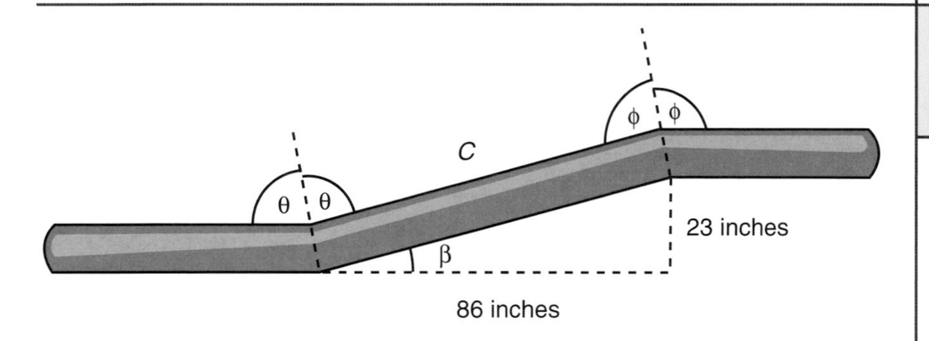

Fig. **7.14.**

The rerouting of an underground pipe

> When a section of pipe is cut at an angle, the cross section is an ellipse. If one of the two resulting pieces is rotated 180 degrees about the axis going through the center of the pipe and then repositioned, the elliptical cross sections match each other, so the two pipes can be joined smoothly [see fig. 7.15].

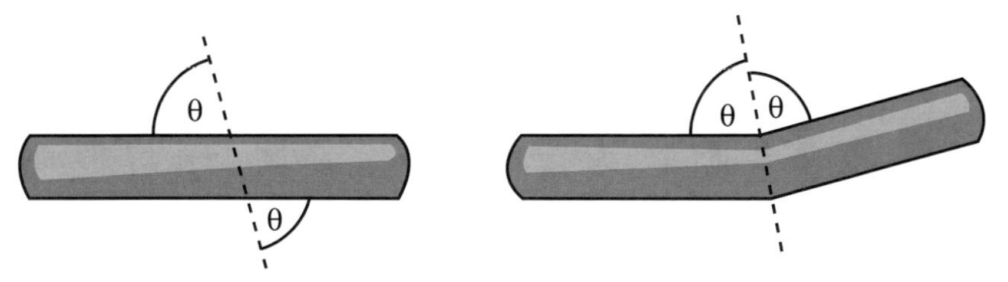

Fig. **7.15.**

Cutting and repositioning a pipe

> A second cut will be needed to send the pipe on in the original direction, as in [figure 7.14]. The angles marked θ are equal in measure, as are the angles marked ϕ. In order to proceed, the pipe fitter needs to find the two angles at which to cut the materials as well as the length *C* of the connecting piece.

To find θ and ϕ, students need to observe that $2\theta + \beta = 180°$, $2\phi + (90° - \beta) = 270°$, and $\tan(\beta) = 23/86$. Working through this kind of problem can help students develop visualization skills and see how what they have learned can be applied in meaningful contexts.

Specify locations and describe spatial relationships using coordinate geometry and other representational systems

Geometric problems can be presented and approached in various ways. For example, many problems from Euclidean geometry, such as showing that the medians of any triangle intersect at a point, can be approached through coordinate geometry. Although it is possible to use a different variable for the coordinates of each vertex (especially if a CAS package is effectively used), a "without loss of generality" argument can be used to lower the level of symbolic complexity. This type of argument relies on conveniently placing a coordinate plane over a general triangle (or other figure), often so that the coordinate axis coincides with a side of the triangle. By making clever choices about naming the coordinates of the vertices, taking care, of course, to be sure that the choices do not introduce unintended conditions, the calculations can be quite reasonable. Once they have obtained a representation like that in figure 7.16, students could determine the equations of two of the medians, find the point at which they intersect, and show that the third median passes through that point. Although proofs of this kind can be difficult for high school students, grappling with them may stimulate growth in students' understanding of geometry, algebraic variables, and generality.

Fig. **7.16.**

A diagram that shows the use of coordinate geometry to prove that the medians of a triangle intersect

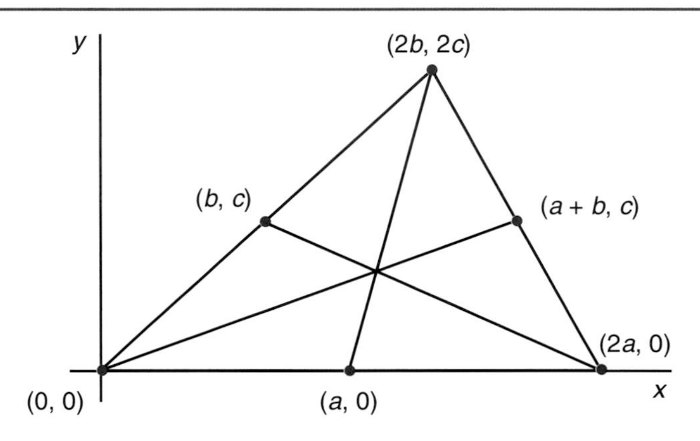

In grades 9–12 students should also explore problems for which using other coordinate systems is helpful. They should have some familiarity with spherical and simple polar coordinate systems, as well as with systems used in navigation. Using rectangular coordinates, for example, students should learn to represent points that lie on a circle of radius 3 centered at the origin as

$$\left(x,\ \pm\sqrt{9-x^2}\right)$$

for $-3 \leq x \leq 3$. With polar coordinates, these pairs are represented more simply as $(3, \theta)$ for $0 \leq \theta \leq 2\pi$, where θ is measured in radians. Students should be able to explain why both of these forms describe the points on a circle. The polar-coordinate representation is simpler in this example and may be more useful for solving certain problems.

Apply transformations and use symmetry to analyze mathematical situations

Middle-grades students should have had experience with such basic geometric transformations as translations, reflections, rotations, and dilations (including contractions). In high school they will learn to represent these transformations with matrices, exploring the properties of the transformations using both graph paper and dynamic geometry tools. For example, students who are familiar with matrix multiplication could be introduced to matrix representations of transformations through tasks such as the one in figure 7.17.

Consider a triangle ABC with vertices $A = (-5, 1)$, $B = (-4, 7)$, and $C = (-8, 5)$.

Reflect the triangle over the line $y = x$ to obtain the triangle $A'B'C'$ as shown.

Determine a matrix M such that $MA = A'$, $MB = B'$, and $MC = C'$, where the points are represented as vectors.

Explore the properties of the matrix M.

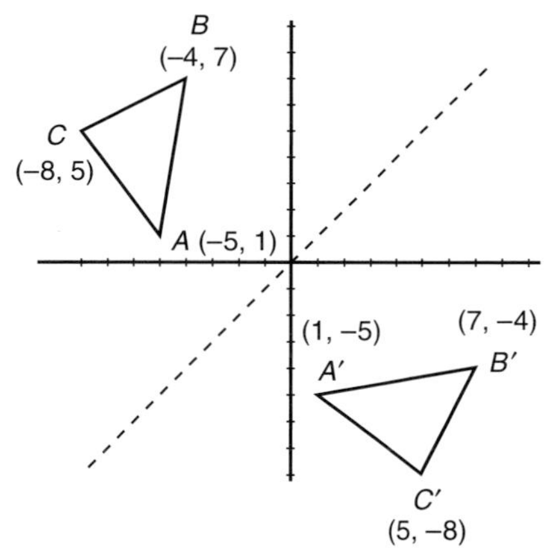

Fig. **7.17.**

Representing a reflection using a matrix (See a similar example in Senk et al. [1996].)

By graphing the triangle and its image, students can see that the first question amounts to determining P, Q, R, and S in the following transformation matrix, where the vertices of the triangles are collected into a 2×3 matrix for convenience:

$$\begin{bmatrix} P & Q \\ R & S \end{bmatrix} \begin{bmatrix} -5 & -4 & -8 \\ 1 & 7 & 5 \end{bmatrix} = \begin{bmatrix} 1 & 7 & 5 \\ -5 & -4 & -8 \end{bmatrix}$$

They might solve the resulting set of equations, showing that the transformation matrix is

$$\begin{bmatrix} 0 & 1 \\ 1 & 0 \end{bmatrix}.$$

Alternatively, they might observe that they need to find an M so that

$$M \cdot \begin{bmatrix} x \\ y \end{bmatrix} = \begin{bmatrix} y \\ x \end{bmatrix}$$

and explore various possibilities until observing that

$$\begin{bmatrix} 0 & 1 \\ 1 & 0 \end{bmatrix} \cdot \begin{bmatrix} x \\ y \end{bmatrix} = \begin{bmatrix} y \\ x \end{bmatrix}.$$

Other transformations with easily accessible matrix representations include reflections over either axis and about the line $y = -x$, rotations about the origin that are multiples of 90 degrees, and dilations from the origin. Students should understand that multiplying transformation matrices corresponds to composing the transformations represented. They should also understand that transformations have many practical applications.

Use visualization, spatial reasoning, and geometric modeling to solve problems

Creating and analyzing perspective drawings, thinking about how lines or angles are formed on a spherical surface, and working to understand orientation and drawings in a three-dimensional rectangular coordinate system all afford opportunities for students to think and reason spatially. With the expanding role of computer graphics in the workplace, students will have increased needs and opportunities to use visualization as a problem-solving tool. Schooling should provide rich mathematical settings in which they can hone their visualization skills. Visualizing a building represented in architectural plans, the shape of a cross section formed when a plane slices through a cone (a conic section) or another solid object, or the shape of the solid swept out when a plane figure is rotated about an axis become easier when students work with physical models, drawings, and software capable of manipulating three-dimensional representations.

Geometric relationships explain procedures used by artists for drawing in perspective (see Smith [1995]), as demonstrated by the following perspective problem adapted from Consortium for Mathematics and Its Applications (1999, pp. 65–67):

> An artist wants to draw a set of evenly spaced telephone poles along the side of a straight road, starting with two telephone poles as shown in figure 7.18a below. Where should the third telephone pole be placed so that it appears as far from the second as the second is from the first?

Fig. **7.18.**

Locating telephone poles so that they appear equidistant in a perspective drawing

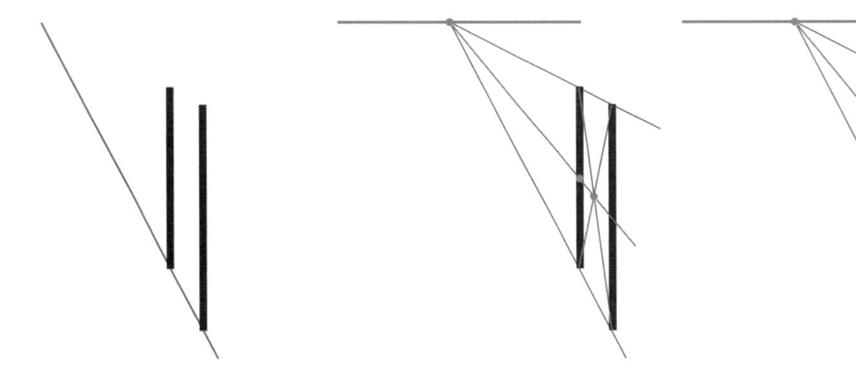

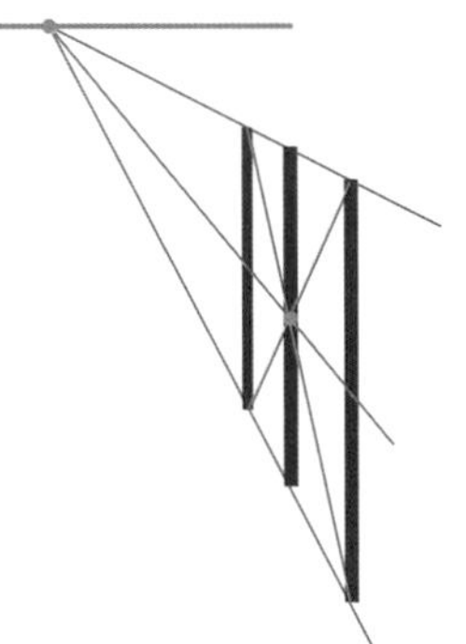

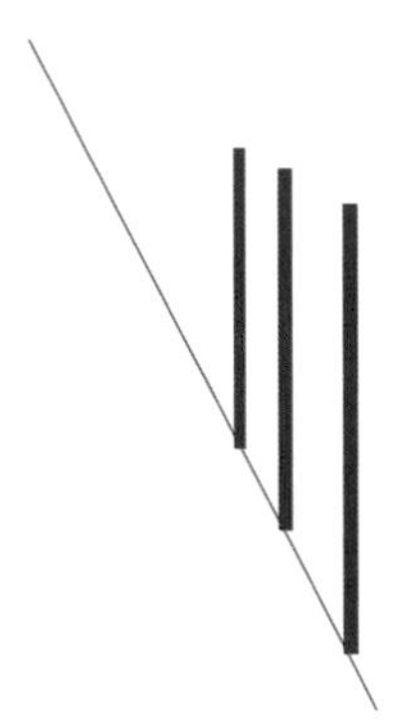

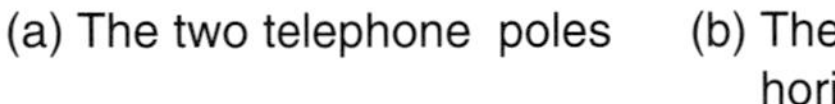

(a) The two telephone poles

(b) The vanishing point, the horizon, and the center of the rectangle

(c) The diagonals of a rectangle determine the location of the third telephone pole.

(d) The three telephone poles in perspective

The intersection of the line through the bottoms of the telephone poles with the line through their tops is the vanishing point for this family of mutually parallel lines in the perspective drawing. Since the tops and bottoms of the two telephone poles are corners of a rectangle and since the type of transformation used in producing perspective drawings preserves intersections, drawing the diagonals of the transformed rectangle locates the image of its center under the transformation. Drawing the line through the center to the vanishing point locates the midpoints of the telephone poles (fig. 7.18b).

The midpoint of the second telephone pole is also the center of the rectangle whose sides are the first and third telephone poles (fig. 7.18c). Thus the line from the top of the first telephone pole through the midpoint of the second pole intersects the ground at the bottom of the third telephone pole (fig. 7.18c); the top of the third pole can be located similarly. Finally, removing the lines used in the construction yields the desired perspective drawing (fig. 7.18d). The process can be continued to locate other telephone poles along the same line.

Although problems formulated for mathematics classes are typically stated with great precision, underspecified problems that students need to formulate clearly for themselves also play an important role. The following problem (Keynes 1998, p. 109) draws on students' knowledge of geometric and trigonometric relationships and on their spatial-visualization skills. The way it is posed compels students to determine what additional information is needed, an important aspect of working problems in real-world contexts.

> You are installing track lighting in an old warehouse that is being remodeled into a restaurant. The lights can adequately illuminate up to 15 feet from the bulbs and, at that distance, illuminate a circle with a 6-foot diameter. Figure out where to place the tracks and the bulbs to provide for maximum illumination of the customer area.

Students might compile a list of questions such as these: Is the ceiling flat or vaulted? What is the height of the ceiling? What is the square footage of the customer area? A discussion of this type furthers students' abilities to solve real problems, which are typically much more open-ended than those found in textbooks.

The following problem comes from discrete mathematics. Vertex-edge graphs can be used to find optimal solutions to problems involving paths, networks, or relationships among a finite number of objects. This example, adapted from Coxford et al. (1998, p. 326), illustrates these ideas.

> Seven small towns in Smith County are connected to one another by dirt roads, as shown in the diagram in figure 7.19. (The diagram depicts only the beginnings, ends, and lengths of the roads. The roads may be straight or curved.) The distances are given in kilometers. The county, which has a limited budget, wants to pave some roads so that people can get from every town to every other town on paved roads, either directly or indirectly, but they want to minimize the total number of kilometers paved. Find a network of paved roads that will fulfill these requirements.

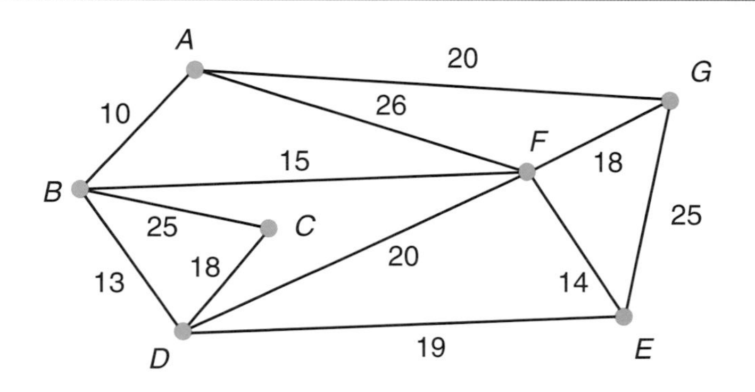

Fig. **7.19.**

A vertex-edge graph depicting the lengths of roads between towns

The diagram in figure 7.19 is a vertex-edge graph modeling the road network. The goal is to find a subnetwork of the given network that is connected, contains no circuits, includes all the cities (vertices), and minimizes the sum of the distances represented by each edge in the network. Such a network is called a *minimal spanning tree.* There are many ways to approach this problem, and students should be encouraged to share as many different approaches as possible with the class. Students can and will approach this problem in many ways, many of which will match the standard, formal solution methods. Students should first be given the opportunity to think about the characteristics of a paved road network that will satisfy the given requirements. This will often lead them to formulate the definition of a minimal spanning tree for themselves. Then they can begin to find a solution systematically. For example, students often proceed as follows: First, choose the shortest edge (*AB*). Then choose the shortest remaining edge (*BD*). Continue in this way, but never choose an edge that closes a circuit. One solution found by using this method is *AB*, *BD*, *EF*, *BF*, *FG*, *CD* (total length = 88).

Another commonly tried method is to start at *A*, then go to the "nearest neighbor," then the nearest neighbor from there, and so on. This leads to *AB*, *BD*, *DC*, and then a dead end, since moving on from *C* creates a circuit. Students might try to resolve this dead-end problem by starting at a different vertex, or they might try to modify their method by finding the nearest neighbor to any vertex that has already been reached (and, as always, not creating a circuit). This latter approach leads to *AB*, *BD*, *DC*, *BF*, *EF*, *FG* (total length = 88). As students try these different methods, they should be encouraged to fully specify their methods as algorithms, compare algorithms with other students, and consider which algorithms produce the desired results and which seem to be easier or more efficient. As it turns out, these two solution methods commonly used by students are the two most commonly used standard algorithms, namely, Kruskal's algorithm and Prim's algorithm.

Students should learn how to formulate and apply other vertex-edge-graph models to solve problems. For example, they can use critical paths to optimally schedule large projects like a school dance or a construction project. Using each edge exactly once (an *Euler path*), students can plan an optimal snow-plowing route or an optimal layout for moving people efficiently through a museum. Using each vertex exactly once (a *Hamiltonian path*), students can find an optimal route for collecting money from ATM machines or an optimal path for a robot to follow in a manufacturing plant. Vertex-coloring methods can be used to solve problems that involve conflict management, such as optimally assigning frequencies to radio stations or scheduling committee meetings.

Measurement Standard for Grades 9–12

Instructional programs from prekindergarten through grade 12 should enable all students to—

Expectations

In grades 9–12 all students should–

Standard	Expectations
Understand measurable attributes of objects and the units, systems, and processes of measurement	• make decisions about units and scales that are appropriate for problem situations involving measurement.
Apply appropriate techniques, tools, and formulas to determine measurements	• analyze precision, accuracy, and approximate error in measurement situations; • understand and use formulas for the area, surface area, and volume of geometric figures, including cones, spheres, and cylinders; • apply informal concepts of successive approximation, upper and lower bounds, and limit in measurement situations; • use unit analysis to check measurement computations.

Measurement

Students should enter grades 9–12 with a good understanding of measurement concepts and well-developed measurement skills. In addition to reading measurements directly from instruments, students should have calculated distances indirectly and used derived measures, such as rates.

Opportunities to use and understand measurement arise naturally during high school in other areas of mathematics, in science, and in technical education. Measuring the number of revolutions per minute of an engine, vast distances in astronomy, or microscopic molecular distances extends students' facility with derived measures and indirect measurement. Calculator- and computer-based measurement instruments facilitate the collection, storage, and analysis of real-time measurement data. Through logarithmic scaling, students can graphically represent a relatively large range of measurements. Insight into formulas for the volume or surface area of a cone or a sphere can be gained by applying methods of successive approximation. These aspects of measurement, along with considerations of precision and error, are important in the students' high school experience.

Opportunities to use and understand measurement arise naturally during high school in other areas of mathematics, in science, and in technical education.

Understand measureable attributes of objects and the units, systems, and processes of measurement

Students should enter high school adept at using rates to express measurements of some attributes. Cars and buses travel at velocities expressed in miles per hour or kilometers per hour, the growth of plants is recorded in centimeters per day, and birth rates are often reported in births per 1000 people. By the time students reach high school, they should be ready to make sound decisions about how quantities should be measured and represented, depending on the situation and the problem under consideration. For example, the velocity of an insect measured as 4 centimeters a second is easier to understand than a velocity of 0.00004 kilometers a second, although they are equal. Students extend their understanding of measurement in the sciences, where many measurements are indirect. For example, they can determine the height of a bridge if they know it takes three seconds for a ball dropped off the bridge to hit the water below.

With the widespread use of calculator and computer technologies for gathering and displaying data, students should understand that selections of scale and viewing window become important choices. For example, the line $f(x) = x$ appears to intersect the two coordinate axes at an angle of 45 degrees only when the horizontal and vertical scales are the same. Likewise, circles viewed on screens where the horizontal and vertical scales differ look like ellipses. The local and global behavior of a function shown in different viewing windows can appear very different; at times, small differences in choices can lead to significant differences in the visual messages. Teachers can help students understand how to make strategic choices about scale and viewing window so that they can solve the problems they are addressing most effectively.

Nonlinear scales help represent some naturally occurring phenomena. For example, human ears have the ability to differentiate among sounds of low intensity that are very close together, but they do not distinguish as well among sounds of high intensity. Consequently, measures of sound intensity are often displayed on a logarithmic scale, tied to equally spaced decibel units as shown in the chart in figure 7.20 (units are in newtons per square meter). Students should be able to compare, for example, the loudness of a whisper (20 decibels) with that of a vacuum cleaner (80 decibels), noting that for each ten-decibel increase, the sound intensity increases by a factor of 10.

Fig. **7.20.**

A chart showing that the decibel scale is logarithmic

			Barely audible whisper ↓						Vacuuming ↓			Listening to Walkman ↓			Jet engines heard from nearby ↓
Decibels	0	10	20	30	40	50	60	70	80	90	100	110	120	130	140
Sound intensity in newtons per m^2	10^{-5}	10^{-4}	10^{-3}	10^{-2}	10^{-1}	10^{0}	10^{1}	10^{2}	10^{3}	10^{4}	10^{5}	10^{6}	10^{7}	10^{8}	10^{9}

Apply appropriate techniques, tools, and formulas to determine measurements

High school students should be able to make reasonable estimates and sensible judgments about the precision and accuracy of the values they report. Teachers can help students understand that measurements of continuous quantities are always approximations. For example, suppose a situation calls for determining the mass of a bar of gold bullion in the shape of a rectangular prism whose length, width, and height are measured as 27.9 centimeters, 10.2 centimeters, and 6.4 centimeters, respectively. Knowing that the density is 19 300 kilograms per cubic meter, students might compute the mass as follows:

$$\begin{aligned}\text{Mass} &= (\text{density})\cdot(\text{volume})\\ &= \left(19300\ \frac{\text{kg}}{\text{m}^3}\right)\cdot\left(\frac{1}{10^6}\ \frac{\text{m}^3}{\text{cm}^3}\right)\cdot\left(1821.312\ \text{cm}^3\right)\\ &= 35.1513216\ \text{kg}\end{aligned}$$

The students need to understand that reporting the mass with this degree of precision would be misleading because it would suggest a degree of accuracy far greater than the actual accuracy of the measurement. Since the lengths of the edges are reported to the nearest tenth of a centimeter, the measurements are precise only to 0.05 centimeter. That is, the edges could actually have measures in the intervals 27.9 ± 0.05, 10.2 ± 0.05, and 6.4 ± 0.05. If students calculate the possible maximum and minimum mass, given these dimensions, they will see that at most one decimal place in accuracy is justified.

As suggested by the example above, units should be reported along with numerical values in measurement computations. The following problem requires both an understanding of derived measurements and facility in unit analysis—keeping track of units during computations:

> While driving through Canada in the late 1990s, a U.S. tourist put 60 liters of gas in his car. The gas cost Can$0.50 a liter (Can$ stands for Canadian dollars). The exchange rate at that time was Can$1.49 for each US$1.00 (United States dollar). The price for a gallon of gasoline in the

United States was US$0.99. The driver wanted to compare prices and decide whether a tank of gas was cheaper in the United States or Canada.

The cost of the gasoline in Canada follows:

$$\frac{\text{Can\$0.50}}{1\,\cancel{L}} \cdot 60\,\cancel{L} = \text{CAN\$30.00}$$

Teachers can help students recognize that in order to compute the cost of the same quantity of gasoline in the United States, it is necessary to convert between both monetary systems and units of volume. Thus, in addition to knowing the exchange rate, it is necessary to know that there are approximately 3.79 liters in each gallon of gas. The cost of 60 liters of gasoline at the U.S. price can then be seen to be

$$\frac{\cancel{\text{US}}\$0.99}{1\,\cancel{\text{gal}}} \cdot \frac{1\,\cancel{\text{gal}}}{3.79\,\cancel{L}} \cdot 60\,\cancel{L} \cdot \frac{\text{Can\$1.49}}{\cancel{\text{US}}\$1.00} \approx \text{Can\$23.35},$$

so the gas is less expensive in the United States. This computation illustrates how unit analysis can be helpful in keeping track of the conversions.

An important measurement idea, which also helps to establish the groundwork for some fundamental ideas of calculus, is that the measurements of some quantities can be determined by sequences of increasingly accurate approximations. For example, suppose that students are exploring ways to find the volumes of three-dimensional solids. Students should know that the volume of any right cylinder is the product of its height and the area of its base. Thus the volume of a right circular cylinder would be $\pi r^2 h$, where r is the radius of its base and h is its height. But how could students determine the volume of a cone?

To illustrate, the right circular cone shown in figure 7.21 has base radius 5 centimeters and height 10 centimeters. Using similar triangles, students should be able to see that slicing the cone parallel to the base at 2-centimeter intervals yields four nearly cylindrical disks and a small cone, each of height 2 centimeters. Each of those five figures would fit completely inside a cylinder that is 2 centimeters high and that has the same radius as its bottom cross section. Hence, the cumulative volume is less than $2(\pi \times 5^2 + \pi \times 4^2 + \pi \times 3^2 + \pi \times 2^2 + \pi \times 1^2)\ \text{cm}^3 = 110\pi\ \text{cm}^3$. At the same time, each of the five figures contains as a subset a cylinder that has the radius of its top cross section. Thus the cumulative volume must be at least $2(\pi \times 4^2 + \pi \times 3^2 + \pi \times 2^2 + \pi \times 1^2 + \pi \times 0^2)\ \text{cm}^3 = 60\pi\ \text{cm}^3$. Repeating the process with 1-centimeter-thick slices would help students see that as they take thinner slices of the cone, the overestimates and underestimates of the volume get increasingly close to each other. (Indeed, the averages of the underestimates and overestimates rapidly approach the actual volume of $(83\ 1/3)\pi\ \text{cm}^3$.) Informally, such experiences serve as an introduction to the idea of approximation by using upper and lower bounds and to the idea of limits. Whether or not these ideas are pursued later in formal coursework, they can introduce powerful ways of thinking about mathematical phenomena and help students establish a basic familiarity with core ideas of calculus.

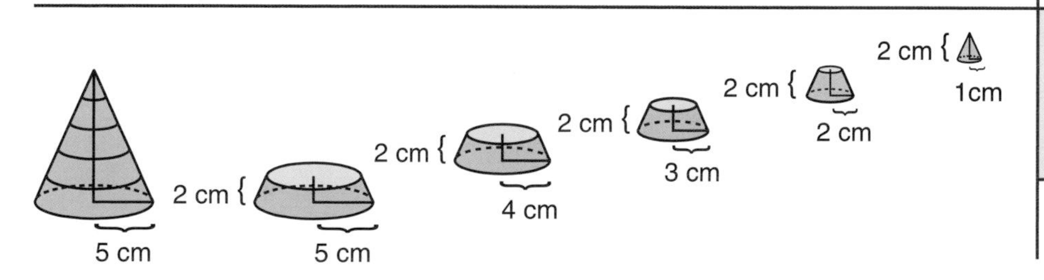

Fig. **7.21.**

Slices of a cone can be used to approximate the volume of the cone by using upper and lower bounds.

Data Analysis and Probability

STANDARD for Grades 9–12

Instructional programs from prekindergarten through grade 12 should enable all students to—

Expectations

In grades 9–12 all students should–

Instructional programs should enable all students to—	In grades 9–12 all students should–
Formulate questions that can be addressed with data and collect, organize, and display relevant data to answer them	• understand the differences among various kinds of studies and which types of inferences can legitimately be drawn from each; • know the characteristics of well-designed studies, including the role of randomization in surveys and experiments; • understand the meaning of measurement data and categorical data, of univariate and bivariate data, and of the term variable; • understand histograms, parallel box plots, and scatterplots and use them to display data; • compute basic statistics and understand the distinction between a statistic and a parameter.
Select and use appropriate statistical methods to analyze data	• for univariate measurement data, be able to display the distribution, describe its shape, and select and calculate summary statistics; • for bivariate measurement data, be able to display a scatterplot, describe its shape, and determine regression coefficients, regression equations, and correlation coefficients using technological tools; • display and discuss bivariate data where at least one variable is categorical; • recognize how linear transformations of univariate data affect shape, center, and spread; • identify trends in bivariate data and find functions that model the data or transform the data so that they can be modeled.
Develop and evaluate inferences and predictions that are based on data	• use simulations to explore the variability of sample statistics from a known population and to construct sampling distributions; • understand how sample statistics reflect the values of population parameters and use sampling distributions as the basis for informal inference; • evaluate published reports that are based on data by examining the design of the study, the appropriateness of the data analysis, and the validity of conclusions; • understand how basic statistical techniques are used to monitor process characteristics in the workplace.
Understand and apply basic concepts of probability	• understand the concepts of sample space and probability distribution and construct sample spaces and distributions in simple cases; • use simulations to construct empirical probability distributions; • compute and interpret the expected value of random variables in simple cases; • understand the concepts of conditional probability and independent events; • understand how to compute the probability of a compound event.

Data Analysis and Probability

Students whose mathematics curriculum has been consistent with the recommendations in *Principles and Standards* should enter high school having designed simple surveys and experiments, gathered data, and graphed and summarized those data in various ways. They should be familiar with basic measures of center and spread, able to describe the shape of data distributions, and able to draw conclusions about a single sample. Students will have computed the probabilities of simple and some compound events and performed simulations, comparing the results of the simulations to predicted probabilities.

In grades 9–12 students should gain a deep understanding of the issues entailed in drawing conclusions in light of variability. They will learn more-sophisticated ways to collect and analyze data and draw conclusions from data in order to answer questions or make informed decisions in workplace and everyday situations. They should learn to ask questions that will help them evaluate the quality of surveys, observational studies, and controlled experiments. They can use their expanding repertoire of algebraic functions, especially linear functions, to model and analyze data, with increasing understanding of what it means for a model to fit data well. In addition, students should begin to understand and use correlation in conjunction with residuals and visual displays to analyze associations between two variables. They should become knowledgeable, analytical, thoughtful consumers of the information and data generated by others.

In grades 9–12 students should gain a deep understanding of the issues entailed in drawing conclusions in light of variability.

As students analyze data in grades 9–12, the natural link between statistics and algebra can be developed further. Students' understandings of graphs and functions can also be applied in work with data.

Basic ideas of probability underlie much of statistical inference. Probability is linked to other topics in high school mathematics, especially counting techniques (Number and Operations), area concepts (Geometry), the binomial theorem, and relationships between functions and the area under their graphs (Algebra). Students should learn to determine the probability of a sample statistic for a known population and to draw simple inferences about a population from randomly generated samples.

Formulate questions that can be addressed with data and collect, organize, and display relevant data to answer them

Students' experiences with surveys and experiments in lower grades should prepare them to consider issues of design. In high school, students should design surveys, observational studies, and experiments that take into consideration questions such as the following: Are the issues and questions clear and unambiguous? What is the population? How should the sample be selected? Is a stratified sample called for? What size should the sample be? Students should understand the concept of bias in surveys and ways to reduce bias, such as using randomization in selecting samples. Similarly, when students design experiments, they should begin to learn how to take into account the nature of the treatments, the selection of the experimental units, and the randomization used to assign treatments to units. Examples of situations students might consider are shown in figure 7.22.

Representation | Connections | Communication | Reasoning & Proof | Problem Solving | Data Analysis & Probability | Measurement | Geometry | Algebra | Number & Operations

Fig. **7.22.**

Three kinds of situations in which statistics are used

Survey

An opinion survey asked people, "Do you use a computer?"

Maria answered Yes because she thought the question meant had she ever used a computer.

Eric answered No because he thought the question was asking whether he used one regularly.

Alex answered No because he only played games on the computer and didn't think this counted as "using" one.

The ambiguity of the question renders the results of the survey useless and makes any conclusions drawn from those results very questionable.

Observational Study

The city of New York has sufficient funds to run one additional commuter train each weekday. Officials must decide whether it would be better to add a train during the morning or the evening rush hour.

Determining the average number of commuters currently riding the train from 6:00 to 9:00 A.M. and the average number riding from 4:00 to 6:00 P.M. on weekdays would provide valuable information about the habits of current riders.

However, conclusions based on this information might not take into account people who currently drive to work to avoid overcrowding on trains.

Experiment

An experiment to determine which brand of tires lasts longest tests one brand on the front wheels and one brand on the rear wheels of cars in rural Montana.

Many factors could explain experimental differences; for example, front-wheel drive or frequent or infrequent braking can systematically cause uneven wear.

To minimize the number of extraneous or confounding factors, it would be better to randomly assign two tires of each kind to the wheels of each car. Also, to broaden the base of applicability, the test should be performed under conditions that involve city as well as rural driving.

There are many reasons to be careful in conducting surveys and analyzing the results.

There are many reasons to be careful in conducting surveys and analyzing the results. In the survey example, the ambiguity of the question about computer usage makes it impossible to interpret the results meaningfully. Students designing surveys must also deal with sampling procedures. The goal of a survey is to generalize from a sample to the population from which it is drawn. Students must understand that a sample is most likely to be representative when it has been randomly chosen from the population.

Nonrandomness in sampling may also limit the conclusions that can be drawn from observational studies. For instance, in the observational study example, it is not certain that the number of people riding trains reflects the number of people who would ride trains if more were available or if scheduling were more convenient. Similarly, it would be inappropriate to draw conclusions about the percentage of the population that ice skates on the basis of observational studies done either in

Florida or in Quebec. Students need to be aware that any conclusions about cause and effect should be made very cautiously in observational studies. They should also know how certain kinds of systematic observations, such as random testing of manufacturing parts taken from an assembly line, can be used for purposes of quality control.

In designed experiments, two or more experimental treatments (or conditions) are compared. In order for such comparisons to be valid, other sources of variation must be controlled. This is not the situation in the tire example, in which the front and rear tires are subjected to different kinds of wear. Another goal in designed experiments is to be able to draw conclusions with broad applicability. For this reason, new tires should be tested on all relevant road conditions. Consider another designed experiment in which the goal is to test the effect of a treatment (such as getting a flu shot) on a response (such as getting the flu) for older people. This is done by comparing the responses of a treatment group, which gets treatment, with those of a control group, which does not. Here, the investigators would randomly choose subjects for their study from the population group to which they want to generalize, say, all males and females aged 65 or older. They would then randomly assign these individuals to the control and treatment groups. Note that interesting issues arise in the choice of subjects (not everyone wants to or is able to participate—could this introduce bias?) and in the concept of a control group (are these seniors then at greater risk of getting the flu?).

Describing center, spread, and shape is essential to the analysis of both univariate and bivariate data.

Select and use appropriate statistical methods to analyze data

Describing center, spread, and shape is essential to the analysis of both univariate and bivariate data. Students should be able to use a variety of summary statistics and graphical displays to analyze these characteristics.

The shape of a distribution of a single measurement variable can be analyzed using graphical displays such as histograms, dotplots, stem-and-leaf plots, or box plots. Students should be able to construct these graphs and select from among them to assist in understanding the data. They should comment on the overall shape of the plot and on points that do not fit the general shape. By examining these characteristics of the plots, students should be better able to explain differences in measures of center (such as mean or median) and spread (such as standard deviation or interquartile range). For example, students should recognize that the statement "the mean score on a test was 50 percent" may cover several situations, including the following: all scores are 50 percent; half the scores are 40 percent and half the scores are 60 percent; half the scores are 0 percent and half the scores are 100 percent; one score is 100 percent and 50 scores are 49 percent. Students should also recognize that the sample mean and median can differ greatly for a skewed distribution. They should understand that for data that are identified by categories—for example, gender, favorite color, or ethnic origin—bar graphs, pie charts, and summary tables often display information about the relative frequency or percent in each category.

Students should learn to apply their knowledge of linear transformations from algebra and geometry to linear transformations of data. They should be able to explain why adding a constant to all observed values in a sample changes the measures of center by that constant but does not change measures of spread or the general shape of the distribution. They should also understand why multiplying each observed value by the same constant multiplies the mean, median, range, and standard deviation by the same factor (see the related discussion in the "Reasoning and Proof" section of this chapter).

The methods used for representing univariate measurement data also can be adapted to represent bivariate data where one variable is categorical and the other is a continuous measurement. The levels of the categorical variable split the measurement variable into groups. Students can use parallel box plots, back-to-back stem-and-leaf, or same-scale histograms to compare the groups. The following problem from Moore (1990, pp. 108–9) illustrates conclusions that can be drawn from such comparisons:

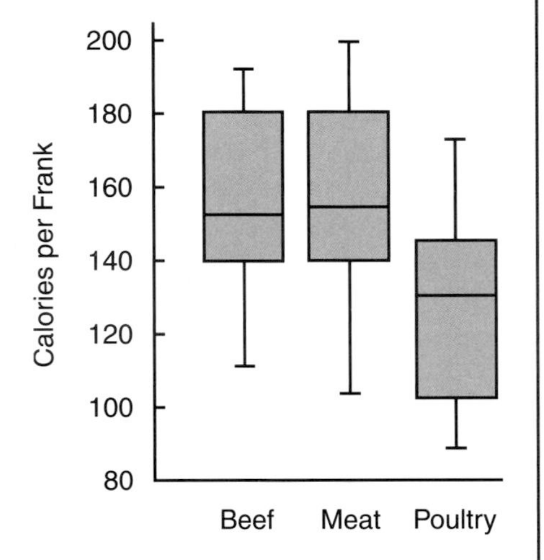

> U.S. Department of Agriculture regulations group hot dogs into three types: beef, meat, and poultry. Do these types differ in the number of calories they contain? The three boxplots to the left display the distribution of calories per hot dog among brands of the three types. The box ends mark the quartiles, the line within the box is the median, and the whiskers extend to the smallest and largest individual observations. We see that beef and meat hot dogs are similar but that poultry hot dogs as a group show considerably fewer calories per hot dog.

Analyses of the relationships between two sets of measurement data are central in high school mathematics. These analyses involve finding functions that "fit" the data well. For instance, students could examine the scatterplot of bivariate measurement data shown in figure 7.23 and consider what type of function (e.g., linear, exponential, quadratic) might be a good model. If the plot of the data seems approximately linear, students should be able to produce lines that fit the data, to compare several such lines, and to discuss what *best fit* might mean. This analysis includes stepping back and making certain that what is being done makes sense practically.

The dashed vertical line segments in figure 7.23 represent *residuals*—the differences between the y-values predicted by the linear model and

Fig. **7.23.**

Fitting a line to the data displayed in a scatterplot (Adapted from Burrill et al. [1999, pp. 14–15, 20])

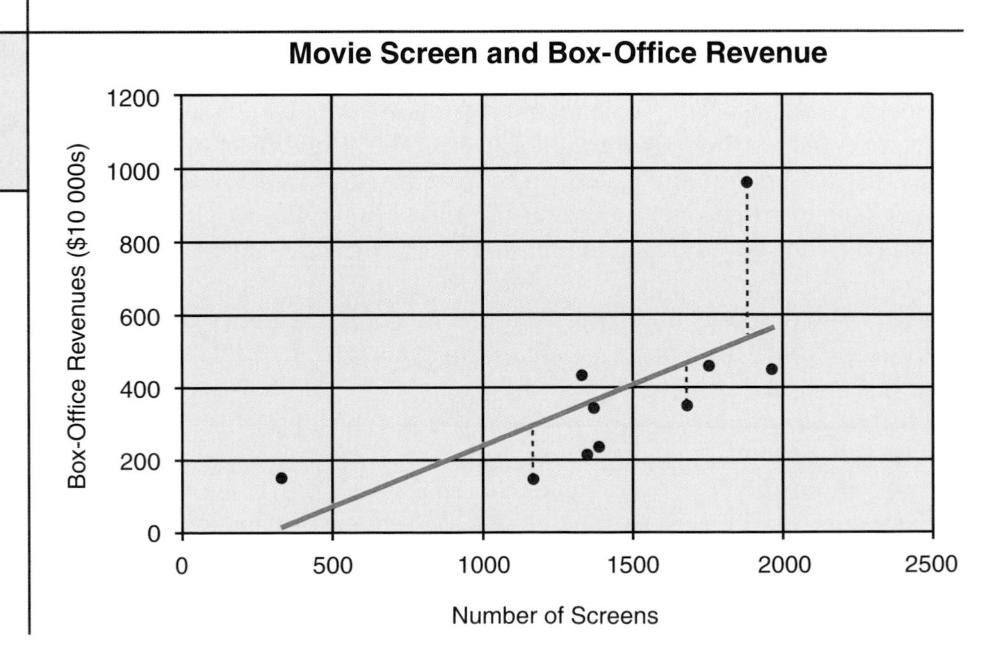

the actual y-values—for three data points. Teachers can help students explore several ways of using residuals to define *best fit*. For example, a candidate for best-fitting line might be chosen to minimize the sum of the absolute values of residuals; another might minimize the sum of squared residuals. Using dynamic software, students can change the position of candidate lines for best fit and see the effects of those changes on squared residuals. The line illustrated in figure 7.23, which minimizes the sum of the squares of the residuals, is called the *least-squares regression line*. Using technology, students should be able to compute the equation of the least-squares regression line and the correlation coefficient, r.

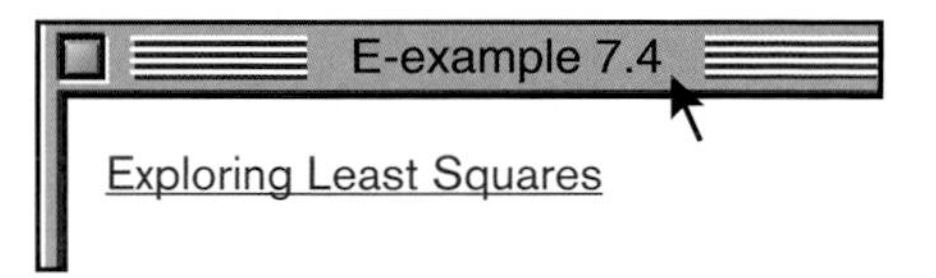

Students should understand that the correlation coefficient r gives information about (1) how tightly packed the data are about the regression line and (2) about the strength of the relationship between the two variables. Students should understand that correlation does not imply a cause-and-effect relationship. For example, the presence of certain kinds of eye problems and the loss of sensitivity in people's feet can be related statistically. However, the correlation may be due to an underlying cause, such as diabetes, for both symptoms rather than to one symptom's causing the other.

Students should understand that correlation does not imply a cause-and-effect relationship.

Develop and evaluate inferences and predictions that are based on data

Once students have determined a model for a data set, they can use the model to make predictions and recognize and explain the limitations of those predictions. For example, the regression line depicted in figure 7.23 has the equation $y = 0.33x - 93.9$, where x represents the number of screens and y represents box-office revenues (in units of \$10 000). To help students understand the meaning of the regression line, its role in making predictions and inferences, and its limitations and possible extensions, teachers might ask questions like the following:

1. Predict the revenue of a movie that is shown on 800 screens nationwide. Of a movie that is shown on 2 500 screens. Discuss the accuracy and limitations of your predictions.
2. Explain the meaning of a slope of 0.33 in the screen-revenue context.
3. Explain why the y-intercept of the regression line does not have meaning in the box-office-revenue context.
4. What other variables might help in predicting box-office revenues?

A *parameter* is a single number that describes some aspect of an entire population, and a *statistic* is an estimate of that value computed from some sample of the population. To understand terms such as *margin of error* in opinion polls, it is necessary to understand how statistics, such as sample proportions, vary when different random samples are chosen from a population. Similarly, sample means computed from measurement data vary according to the random sample chosen, so it is important to understand the distribution of sample means in order to assess how well a specific sample mean estimates the population mean.

Understanding how to draw inferences about a population from random samples requires understanding how those samples might be distributed. Such an understanding can be developed with the aid of simulations. Consider the following situation:

Suppose that 65% of a city's registered voters support Mr. Blake for mayor. How unusual would it be to obtain a random sample of 20 registered voters in which at most 8 support Mr. Blake?

Here the parameter for the population is known: 65 percent of all registered voters support Mr. Blake. The question is, How likely is a random sample with a very different proportion (at most 8 out of 20, or 40%) of supporters? The probability of such a sample can be approximated with a simulation. Figure 7.24 shows the results of drawing 100 random samples of size 20 from a population in which 65 percent support Mr. Blake.

Fig. **7.24.**

The results of a simulation of drawing 100 random samples of size 20 from a population in which 65 percent support Mr. Blake

Number of registered voters supporting Mr. Blake	0	1	2	3	4	5	6	7	8	9	10	11	12	13	14	15	16	17	18	19	20
Number of random samples of size 20	0	0	0	0	0	1	0	0	1	4	8	8	13	22	24	8	4	6	1	0	0

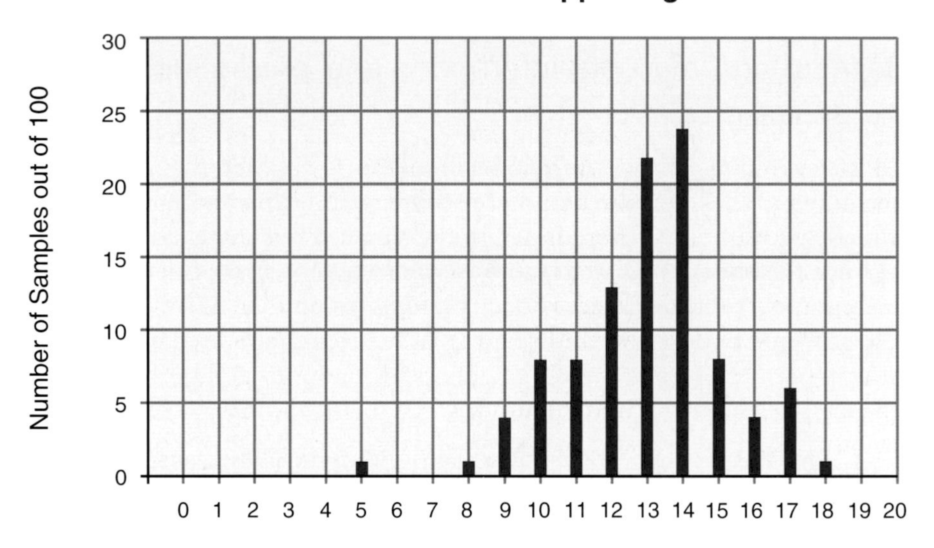

Only 2 percent of the samples had 8 or fewer registered voters supporting Mr. Blake. The value 8 occurs well out on the left tail of the histogram. One can reasonably conclude that a sample outcome of 8 or fewer supporters out of 20 randomly selected voters is a rare event when sampling from this population. This kind of exercise can be used to develop the concept of hypothesis testing for a single proportion or mean.

In the situation just described, a parameter of the population was known and the probability of a particular sample characteristic was estimated in order to understand how sampling distributions work. However, in applications of this idea in real situations, the information about a population is unknown and a sample is used to project what that information might be without having to check all the individuals in the population. For example, suppose that the proportion of registered voters supporting Mr. Blake was unknown (a realistic situation) and that a pollster wanted to find out what that proportion might be. If

the pollster surveyed a sample of 20 voters and found that 65 percent of them support the candidate, is it reasonable to expect that about 65 percent of all voters support the candidate? What if the sample was 200 voters? 2000 voters? As indicated above, the proportion of voters who supported Mr. Blake could vary substantially from sample to sample in samples of 20. There is much less variation in samples of 200. By performing simulations with samples of different sizes, students can see that as sample size increases, variation decreases. In this way, they can develop the intuitive underpinnings for understanding confidence intervals.

A similar kind of reasoning about the relationship between the characteristics of a sample and the population from which it is drawn lies behind the use of sampling for monitoring process control and quality in the workplace.

Understand and apply basic concepts of probability

In high school, students can apply the concepts of probability to predict the likelihood of an event by constructing probability distributions for simple sample spaces. Students should be able to describe sample spaces such as the set of possible outcomes when four coins are tossed and the set of possibilities for the sum of the values on the faces that are down when two tetrahedral dice are rolled.

High school students should learn to identify mutually exclusive, joint, and conditional events by drawing on their knowledge of combinations, permutations, and counting to compute the probabilities associated with such events. They can use their understandings to address questions such as those in following series of examples.

> The diagram at the right shows the results of a two-question survey administered to 80 randomly selected students at Highcrest High School.
>
> - Of the 2100 students in the school, how many would you expect to play a musical instrument?

Do you play on a sports team?	Do you play a musical instrument? Yes	No
Yes	14	32
No	20	14

- Estimate the probability that an arbitrary student at the school plays on a sports team and plays a musical instrument. How is this related to estimates of the separate probabilities that a student plays a musical instrument and that he or she plays on a sports team?
- Estimate the probability that a student who plays on a sports team also plays a musical instrument.

High school students should learn to compute expected values. They can use their understanding of probability distributions and expected value to decide if the following game is fair:

> You pay 5 chips to play a game. You roll two tetrahedral dice with faces numbered 1, 2, 3, and 5, and you win the sum of the values on the faces that are not showing.

Fig. **7.25.**

The sample space for the roll of two tetrahedral dice

First die

Second die	1	2	3	5
1	2	3	4	6
2	3	4	5	7
3	4	5	6	8
5	6	7	8	10

Teachers can ask students to discuss whether they think the game is fair and perhaps have the students play the game a number of times to see if there are any trends in the results they obtain. They can then have the students analyze the game. First, students need to delineate the sample space. The outcomes are indicated in figure 7.25. The numbers on the first die are indicated in the top row. The numbers on the second die are indicated in the first column. The sums are given in the interior of the table. Since all outcomes are equally likely, each cell in the table has a probability of 1/16 of occurring.

Students can determine that the probability of a sum of 2 is 1/16; of a 3, 1/8; of a 4, 3/16; of a 5, 1/8; of a 6, 3/16; of a 7, 1/8; of an 8, 1/8; of a 10, 1/16. The expected value of a player's "income" in chips from rolling the dice is

$$\left[2\left(\frac{1}{16}\right)+3\left(\frac{1}{8}\right)+4\left(\frac{3}{16}\right)+5\left(\frac{1}{8}\right)+6\left(\frac{3}{16}\right)+7\left(\frac{1}{8}\right)+8\left(\frac{1}{8}\right)+10\left(\frac{1}{16}\right)\right]=5.5 \text{ chips.}$$

If a player pays a five-chip fee to play the game, on average, the player will win 0.5 chips. The game is not statistically fair, since the player can expect to win.

Students can also use the sample space to answer conditional probability questions such as "Given that the sum is even, what is the probability that the sum is a 6?" Since ten of the sums in the sample space are even and three of those are 6s, the probability of a 6 given that the sum is even is 3/10.

The following situation, adapted from Coxford et al. (1998, p. 469), could give rise to a very rich classroom discussion of compound events.

> In a trial in Sweden, a parking officer testified to having noted the position of the valve stems on the tires on one side of a car. Returning later, the officer noted that the valve stems were still in the same position. The officer noted the position of the valve stems to the nearest "hour." For example, in figure 7.26 the valve stems are at 10:00 and at 3:00. The officer issued a ticket for overtime parking. However, the owner of the car claimed he had moved the car and returned to the same parking place.
>
> The judge who presided over the trial made the assumption that the wheels move independently and the odds of the two valve stems returning to their previous "clock" positions were calculated as 144 to 1. The driver was declared to be innocent because such odds were considered insufficient—had all four valve stems been found to have returned to their previous positions, the driver would have been declared guilty (Zeisel 1968).

Fig. **7.26.**

A diagram of tires with valves at the 10:00 and 3:00 positions

Given the assumption that the wheels move independently, students could be asked to assess the probability that if the car is moved, two (or four) valve stems would return to the same position. They could do so by a direct probability computation, or they might design a simulation, either by programming or by using spinners, to estimate this probability. But is it reasonable to assume that two front and rear wheels or all four wheels move independently? This issue might be resolved empirically. The students might drive a car around the block to see if its wheels do rotate independently of one another and decide if the judge's assumption was justified. They might consider whether it would be more reasonable to assume that all four wheels move as a unit and ask related questions: Under what circumstances might all four wheels travel the same distance? Would all the wheels travel the same distance if the car was driven around the block? Would any differences be large enough to show up as differences in "clock" position? In this way, students can learn about the role of assumptions in modeling, in addition to learning about the computation of probabilities.

Students could also explore the effect of more-precise measurements on the resulting probabilities. They could calculate the probabilities if, say, instead of recording markings to the nearest hour on the clockface, the markings had been recorded to the nearest half or quarter hour. This line of thinking could raise the issue of continuous distributions and the idea of calculating probabilities involving an interval of values rather than a finite number of values. Some related questions are, How could a practical method of obtaining more-precise measurements be devised? How could a parking officer realistically measure tire-marking positions to the nearest clock half-hour? How could measurement errors be minimized? These could begin a discussion of operational definitions and measurement processes.

Students should be able to investigate the following question by using a simulation to obtain an approximate answer:

> How likely is it that at most 25 of the 50 people receiving a promotion are women when all the people in the applicant pool from which the promotions are made are well qualified and 65% of the applicant pool is female?

Those students who pursue the study of probability will be able to find an exact solution by using the binomial distribution. Either way, students are likely to find the result rather surprising.

Problem Solving STANDARD for Grades 9–12

Instructional programs from prekindergarten through grade 12 should enable all students to—

Build new mathematical knowledge through problem solving

Solve problems that arise in mathematics and in other contexts

Apply and adapt a variety of appropriate strategies to solve problems

Monitor and reflect on the process of mathematical problem solving

To meet new challenges in work, school, and life, students will have to adapt and extend whatever mathematics they know. Doing so effectively lies at the heart of problem solving. A problem-solving disposition includes the confidence and willingness to take on new and difficult tasks. Successful problem solvers are resourceful, seeking out information to help solve problems and making effective use of what they know. Their knowledge of strategies gives them options. If the first approach to a problem fails, they can consider a second or a third. If those approaches fail, they know how to reconsider the problem, break it down, and look at it from different perspectives—all of which can help them understand the problem better or make progress toward its solution. Part of being a good problem solver is being a good planner, but good problem solvers do not adhere blindly to plans. Instead, they monitor progress and consider and make adjustments when things are not going as well as they should (Schoenfeld 1985).

In high school, students' repertoires of problem-solving strategies expand significantly because students are capable of employing more-complex methods and their abilities to reflect on their knowledge and act accordingly have grown. Thus, students should emerge from high school with the disposition, knowledge, and strategies to deal with the new challenges they will encounter.

As in the earlier grades, problems and problem solving play an essential role in students' learning of mathematical content and in helping students make connections across mathematical content areas. Much of school mathematics can be seen as the codification of answers to sets of interesting problems. Accordingly, much of the mathematics that students encounter can be introduced by posing interesting problems on which students can make legitimate progress. (See, for example, the use of such an example in Mr. Robinson's class, as described in the "Connections" section of this chapter.) Approaching the content in this way does more than motivate students. It reveals mathematics as a sense-making discipline rather than one in which rules for working exercises are given by the teacher to be memorized and used by students.

What should problem solving look like in grades 9 through 12?

Problem solving plays a dual role in the high school curriculum. On the one hand, solving problems that have been strategically chosen and carefully sequenced is a fundamental vehicle for learning mathematical content (see the "counting rectangles" problem later in this section). In addition to carefully designing problems, teachers should seize unexpected opportunities (see the discussion of Ms. Rodriguez's class later in this section) to use problems to engage students in important mathematical ideas and to develop a deep understanding of those ideas through such engagement. Most mathematical concepts or generalizations can be effectively introduced using a problem situation that helps students see important aspects of the idea to be generalized. For example, rather than begin the consideration of the volume of a sphere by reminding students of the formula or technique for computing the volume, a teacher might begin by posing a question such as "How might you find the volume of a sphere whose radius is ten centimeters?" As students consider possible approaches, they can come to appreciate the difficulties inherent in what appears to be an easy question. In other instances, their proposed solutions may either directly lead to the desired conclusion or serve as a springboard to class discussions of the idea.

On the other hand, a major goal of high school mathematics is to equip students with knowledge and tools that enable them to formulate, approach, and solve problems beyond those that they have studied. High school students should have significant opportunities to develop a broad repertoire of problem-solving (or *heuristic*) strategies. They should have opportunities to formulate and refine problems because problems that occur in real settings do not often arrive neatly packaged. Students need experience in identifying problems and articulating them clearly enough to determine when they have arrived at solutions. The curriculum should include problems for which students know the goal to be achieved but for which they need to specify—or perhaps gather from other sources—the kinds of information needed to achieve it. Recall, for example, the "track lighting" problem discussed in the "Geometry" section of this chapter. In addressing the problem, students would need to know about the geometry of the floor plan, about the structure and height of the ceiling, about the positioning of the "customer area," and about the areas that should receive priority lighting.

Most mathematical concepts or generalizations can be effectively introduced using a problem situation.

The following problem serves multiple purposes. It gives students an opportunity to build their content knowledge during the problem-solving process, to learn or practice some heuristic strategies, and to make connections among various ways of thinking about the same mathematical content. It might be posed during a unit on combinatorics or as part of an ongoing series of problems that are given out of context so that students have to determine the relevant approaches for themselves.

> How many rectangles are there on a standard 8×8 checkerboard? Count only those rectangles (including squares) whose sides lie on grid lines. For example, there are nine rectangles on a 2×2 board, as shown in figure 7.27.

There are numerous ways to approach and solve this problem. Students should be given significant latitude to explore it rather than be led

Fig. **7.27.**

Rectangles on a 2×2 "checkerboard"

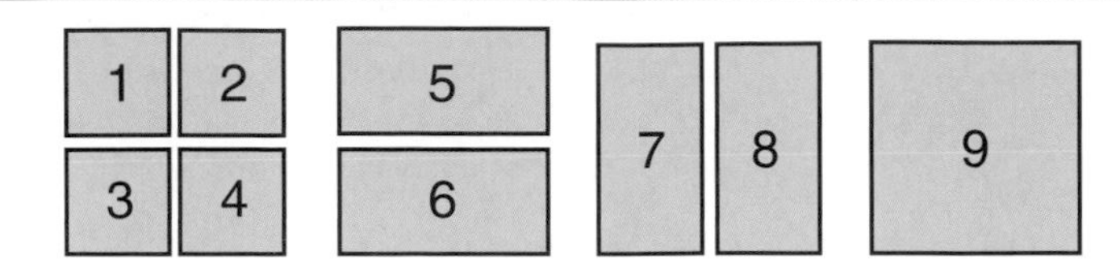

directly to the solution method or methods that the teacher has in mind. With planning, a teacher can use such explorations to develop interesting mathematics and to make connections that might otherwise be overlooked.

Often students begin this problem by trying to count the number of rectangles directly, but it is hard to keep track of which rectangles have been counted, so they usually lose count. This situation presents an opportunity to discuss the need to be systematic—to find an approach that will identify all rectangles on the board and count each only once.

Typically, some students continue to count while others look for alternatives. For problems such as this one, it is often useful to consider the heuristic strategy "try an easier related problem." But which related problem should they try, and what is to be learned from it? Some students will want to count rectangles on the 3×3 and 4×4 boards, looking for patterns. The results for the 1×1, 2×2, 3×3, and 4×4 boards are 1, 9, 36, and 100, respectively. This result suggests that the number of rectangles on an $n \times n$ grid is $(1 + 2 + \cdots + n)^2$—a very nice observation that leaves unanswered the question of *why* the sum should be what it seems to be. Some students might focus on developing a systematic way to count the rectangles on the smaller grids, and then return to the 8×8 grid to apply what they have learned. Others will try to show that the generalization holds by using the result from the $n \times n$ square as a stepping-stone to counting the rectangles in the $(n + 1) \times (n + 1)$ square. They discover that it is difficult and may need to turn their attention elsewhere. An important lesson about problem solving is that not all approaches work and that after consideration, some need to be abandoned.

The heuristic "trying an easier related problem" can be implemented in other ways in solving this problem. Instead of working with an $n \times n$ board, for example, some students may look for ways to count rectangles on a 1×8 board. That the 1×8 board has $1 + 2 + 3 + 4 + 5 + 6 + 7 + 8$ different subrectangles can be shown in a number of ways, and students can compare notes on how efficient or compelling their methods are. When they work on the 2×8 board, they may notice that the pattern of rectangles found in the 1×8 example can be found in the top row, in the second row, and extending across both rows. This strategy has a natural extension to 3×8 boards, and so on. Students should reflect on the value of strategies that can easily be generalized to other examples. Teachers can ask them to talk about and demonstrate that their strategy works in all such instances and convince others of its validity. As the students do so, they are exploring patterns systematically and verifying algebraically that the patterns hold.

Simple questions such as "What determines a rectangle?"—a version of Pólya's (1957) heuristic strategy "look at the unknown," tailored to this situation—can help students reformulate the problem. Some students may recognize that it is possible to characterize a rectangle on a grid by specifying two opposite corners. They could then count the number of rectangles that have a fixed upper left corner on the 8×8 grid as follows: Pick any corner on the grid. If there are m grid lines below that corner and n grid lines to the right of it, the number of rectangles on the grid with that point as an upper left corner is $m \times n$. The number of rectangles can then be indicated, as shown in figure 7.28, where the product written in each square of the grid is the number of rectangles on the whole grid that have the same upper left corner as the square. The task becomes to add the 64 numbers in the grid. The first column sums to $8(8 + 7 + \cdots + 1)$, the second column to $7(8 + 7 + \cdots + 1)$, and so on. Hence the sum of the columns is $(8 + 7 + \cdots + 1)(8 + 7 + \cdots + 1) = (8 + 7 + \cdots + 1)^2$. This result confirms the conjecture that some students had made earlier, when they examined the 1×1, 2×2, 3×3, and 4×4 grids, and presents an opportunity to see how one approach can shed light on a result obtained by another method. Also, once the pattern is apparent, the students should be encouraged to abstract and represent their results for more-general cases. They should, for instance, be able to conjecture that the number of rectangles on an $n \times m$ board is $(1 + 2 + \cdots + n)(1 + 2 + \cdots + m)$. From other work they might have done adding sequences of consecutive integers, they should be able to represent this as $[(n)(n + 1)(m)(m + 1)/4]$.

8 × 8	7 × 8	6 × 8	5 × 8	4 × 8	3 × 8	2 × 8	1 × 8
8 × 7	7 × 7	6 × 7	5 × 7	4 × 7	3 × 7	2 × 7	1 × 7
8 × 6	7 × 6	6 × 6	5 × 6	4 × 6	3 × 6	2 × 6	1 × 6
8 × 5	7 × 5	6 × 5	5 × 5	4 × 5	3 × 5	2 × 5	1 × 5
8 × 4	7 × 4	6 × 4	5 × 4	4 × 4	3 × 4	2 × 4	1 × 4
8 × 3	7 × 3	6 × 3	5 × 3	4 × 3	3 × 3	2 × 3	1 × 3
8 × 2	7 × 2	6 × 2	5 × 2	4 × 2	3 × 2	2 × 2	1 × 2
8 × 1	7 × 1	6 × 1	5 × 1	4 × 1	3 × 1	2 × 1	1 × 1

Fig. **7.28.**

The number of rectangles in an 8×8 grid

Other heuristics may also prove useful. Some students may notice that each rectangle in the 8×8 grid is determined by the lines on which its top and bottom borders lie (there are $\binom{9}{2}$ choices) and the lines on which its left and right borders lie (also $\binom{9}{2}$ choices). So there are $\binom{9}{2} \cdot \binom{9}{2}$ rectangles in the 8×8 grid. This approach can be generalized to show that there are $\binom{n+1}{2} \cdot \binom{m+1}{2}$ rectangles in an $n \times m$ grid. Thus, this problem provides students with an opportunity to review counting techniques and show their power, as well as to use their prior knowledge in other approaches to solving the problem. Moreover, just as the class has not finished when students have found one solution—the variety of solutions is important—it has not finished when a variety of solutions have been found. A teacher and a class with a problem-solving disposition will be quick to formulate interesting extensions such as, What would an analogous three-dimensional problem look like?

In addition to using carefully chosen problems with particular curricular purposes in mind, teachers with problem-solving dispositions can take advantage of events that occur in the classroom to promote further understanding through problem solving. Such an opportunity arises in the following story, based on Zbiek and Glass (forthcoming):

E-example 7.5

Linking Representations of a Linear Function

> Ms. Rodriguez's class was studying classes of functions. The focus of the lesson was on exploring the properties of the graphs of quadratic functions. To ease the students into the use of dynamic graphing technology, the lesson began with parametric explorations of a more familiar class of functions, linear functions of the form $f(x) = mx + b$. Ms. Rodriguez introduced students to the day's goal:
>
> *Ms. R.*: Today, we will learn about a new family of functions called *quadratic functions*. But first, you will need to become familiar with a new type of computer representation. So we will begin by exploring a familiar family, $f(x) = mx + b$. You will be using a computer sketch with sliders to change the values of m and b. I think you'll like using the sliders—they'll save you a lot of work and let you concentrate on big ideas.
>
> The computer software enabled the students to change the value of m or the value of b and observe the effects of these changes on the graph of $f(x) = mx + b$. Ms. Rodriguez demonstrated the use of the top slider to control the value of m and the bottom slider to control the value of b (see fig. 7.29 for a view of the slider screen). She then handed out a worksheet to structure the students' activities.
>
> As often happens, some students were exploring the graphing software without following the instructions on the teacher's worksheet. Shelly was one of those students. Soon she was calling for Ms. Rodriguez's attention:
>
> *Shelly*: Ms. R., Ms. R.! I found something really awesome!
>
> *Ms. R.*: That's great, Shelly. But don't tell others what you've seen yet. I want the others to have a chance to make their own discoveries. Please keep track of your work for when we compare notes.
>
> As she approached Shelly's desk, Ms. Rodriguez expected the trace function on Shelly's computer to show a set of parallel lines (if Shelly had used the slider for b) or a set of lines that pass through the same point on the y-axis (if Shelly had used the slider for m). What she saw was different (see fig. 7.30). Ms. Rodriguez was puzzled.
>
> *Ms. R.*: Shelly, could you show me how you got the computer screen to that point?
>
> *Shelly*: I don't really know, but for some reason, when I moved m, b also moved.
>
> *Ms. R.*: Can you do it again?

Fig. **7.29.**
A slider screen

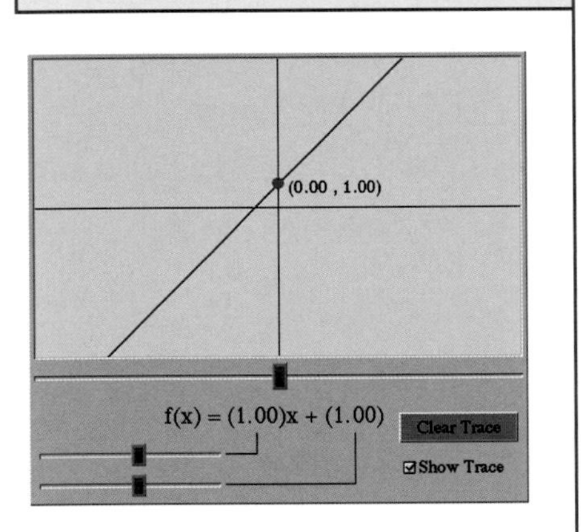

Fig. **7.30.**
Shelly's screen

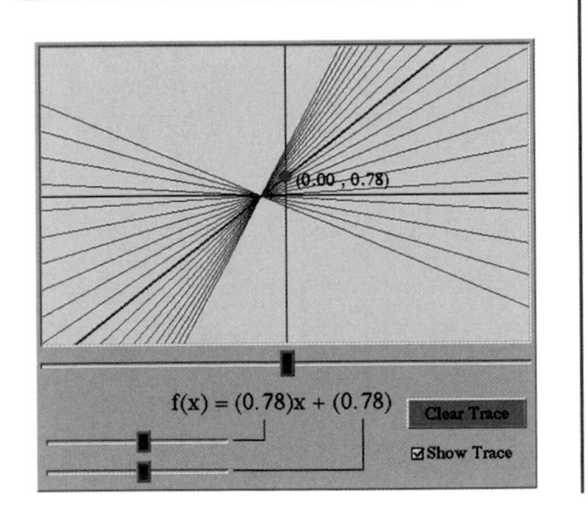

Shelly demonstrated, clicking on m and b in such a way that both sliders moved at the same time and rate. Since the m and b scales were the same, as she moved m, b moved the same amount.

Ms. R.: What's the really awesome thing you are seeing?

Shelly: Look! They all intersect at the same point.

Ms. R.: That is really interesting!

Ms. Rodriguez found this observation intriguing and decided that exploring the situation might lead to some interesting insights. She asked Shelly to show the class how she created the pattern. Soon Michael, intrigued by Shelly's result, tried her technique on another function, with the result shown in figure 7.31.

Fig. **7.31**.

Michael's screen

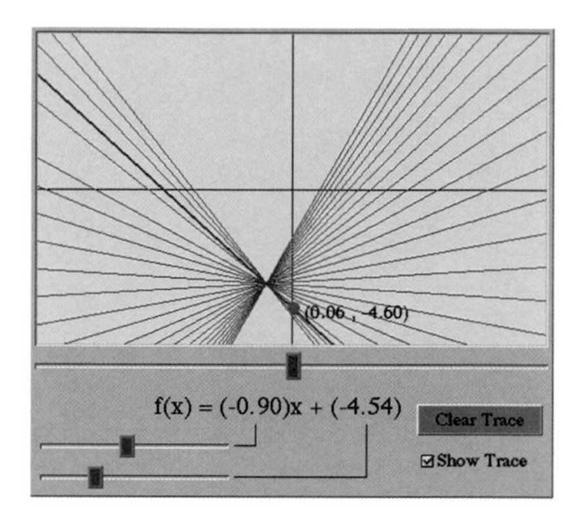

Michael: Look at what happened! I got the same thing as Shelly, but I started with a different function.

At Ms. Rodriguez's request, Michael demonstrated his finding for the class.

Ms. R.: Tell us what you mean when you say you "got the same thing."

Michael: All my lines crossed at one point.

Jerod: I tried a different one, and mine crossed at one point!

Ms. Rodriguez saw this development as an opportunity for group problem solving and knew that she could entice her students with a challenge. She asked, "Can anyone come up with a similar picture for another linear function?" The room buzzed with activity as students tried Shelly's technique. Some students helped one another learn the technique while Ms. Rodriguez helped others. Every time the students applied Shelly's technique to a linear function, their set of equations passed through one point. As the students compared their results, they began to notice that the points of intersection always had the x-value -1, a fact that rippled rapidly through the

classroom. Ms. Rodriguez brought the class to a close with a discussion of what had been found. The class was sure that Shelly's method produced the same result for all linear functions, and the students had mixed opinions about whether it would work for other functions. A student formulated a statement of the class's conjecture and wrote it on the board as follows:

> Applying Shelly's technique to a linear function results in a set of linear functions, all of which pass through the same point. The point of intersection has an x-value of -1.

The next day, Ms. Rodriguez began the class by asking each student to choose a linear function and apply Shelly's technique to it. She told the students that each person would be responsible for showing why Shelly's method did or did not work for his or her function. As the students worked on symbolizing what they had done, questions arose such as, What should I write down? I know I want to show that they all intersect in the same place, but how do I show that? Ms. Rodriguez suggested that students work together in small groups. At some point, each group realized that the solution involved finding the point of intersection of the graphs of two equations—the one they started with and a second one. Karen produced the following argument:

> My equation was $y = 3x + 5$. I moved the sliders about two units to the right. The equation for the new line had m and b two units bigger than the original. So the new line was $y = 5x + 7$. I found the intersection of those two lines and got $2x + 2 = 0$, so the intersection is at $x = -1$ and $y = 2$. [Karen wrote the following on the chalkboard.]

$$
\begin{aligned}
3x + 5 &= 5x + 7 \\
0 &= 2x + 2 \\
-2 &= 2x \\
-1 &= x
\end{aligned}
$$

Pao announced excitedly that Karen's approach worked even for an equation as complicated as $y = -3.23x + 4.577$. After most of the students had found that their equations intersect at a point with an x-value of -1, Ms. Rodriguez led the class in a discussion of exactly what they knew. Mike argued that since they had chosen their equations randomly, the result must be true in general. Pete said that even though they had looked at several equations, their results were not sufficient to prove the conjecture for every equation. Pat added that they also hadn't tried every possible movement of m.

The discussion of this problem continued until the class generated a general proof. With Ms. Rodriguez's guidance, the students worked to represent the result of the simultaneous slider shift of a linear function written in the form $y = mx + b$ as $y = (m + k)x + (b + k)$. The students determined the point of intersection by solving the system of equations:

$$
\begin{aligned}
y &= mx + b \\
y &= (m + k)x + (b + k)
\end{aligned}
$$

They then confirmed that every equation of the form $y = (m + k)x + (b + k)$ passes through the point $(-1, b - m)$.

Ms. Rodriguez may well have felt very satisfied at the end of this episode, since her class had had an opportunity to explore a striking graphical pattern observed by a student. The exploration had included considering whether it would always work and if so, why. Pursuing the answer to that question helped her students see the linkage between algebraic and graphical representations of functions and understand how algebraic arguments can be used to establish the truth of a mathematical conjecture. It gave them an opportunity to review and apply previously learned knowledge of solving systems of equations. She will have to revise her instructional plan for the coming weeks. However, this experience has helped reinforce her students' dispositions toward posing and solving problems as well as their algebraic fluency. Indeed, the situation may have helped her students realize that mathematical situations can be interesting and worthy of exploration. Thus problem posing and problem solving led to a deeper understanding of both content and process.

Teaching is itself a problem-solving activity.

What should be the teacher's role in developing problem solving in grades 9 through 12?

As Halmos (1980) writes, problem solving is the "heart of mathematics." Successful problem solving requires knowledge of mathematical content, knowledge of problem-solving strategies, effective self-monitoring, and a productive disposition to pose and solve problems. Teaching problem solving requires even more of teachers, since they must be able to foster such knowledge and attitudes in their students. A significant part of a teacher's responsibility consists of planning problems that will give students the opportunity to learn important content through their explorations of the problems and to learn and practice a wide range of heuristic strategies. The teacher must be courageous, for even well-planned lessons can veer into uncharted territory. Students may make novel suggestions as they try to solve problems; they may make observations that give rise to new conjectures or explorations; they may suggest generalizations whose validity may be unknown to the teacher. Teachers must exercise judgment in deciding what responses to pursue and recognize the potential for both productive learning and improved attitudes when students generate new ideas, but they must also acknowledge that not all responses lead to fruitful discussions and that time constraints do not allow them to pursue every interesting idea. It is the teacher's job to make the tough calls. The teacher must also be reflective in order to create an environment in which students are inclined to reflect on their work as they engage in it (Thompson 1992). In short, teaching is itself a problem-solving activity. Effective teachers of problem solving must themselves have the knowledge and dispositions of effective problem solvers.

Reasoning and Proof Standard for Grades 9–12

Instructional programs from prekindergarten through grade 12 should enable all students to—

- Recognize reasoning and proof as fundamental aspects of mathematics
- Make and investigate mathematical conjectures
- Develop and evaluate mathematical arguments and proofs
- Select and use various types of reasoning and methods of proof

Mathematics should make sense to students; they should see it as reasoned and reasonable. Their experience in school should help them recognize that seeking and finding explanations for the patterns they observe and the procedures they use help them develop deeper understandings of mathematics. As illustrated throughout this chapter, opportunities for mathematical reasoning and proof pervade the high school curriculum. Students should develop an appreciation of mathematical justification in the study of all mathematical content. In high school, their standards for accepting explanations should become more stringent, and they should develop a repertoire of increasingly sophisticated methods of reasoning and proof.

What should reasoning and proof look like in grades 9 through 12?

Reasoning and proof are not special activities reserved for special times or special topics in the curriculum but should be a natural, ongoing part of classroom discussions, no matter what topic is being studied. In mathematically productive classroom environments, students should expect to explain and justify their conclusions. When questions such as, What are you doing? or Why does that make sense? are the norm in a mathematics classroom, students are able to clarify their thinking, to learn new ways to look at and think about situations, and to develop standards for high-quality mathematical reasoning (Collins et al. 1989).

Consider the following hypothetical classroom scenario:

> Mr. Hamilton's class at Manorville High School has established e-mail contact with a high school class in Osaka, Japan. The two groups of students begin by gathering and sharing information about their classes. They exchange data about the number of people in each student's family, about how far each student lives from the school, and about the number and type of pets each student has. The Japanese students have heard that houses and apartments in North America are very large, and they want to compare the living areas of the students in Manorville with their own. Each class makes a list of the floor areas in their families' houses or apartments. They compute the mean, the median, the mode, the range, and the standard deviation for their data and share them electronically.

When the data arrive from their Japanese friends, Mr. Hamilton's students realize that they will have to do a bit more work before a comparison can be made. It has not occurred to them that the information from the Japanese students would be reported in square meters whereas all their measurements are recorded as square feet. At first Angela thinks that they will need to get the original data about the Japanese room sizes, convert the measurements to feet, and recompute the size of each living area before they can determine the statistics. Shanika points out that a spreadsheet would do the conversions quickly but thinks that they will still have to ask for the original data. Mr. Hamilton suggests that before asking the Japanese students to enter all those extra data, they might work with the summary statistics they have and see if they can find a way to compare them directly with their own.

Reasoning and proof are not special activities reserved for special times in the curriculum.

The class decides to focus first on the mode. They know that the mode of the areas corresponds to the measurement of an actual living area. They think that if they figure out how to convert that value from square meters to square feet, they might be able to get started on the other statistics. Jacob observes that a meter is roughly equal to 3.3 feet. He proposes that they multiply the mode of the Japanese data by 3.3 to convert square meters to square feet. Mr. Hamilton asks the class if they agree. Several students nod, but Shanika objects. She points out that they aren't thinking about the fact that the mode is given in square meters and that 1 square meter is equal to about 10.9 square feet. Angela says she doesn't understand where that number comes from so Shanika draws her a diagram (see fig. 7.32) and explains. "See, suppose this is 1 meter by 1 meter. We could make it 3.3 feet by 3.3 feet, then when we multiply to find the area, we get about 10.9 square feet." Her reasoning makes sense to the rest of the class, so they convert the mode value.

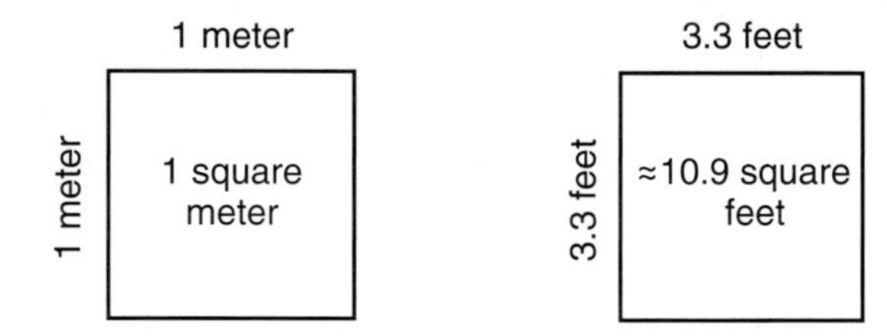

Fig. **7.32.**
Shanika's diagram for explaining the conversion from square meters to square feet

Mr. Hamilton asks them to think about which of the other statistics could be converted in the same way. Angela says that they can use the same method to convert the median. At that point, two more students join the conversation.

Chuen: You can do it with the average of two numbers. The first number would get multiplied by 10.9, and so would the second, then you average them. But if you just factor out the 10.9, you get the average of the two numbers you had times 10.9.

Robert: If it works when we average two numbers that way, it ought to work when we average more numbers.

Chuen tries the mean of three numbers and announces that the method works in that case also. Mr. Hamilton asks the class to show it would always work. With some help, they argue that for any values x_1, x_2, and x_3,

$$10.9x_1 + 10.9x_2 + 10.9x_3 = 10.9(x_1 + x_2 + x_3).$$

Shanika then observes, "It doesn't matter how many x's you have. You can always factor out the 10.9."

The discussion of standard deviation is similar. Damon says he tried the method with some simple numbers and it worked, so he thought it should be true. "OK," says Mr. Hamilton. "So we think it's likely to be true. But how do we know it will be? Are there any hints in anything we have written down?" Damon then suggests that they write out the formula for standard deviation and replace all the values of x with $10.9x$. It takes a while to work through the details, but the class ultimately shows that if every number in a data set is multiplied by a constant, the standard deviation of the resulting data set equals the standard deviation of the original data set multiplied by the same constant.

An important point in this example is that reasoning and proof enabled students to abstract and codify their observations. Chuen's initial observation was that if each of two numbers is multiplied by 10.9, the mean of the resulting numbers is 10.9 times the mean of the original numbers. The reasoning he used ultimately produced the argument that if every number in a data set is multiplied by a constant, the mean of the resulting data set equals the mean of the original data set multiplied by that constant. The fact that this argument could be made algebraically furnished a mechanism for making a similar argument about the standard deviation. In this way, results with similar justifications can emerge.

Sometimes developing a proof is a natural way of thinking through a problem.

Sometimes developing a proof is a natural way of thinking through a problem. For example, a teacher posed the problem of finding four consecutive integers whose sum is 44. The students tried the task and decided it was impossible. The teacher responded, "OK, so you couldn't find the integers. How do you know that someone else won't be able to find them?" The students worked quietly for a few minutes, and one student offered, "Look, if you call the first number n, the next three are $n + 1$, $n + 2$, and $n + 3$. Add those four numbers and set them equal to 44. You get $4n + 6 = 44$, and the solution to that equation is $n = 9\ 1/2$. So no whole number does it." Here the proof works nicely to explain why something is impossible.

The habit of asking why is essential for students to develop sound mathematical reasoning. In one class, imagine a student wants to divide an 8 1/2 inch × 11 inch sheet of paper into three columns of equal width. The student is ready to measure off lengths of 2 5/6 inches, but the teacher says, "Let me show you a carpenter's trick." He places a 12-inch ruler at an angle on the page so that the 0-inch and 12-inch marks on the ruler are on the left- and right-hand edges, respectively, and makes marks at the 4- and 8-inch points on the ruler. He then repeats the procedure, with the ruler farther down the page. Drawing lines through the 4-inch marks and the 8-inch marks divides the page neatly into three equal parts. The teacher then says, "Carpenters use this trick to divide boards into thirds (see fig. 7.33). My questions to you are,

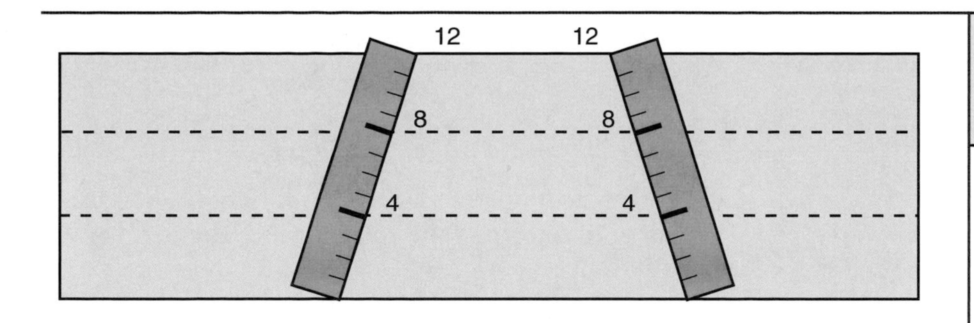

Fig. **7.33.**

A carpenter's method for trisecting a board

Why does it work? Can you find similar procedures to divide a board into four, five, or any number of equal parts?

The repertoire of proof techniques that students understand and use should expand through the high school years. For example, they should be able to make direct arguments to establish the validity of a conjecture. Such reasoning has long been at the heart of Euclidean geometry, but it should be used in all content areas: consider, for example, the number-theory arguments discussed in the "Number and Operations" section of this chapter or the arguments given about the effect on some statistics of a data set of multiplying every element in the data set by a given constant. Students should understand that having many examples consistent with a conjecture may suggest that the conjecture is true but does not prove it, whereas one counterexample demonstrates that a conjecture is false. Students should see the power of deductive proofs in establishing results. They should be able to produce logical arguments and present formal proofs that effectively explain their reasoning, whether in paragraph, two-column, or some other form of proof.

Because conjectures in some situations are not conducive to direct means of verification, students should also have some experience with indirect proofs. And since iterative and recursive methods are increasingly common (see, e.g., the "drug dosage" problem discussed in the "Algebra" section in this chapter), students should learn that certain types of results are proved using the technique of mathematical induction.

Students should reason in a wide range of mathematical and applied settings. Spatial reasoning gives insight into geometric results, especially in two- and three-dimensional geometry. Probabilistic reasoning is helpful in analyzing the likelihood that an event will occur. Statistical reasoning allows students to assess risks and make generalizations about a population by using representative samples drawn from that population. Algebra is conducive to symbolic reasoning. Students who can use many types of reasoning and forms of argument will have resources for more-effective reasoning in everyday situations.

What should be the teacher's role in developing reasoning and proof in grades 9 through 12?

To help students develop productive habits of thinking and reasoning, teachers themselves need to understand mathematics well (Borko and Putnam 1996). Through the classroom environments they create, mathematics teachers should convey the importance of knowing the reasons for mathematical patterns and truths. In order to evaluate the validity of proposed explanations, students must develop enough confidence in their reasoning abilities to question others' mathematical arguments as well as their own. In this way, they rely more on logic than on external authority to determine the soundness of a mathematical argument.

As in other grades, teachers of mathematics in high school should strive to create a climate of discussing, questioning, and listening in their classes. Teachers should expect their students to seek, formulate, and critique explanations so that classes become communities of inquiry. Teachers should also help students discuss the logical structure of their own arguments. Critiquing arguments and discussing conjectures are delicate matters: plausible guesses should be discussed even if they turn out to be wrong. Teachers should make clear that the ideas are at stake, not the students who suggest them. With guidance, students should develop high standards for accepting explanations, and they should understand that they have both the right and the responsibility to develop and defend their own arguments. Such expectations were visible in the Manorville High School episode: Informal reasoning and a few calculations suggested to students how a summary statistic given in one unit seemed to be related to the same statistic given in a different unit. In that classroom, however, informal reasoning and supporting examples were a starting point rather than an end point. In a supportive environment, students were encouraged to furnish a carefully reasoned argument for verifying their conjecture that would meet the standards of the broader mathematics community.

Communication Standard for Grades 9–12

Instructional programs from prekindergarten through grade 12 should enable all students to—

Organize and consolidate their mathematical thinking through communication

Communicate their mathematical thinking coherently and clearly to peers, teachers, and others

Analyze and evaluate the mathematical thinking and strategies of others

Use the language of mathematics to express mathematical ideas precisely

Changes in the workplace increasingly demand teamwork, collaboration, and communication. Similarly, college-level mathematics courses are increasingly emphasizing the ability to convey ideas clearly, both orally and in writing. To be prepared for the future, high school students must be able to exchange mathematical ideas effectively with others.

However, there are more-immediate reasons for emphasizing mathematical communication in high school mathematics. Interacting with others offers opportunities for exchanging and reflecting on ideas; hence, communication is a fundamental element of mathematics learning. For that reason, it plays a central role in all the classroom episodes in this chapter. Sharing ideas and building on the work of others were essential ingredients in the ability of Ms. Rodriguez's class to provide an analytic explanation of a surprising visual result (see the discussion in the "Problem Solving" section). Communication was also important in the episode about converting the unit of measurement for various statistics in the "Reasoning and Proof" section where informal observations led to discussions of specific cases that were ultimately abstracted and proved as general results. Making connections among the various geometric examples in the upcoming "Connections" section depends heavily on the exchange of information. Students' written work is valuable for assessment, as readers will see in the discussion in the "Representation" section, where the students' incorrect distance-versus-time graph gives the teacher insights into their misconceptions. In all these examples, the act of formulating ideas to share information or arguments to convince others is an important part of learning. When ideas are exchanged and subjected to thoughtful critiques, they are often refined and improved (Borasi 1992; Moschkovich 1998). In the process, students sharpen their skills in critiquing and following others' logic. As students develop clearer and more-coherent communication (using verbal explanations and appropriate mathematical notation and representations), they will become better mathematical thinkers.

What does communication look like in grades 9 through 12?

In high school, there should be substantial growth in students' abilities to structure logical chains of thought, express themselves coherently and clearly, listen to the ideas of others, and think about their audience when they write or speak. The relationships students wish to express symbolically and with graphs, as well as the notation and representations for expressing them, should become increasingly sophisticated. Consequently, communication in grades 9–12 can be distinguished from that in lower grades by higher standards for oral and written exposition and by greater mathematical sophistication.

High school students should be good critics and good self-critics. They should be able to generate explanations, formulate questions, and write arguments that teachers, coworkers, or mathematicians would consider logically correct and coherent. Whether they are making their points using spreadsheets, geometric diagrams, natural language, or algebraic symbols, they should use mathematical language and symbols correctly and appropriately. Students also should be good collaborators who work effectively with others.

Proofs should be a significant part of high school students' mathematical experience, as well as an accepted method of communication. The following episode, drawn from real experience, illustrates the part that mathematical communication plays in the development of increased understanding.

To be prepared for the future, high school students must be able to exchange mathematical ideas effectively with others.

> Marta and Nancy attended the biweekly mathematics competition in their school district's mathematics league and encountered the following problem:
>
> > A string is stretched corner to corner on a floor tiled with square tiles. If the floor is 28 tiles long and 35 tiles wide, over how many tiles does the string pass?
>
> They didn't solve the problem, but they reported to their faculty sponsor, Ms. Koech, that they had heard a fellow contestant say that he got the answer by adding the two sides and subtracting their greatest common divisor. Not content with a formula "dropped from the sky," Ms. Koech encouraged the girls to see if this method always works and if so, why. Over a period of days the girls developed some intuitions about the problem and tried various ways of sharing them with their teacher. Marta, who had a broken leg, demonstrated one emerging insight by dragging her cast along the tiles on the classroom floor.
>
> *Marta:* See? The string crosses everything!
>
> *Ms. Koech*: I'm not sure I know what you mean.
>
> *Nancy*: Watch this part of the floor (*she outlines a 3 × 2 tile pattern like that shown and points to the points of intersection of the imagined string and the edges of the tiles*). Here it crosses. Here it crosses. It hits the lines five times—and that comes from the 2 and the 3! (*Nancy has counted the number of points of intersection of the string with the vertical and horizontal line segments and notices*

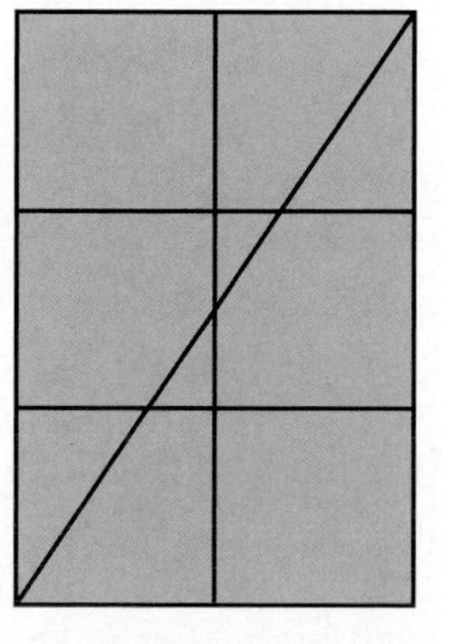

that the total is the same as the sum of 2 and 3.)

Ms. Koech: That's interesting! Have you seen that happen in other cases?

Marta: Yes! It works for most small numbers.

Ms. Koech: What do you mean—most?

Marta: Well, it didn't work for this one. *(Marta shows Ms. Koech a drawing of a 3 × 6 tiling pattern for which there are seven points of intersection.)*

Nancy: Why is this one different?

Ms. Koech suggested that the students pursue Nancy's question. Over a period of days they determined that the relationship breaks down when the string passes through a corner of a tile "inside" the pattern. Gradually and after pursuing some approaches that led nowhere, they realized that whenever the number of intersections was the same as the sum of the dimensions of the rectangle, the dimensions of the room are relatively prime. Thinking back to Marta's cast clicking over the tiles—each time it clicked, she touched another tile—they realized that they could trace along the string and see how many times the string "exits" a tile. Except at the start of the path, the string exits a tile each time it crosses either a vertical or a horizontal line segment. Moreover, the string has to cross every vertical and horizontal line segment as it traverses the tiled room from one corner to the other. So the number of tiles the string crosses can be determined by the number of horizontal and vertical line segments in the tile configuration. After a number of attempts—first for specific cases such as the 3 × 2 and 3 × 4 configurations and then in general—the students were able to show that if n and m (the dimensions of the rectangle) are relatively prime, the string passes over $(m + n - 1)$ tiles.

With this result established, they returned to the general problem. A close look at the 9 × 6 tile configuration (see fig. 7.34) shows that the area the string passes through can be considered as three 3 × 2 configurations. Thus the string passes over three times as many tiles in the 9 × 6 configuration as it passes over in the 3 × 2 configuration. This happens to be $3(2 + 3 - 1) = 6 + 9 - \gcd(6, 9)$ tiles, which conforms with the method they had heard from their fellow contestant. From this analysis, the generalization to $m + n - \gcd(m, n)$ was a natural next step. The students wrote up their results for a regional student mathematics publication.

Fig. **7.34.**

A 9 × 6 tile configuration

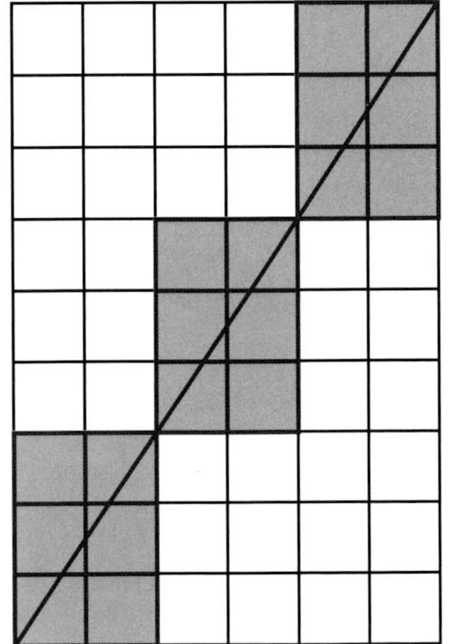

A formal explanation follows: Suppose n and m have a gcd of g. The smaller, relatively prime grid is $(n/g) \times (m/g)$. The students discovered that the number of grid lines crossed there is $(n/g) + (m/g) - 1$, which indicates that when a string passes through g of them, it passes through $g[(n/g) + (m/g) - 1] = n + m - g$ grid lines.

In this example, communication serves at least two purposes. First, it is motivational: Marta and Nancy kept working at the problem in part because it was a collaborative effort, and they were discussing their work. Second, repeated attempts to explain their reasoning to each other and to their teacher helped Marta and Nancy clarify their thinking and focus on essential elements of the problem. Notice that the teacher offered some key observations and questions: "I'm not sure what you mean." "Have you seen this happen in other cases?" "What do you mean, most?" Nancy then asked the question that launched the girls' exploration: "Why is this one different?" This focus was a major factor in their success in eventually solving the problem.

What should be the teacher's role in developing communication in grades 9 through 12?

High school teachers can help students use oral communication to learn and to share mathematics by creating a climate in which all students feel safe in venturing comments, conjectures, and explanations. Teachers must help students clarify their statements, focus carefully on problem conditions and mathematical explanations, and refine their ideas. Orchestrating classroom conversations so that the appropriate level of discourse and mathematical argumentation is maintained requires that teachers know mathematics well and have a clear sense of their mathematical goals for students. Teachers should help students become more precise in written mathematics and have them learn to read increasingly technical text. In both written and oral communication, teachers need to attend to their students and carefully interpret what the students know from what they say or write.

Teachers should help students learn to read increasingly technical text.

Communication can be used in many ways as a vehicle for assessment and learning. Early in a mathematics course or when a new topic is being introduced, teachers can request information about the students' knowledge of the topic. At the beginning of a tenth-grade unit on circles, for example, a teacher might ask students to tell her everything they know about circles. The teacher could then compile the responses, distribute copies of them the next day, and ask the students to agree or disagree with each of the statements. Students should justify the stance they took. The teacher could look at students' incorrect observations and design a lesson to address those misconceptions. In this way, the students' knowledge becomes a starting point for instruction, and the teacher can establish the idea that the students are expected to have reasons for their mathematical opinions. Activities such as this can serve to establish a

classroom climate conducive to the respectful exchange of ideas. More generally, having students present work to the class at the chalkboard, overhead projector, or flip chart—and having the class respond to what is presented rather than having only the teacher judge the correctness of what has been said—can be a valuable way to foster classroom communication. Over time, such activities help students sharpen their ideas through attempts to communicate orally and in writing.

There are various ways that teachers can help students communicate effectively using written mathematics. Problems that require explanations can be assigned regularly, and the class can discuss and compare the adequacy of those explanations. The following exercises can help students sharpen their mathematical writing skills:

- Imagine you are talking to a student in your class on the telephone and want the student to draw some figures. The other student cannot see the figures. Write a set of directions so that the other student can draw the figures exactly as shown [in figure 7.35]. (California State Department of Education 1989, p. 7)
- Suppose you are hired as a consultant to help a business choose between two options (e.g., which taxicab company is the better one to use or which telephone plan is a better buy). Write a memo saying which option is better and why. (For a similar activity, see Balanced Assessment for the Mathematics Curriculum [2000, pp. 16–17].)
- Make a design plan for a dog house. The dog house will be made out of wood cut from a 4' by 8' sheet of plywood, and its volume should be reasonably large. Explain why you made the choices you have made in putting together the design plan (adapted from National Research Council [1989, p. 32]).

Such exercises also serve as good assessment devices that can help teachers understand students' thinking.

Fig. **7.35.**

Figures to be described orally so they can be reproduced

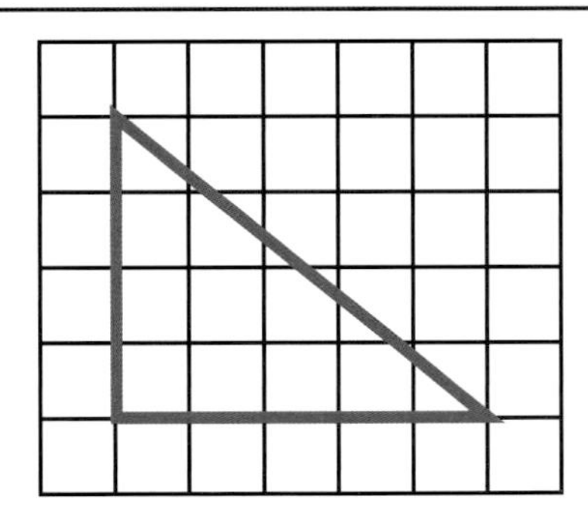

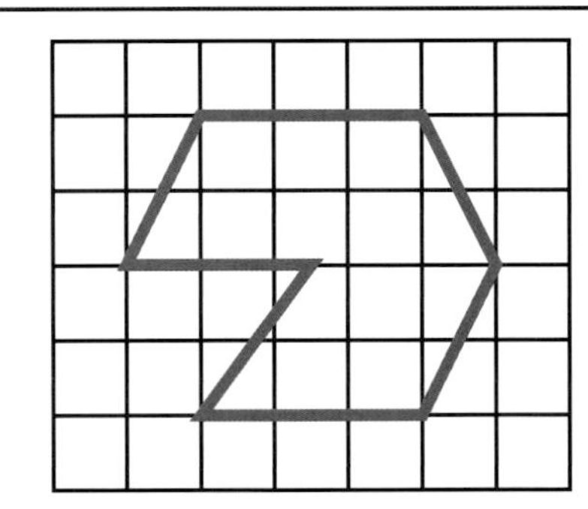

Writing is a valuable way of reflecting on and solidifying what one knows, and several kinds of exercises can serve this purpose. For example, teachers can ask students to write down what they have learned about a particular topic or to put together a study guide for a student who was absent and needs to know what is important about the topic. Students who have done a major project or worked on a substantial long-range problem can be asked to compare some of their early work with later work and explain how the later work reflects greater understanding. In these ways, teachers can help students develop skills in mathematical communication that will serve them well both inside and outside the classroom. Using these skills will in turn help students develop deeper understandings of the mathematical ideas about which they speak, hear, read, and write.

Connections Standard for Grades 9–12

Instructional programs from prekindergarten through grade 12 should enable all students to—

Recognize and use connections among mathematical ideas

Understand how mathematical ideas interconnect and build on one another to produce a coherent whole

Recognize and apply mathematics in contexts outside of mathematics

When students can see the connections across different mathematical content areas, they develop a view of mathematics as an integrated whole. As they build on their previous mathematical understandings while learning new concepts, students become increasingly aware of the connections among various mathematical topics. As students' knowledge of mathematics, their ability to use a wide range of mathematical representations, and their access to sophisticated technology and software increase, the connections they make with other academic disciplines, especially the sciences and social sciences, give them greater mathematical power.

What should connections look like in grades 9 through 12?

Students in grades 9–12 should develop an increased capacity to link mathematical ideas and a deeper understanding of how more than one approach to the same problem can lead to equivalent results, even though the approaches might look quite different. (See, e.g., the "counting rectangles" problem in the "Problem Solving" section in this chapter.) Students can use insights gained in one context to prove or disprove conjectures generated in another, and by linking mathematical ideas, they can develop robust understandings of problems.

The following hypothetical example highlights the connections among what would appear to be very different representations of, and approaches to, a mathematical problem.

> The students in Mr. Robinson's tenth-grade mathematics class suspect they are in for some interesting problem solving when he starts class with this story: "I have a dilemma. As you may know, I have a faithful dog and a yard shaped like a right triangle. When I go away for short periods of time, I want Fido to guard the yard. Because I don't want him to get loose, I want to put him on a leash and secure the leash somewhere on the lot. I want to use the shortest leash possible, but wherever I secure the leash, I need to make sure the dog can reach every corner of the lot. Where should I secure the leash?"
>
> After Mr. Robinson responds to the usual array of questions and comments (such as "Do you really have a dog?" "Only a math teacher would have a triangle-shaped lot—or notice that the lot was

triangular!" "What type of dog is it?"), he asks the students to work in groups of three. All their usual tools, including compass, straightedge, calculator, and computer with geometry software, are available. They are to come up with a plan to solve the problem.

Jennifer dives into the problem right away, saying, "Let's make a sketch using the computer." With her group's agreement, she produces the sketch in figure 7.36.

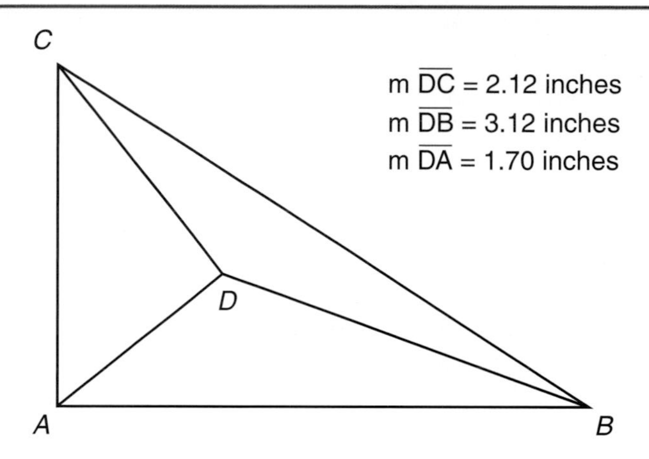

Fig. **7.36.**

Jennifer's computer-drawn sketch of the "dog in the yard" problem

As Mr. Robinson circulates around the room, he observes each group long enough to monitor its progress. On his first pass, Jennifer's group seems to be experimenting somewhat randomly with dragging the point D to various places, but on his second pass, their work seems more systematic. To assess what members of the group understand, he asks how they are doing:

Mr. R: Joe, can you bring me up-to-date on the progress of your group?

Joe: We're trying to find out where to put the point.

Jeff: We don't want the point too close to the corners of the triangle.

Jennifer: I get it! We want all the lengths to be equal! They all work against each other.

When students can see the connections across content areas, they develop a view of mathematics as an integrated whole.

Before moving on to work with other groups, Mr. Robinson works with the members of Jennifer's group on clarifying their ideas, using more-standard mathematical language, and checking with one another for shared understanding. Jennifer clarifies her idea, and the group decides that it seems reasonable. They set a goal of finding the position for D that results in the line segments DA, DB, and DC all being the same length. When Mr. Robinson returns, the group has concluded that point D has to be the midpoint of the hypotenuse, otherwise, they say, it could not be equidistant from B and C. (Mr. Robinson notes to himself that the group's conclusion is not adequately justified, but he decides not to intervene at this point; the work they will do later in creating a proof will ensure that they examine this reasoning.)

Mr. R: What else would you need to know?

Jeff: We're not sure yet whether D is the same distance from all three vertices.

Jennifer: It has to be! At least I think it is. It looks like it's the center of a circle.

Small-group conversations continue until several groups have made observations and conjectures similar to those made in Jennifer's group. Mr. Robinson pulls the class back together to discuss the problem. When the students converge on a conjecture, he writes it on the board as follows:

> *Conjecture*: The midpoint of the hypotenuse of a right triangle is equidistant from the three vertices of the triangle.

He then asks the students to return to their groups and work toward providing either a proof or a counterexample. The groups continue to work on the problem, settling on proofs and selecting group members to present them on the overhead projector. As always, Mr. Robinson emphasizes the fact that there might be a number of different ways to prove the conjecture.

Remembering Mr. Robinson's mantra about placing the coordinate system to "make things eeeasy," one group places the coordinates as shown in figure 7.37a, yielding a common distance of $\sqrt{a^2+b^2}$. Alfonse, who is explaining this solution, proudly remarks that it reminds him of the Pythagorean theorem. Mr. Robinson builds on that observation, noting to the class that if the students drop a perpendicular from M to AC, each of the two right triangles that result has legs of length a and b; thus the length of the hypotenuses, MC and MA, are indeed $\sqrt{a^2+b^2}$.

Jennifer's group returns to her earlier comment about the three points A, B, and C being on a circle. After lengthy conversations with, and questions from, Mr. Robinson, that group produces a second proof based on the properties of inscribed angles (fig. 7.37b).

Fig. **7.37.**

Diagrams corresponding to four proofs of the midpoint-of-hypotenuse theorem

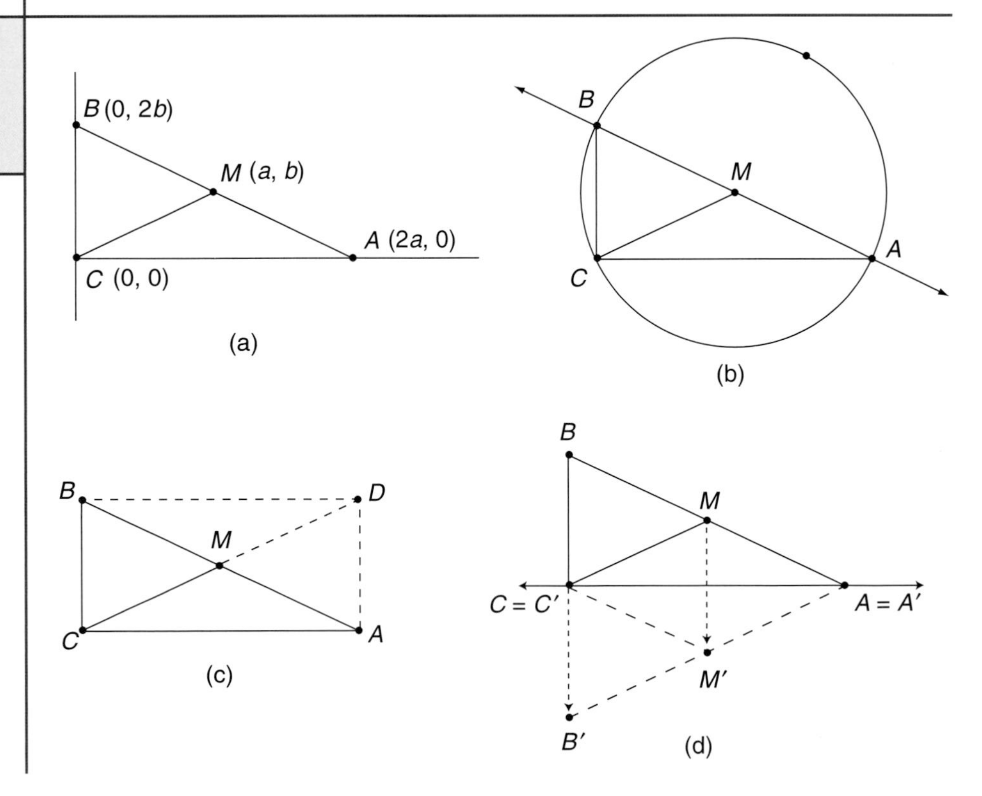

Pedro presents his group's solution showing how they constructed a rectangle that includes the three vertices of the right triangle (fig. 7.37c) and reasoned about the properties of the diagonals of a rectangle. Anna presents a solution using transformational geometry (figure 7.37d). Since M and M' are the midpoints of $\overline{AB}$ and $\overline{A'B'}$, respectively, the triangle MAM' is similar to the triangle BAB', with each of the sides of the smaller triangle half the length of the corresponding side of the larger triangle. The same relationship holds for triangles BMC and BAB'. Using this fact and the fact that BAB′ is isosceles (since $\overline{BA}$ reflects onto $\overline{B'A}$), Anna shows that triangle MAM' is congruent to triangle CMB, from which it follows that CM and MA are the same length.

Mr. Robinson congratulates the students on the quality of their work and on the variety of approaches they used. He points out that some basic mathematical ideas such as congruence were actually part of the mathematics in a number of their solutions and that some of their thinking, such as Alfonse's comment about the Pythagorean theorem, highlighted connections to other mathematical ideas. Taking a step backward to reflect, the students begin to see how different approaches—using coordinate geometry, Euclidean geometry, and transformational geometry—are all connected. Mr. Robinson notes that it is good to have all these ways of thinking in their mathematical "tool kit." Any one of them might be the key to solving the next problem they encounter.

Although the students learned a great deal from working on the problem, the class was not yet finished with it. Mr. Robinson had selected this problem for the class to work on because it supports a number of interesting explorations and because the students would be exploring the properties of triangles and circles as they worked on it. And, indeed, as the students worked on the problem, they remarked that they were "seeing circles everywhere." (The following discussion is inspired by Goldenberg, Lewis, and O'Keefe [1992].)

Fig. **7.38.**

Dynamic representations of right triangles (from Goldenberg, Lewis, and O'Keefe [1992, p. 257])

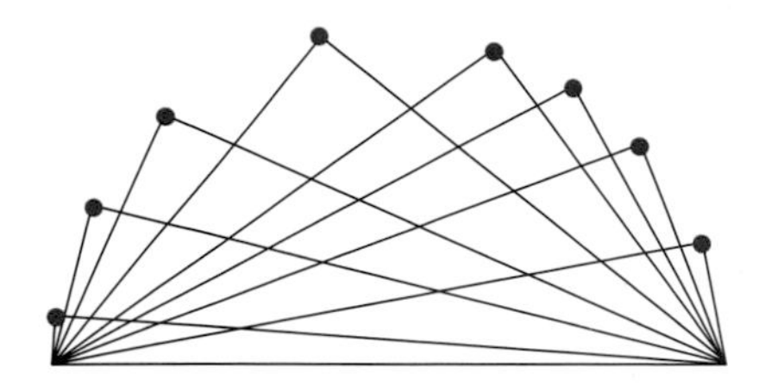

(a) A trace of the locus of the vertices of right angles with a common hypotenuse

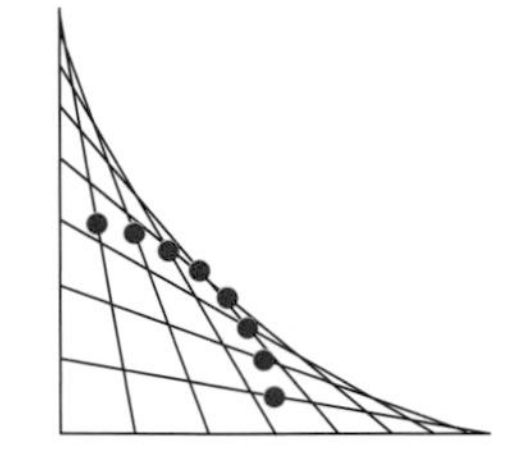

(b) A trace of the locus of midpoints of hypotenuses of fixed length in triangles with a common right angle

One group decides to look at the set of all the right triangles they can find, given a fixed hypotenuse. A group member starts by constructing a right triangle with the given hypotenuse and then dragging the right angle (fig. 7.38a). Another group decides to fix the position of the right angle and look at the set of right triangles whose hypotenuses are the given length (fig. 7.38b). They observe that the plot of the midpoints of the hypotenuses of the right triangles appears to trace out the arc of a circle. At first the students are ready to dismiss the circular pattern as a coincidence. But Mr. Robinson, seeing the potential for making a connection, asks questions such as, "Why do you think you get that pattern?" and "Does the circle in your pattern have anything to do with the circle in Jennifer's group's solution?" As the groups begin to understand Mr. Robinson's questions, they begin to see the connections among the circles in their new drawings, the definition of a circle, and the fact that their problem deals with points that are equally distant from a third point.

Mr. Robinson adds a final challenge for homework: can the students connect this problem (or problems related to it) to real-world situations or to other mathematics? The students create posters illustrating the mathematical connections they see. Most of the posters depict situations similar to the original problem in which something, for some reason, needs to be positioned the same distance from the vertices of a right triangle. One group, however, creates an experiment that they demonstrate for the class in one of the dark, windowless rooms in the building. They put on the floor a large sheet of white chart paper with a right triangle drawn on it, place candles (all of the same height) at each vertex, and stand an object shorter than the candles inside the triangle. The class watches the shadows of the object change as one of the group members moves it around inside the triangle. The three shadows are of equal length only when the object is placed at the midpoint of the hypotenuse—a phenomenon that delights both Mr. Robinson and his students. This activity concludes the discussion of right triangles, but it is far from the end of the class's work. Mr. Robinson reminds the students of the problem that started their discussion and asks them how the problem might be extended. "After all," he says, "not all backyards have right angles or are triangular in shape." This comment sets the stage for abstracting and generalizing some of their work—and for making more connections.

What should be the teacher's role in developing connections in grades 9 through 12?

The story of Mr. Robinson's classroom indicates many of the ways in which teachers can help students seek and make use of mathematical connections. Problem selection is especially important because students are unlikely to learn to make connections unless they are working on problems or situations that have the potential for suggesting such linkages. Teachers need to take special initiatives to find such integrative problems when instructional materials focus largely on content areas and when curricular arrangements separate the study of content areas such as geometry, algebra, and statistics. Even when curricula offer problems that cut across traditional content boundaries, teachers will need to develop expertise in making mathematical connections and in helping students develop their own capacity for doing so.

One essential aspect of helping students make connections is establishing a classroom climate that encourages students to pursue mathematical ideas in addition to solving the problem at hand. Mr. Robinson started with a problem that allowed for multiple approaches and solutions. While the students worked the problem, they were encouraged to pursue various leads. Incorrect statements weren't simply judged wrong and dismissed; Mr. Robinson helped the students find the kernels of correct ideas in what they had said, and those ideas sometimes led to new solutions and connections. The students were encouraged to reflect on and compare their solutions as a means of making connections. When they had done just about everything they were able to do with the given problem, they were encouraged to generalize what they had done. Rich problems, a climate that supports mathematical thinking, and access to a wide variety of mathematical tools all contribute to students' ability to see mathematics as a connected whole.

Rich problems, a climate that supports mathematical thinking, and access to mathematical tools contribute to students' seeing connections.

Representation Standard for Grades 9–12

Instructional programs from prekindergarten through grade 12 should enable all students to—

Create and use representations to organize, record, and communicate mathematical ideas

Select, apply, and translate among mathematical representations to solve problems

Use representations to model and interpret physical, social, and mathematical phenomena

If mathematics is the "science of patterns" (Steen 1988), representations are the means by which those patterns are recorded and analyzed. As students become mathematically sophisticated, they develop an increasingly large repertoire of mathematical representations and the knowledge of how to use them productively. This knowledge includes choosing specific representations in order to gain particular insights or achieve particular ends.

The importance of representations can be seen in every section of this chapter. If large or small numbers are expressed in scientific notation, their magnitudes are easier to compare and they can more readily be used in computations. Representation is pervasive in algebra. Graphs convey particular kinds of information visually, whereas symbolic expressions may be easier to manipulate, analyze, and transform. Mathematical modeling requires representations, as illustrated in the "drug dosage" problem and in the "pipe offset" problem. The use of matrices to represent transformations in the plane illustrates how geometric operations can be represented visually yet also be amenable to symbolic representation and manipulation in a way that helps students understand them. The various methods for representing data sets further demonstrate the centrality of this topic.

A wide variety of representations can be seen in the examples in this chapter. By using various representations for the "counting rectangles" problem in the "Problem Solving" section, students could find different solutions and compare them. The use of algebraic symbolism to explain a striking graphical phenomenon is central to the "string traversing tiles" task in the "Communication" section. Representations facilitate reasoning and are the tools of proof: they are used to examine statistical relationships and to establish the validity of a builder's shortcut. They are at the core of communication and support the development of understanding in Marta's and Nancy's work on the "string traversing tiles" problem. Although at one level the story of Mr. Robinson's class is about connections, at another level it is about representation: one group of students places coordinates that "make things eeeasy," the class gains insights from dynamic representations of geometric objects, and the students produce proofs in coordinate and Euclidean geometry. A major lesson of that story is that different representations support different ways of thinking about and manipulating mathematical objects. An object can be better understood when viewed through multiple lenses.

What should representation look like in grades 9 through 12?

In grades 9–12, students' knowledge and use of representations should expand in scope and complexity. As they study new content, for example, students will encounter many new representations for mathematical concepts. They will need to be able to convert flexibly among these representations. Much of the power of mathematics comes from being able to view and operate on objects from different perspectives.

In elementary school, students most often use representations to reason about objects and actions they can perceive directly. In the middle grades, students increasingly create and use mathematical representations for objects that are not perceived directly, such as rational numbers or rates. By high school, students are working with such increasingly abstract entities as functions, matrices, and equations. Using various representations of these objects, students should be able to recognize common mathematical structures across different contexts. For example, the sum of the first n odd natural numbers, the areas of square gardens, and the distance traveled by a vehicle that starts at rest and accelerates at a constant rate can be represented by functions of the form $f(x) = ax^2$. The fact that these situations can be represented by the same class of functions implies that they are alike in some fundamental mathematical way. Students are ready in high school to see similarity in the underlying structure of mathematical objects that appear contextually different but whose representations look quite similar.

High school students should be able to create and interpret models of more-complex phenomena, drawn from a wider range of contexts, by identifying essential features of a situation and by finding representations that capture mathematical relationships among those features. They should recognize, for example, that phenomena with periodic features often are best modeled by trigonometric functions and that population growth tends to be exponential, or logistic. They will learn

Different representations support different ways of thinking about and manipulating mathematical objects.

to describe some real-world phenomena with iterative and recursive representations.

Consider the graph of the concentration of CO_2 in the atmosphere as a function of time and latitude during the period from 1986 through 1991 (see fig. 7.39) (Sarmiento 1993). Teachers might use an example such as this to help students understand and interpret several aspects of representation. Students could discuss the trends in the change in concentration of CO_2 as a function of time as well as latitude. Doing so would draw on their knowledge about classes of functions and their ability to interpret three-dimensional graphs. They should be able to see a roughly linear increase across time, coupled with a sinusoidal fluctuation with the seasons. Focusing on the change in the character of the graph as a function of latitude, students should note that the amplitude of the sinusoidal function lessens from north to south. Students can test whether the trends they observe in the graph correspond to recent theoretical work on CO_2 concentration in the atmosphere. For example, the author of the article attributes the sinusoidal fluctuation to seasonal variations in the amount of photosynthesis taking place in the terrestrial biosphere. Students could discuss the differences in amplitude across seasons in the Northern and Southern Hemispheres.

Fig. **7.39.**

A three-dimensional graph of the concentration of CO_2 in the atmosphere as a function of time and latitude (Adapted from Sarmiento [1993])

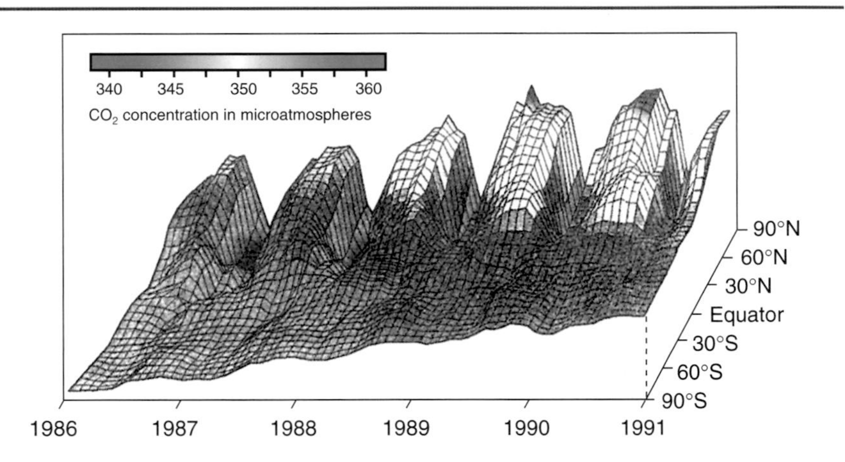

Electronic technologies provide access to problems and methods that until recently were difficult to explore meaningfully in high school. In order to use the technologies effectively, students will need to become familiar with the representations commonly used in technological settings. For example, solving equations or multiplying matrices using a computer algebra system calls for learning how to input and interpret information in formats used by the system. Many software tools that students might use include special icons and symbols that carry particular meaning or are needed to operate the tool; students will need to learn about these representations and distinguish them from the mathematical objects they are manipulating.

What should be the teacher's role in developing representation in grades 9 through 12?

An important part of learning mathematics is learning to use the language, conventions, and representations of mathematics. Teachers should introduce students to conventional mathematical representations and help them use those representations effectively by building on the students' personal and idiosyncratic representations when necessary. It is important for teachers to highlight ways in which different representations of the same objects can convey different information and to emphasize the importance of selecting representations suited to the particular mathematical tasks at hand (Yerushalmy and Schwartz 1993; Moschkovich, Schoenfeld, and Arcavi 1993). For example, tables of values are often useful for quick reference, but they provide little information about the nature of the function represented. Consider the table in the "Algebra" section in this chapter that gives the number of minutes of daylight in Chicago every other day for the year 2000. The values in the table suggest that the function is initially increasing and then becomes decreasing. Knowledge of the context of a graph of those values suggests that the behavior is actually periodic. Similarly, algebraic and graphical representations of functions may provide different information. Some global properties of functions, such as asymptotic behavior or the rate of growth of a function, are often most readily apparent from graphs. But information about specific aspects of a function—the exact value of $f(\pi)$ or exact values of x where $f(x)$ has a maximum or a minimum—may best be determined using an algebraic representation of the function. Suppose $g(x)$ is given by the equation $g(x) = f(x) + 1$, for all x. The analytic definitions of $f(x)$ and $g(x)$ may offer the most-effective ways of computing specific values of $f(x)$ and $g(x)$, but graphing the function reveals that the "shape" of $g(x)$ is precisely the same as that of $f(x)$—that the graph of $g(x)$ is obtained by translating the graph of $f(x)$ one unit upward.

As in all instruction, what matters is what the student sees, hears, and understands. Often, students interpret what teachers may consider wonderfully lucid presentations in ways that are very different from those their teachers intended (Confrey 1990; Smith, diSessa, and Roschelle 1993). Or they may invent representations of content that are idiosyncratic and have personal meaning but do not look at all like conventional mathematical representations (Confrey, 1991; Hall et al. 1989). Part of the teacher's role is to help students connect their personal images to more-conventional representations. One very useful window into students' thinking is student-generated representations. To illustrate this point, consider the following problem (adapted from Hughes-Hallett et al. [1994, p. 6]) that might be presented to a tenth-grade class:

> A flight from SeaTac Airport near Seattle, Washington, to LAX Airport in Los Angeles has to circle LAX several times before being allowed to land. Plot a graph of the distance of the plane from Seattle against time from the moment of takeoff until landing.

Students could work individually or in pairs to produce distance-versus-time graphs for this problem, and teachers could ask them to present and defend those graphs to their classmates. Graphs produced by this class, or perhaps by students in other classes, could be handed out for careful critique and comment. When they perform critiques, students get a considerable amount of practice in communicating mathematics as well as in constructing and improving on representations, and the teacher gets information that can be helpful in assessment. One representation of the flight that a student might produce is shown in figure 7.40.

Part of the teacher's role is to help students connect their personal images to more-conventional representations.

Fig. **7.40.**

A representation that a student might produce of an airplane's distance from its take-off point against the time from takeoff to landing

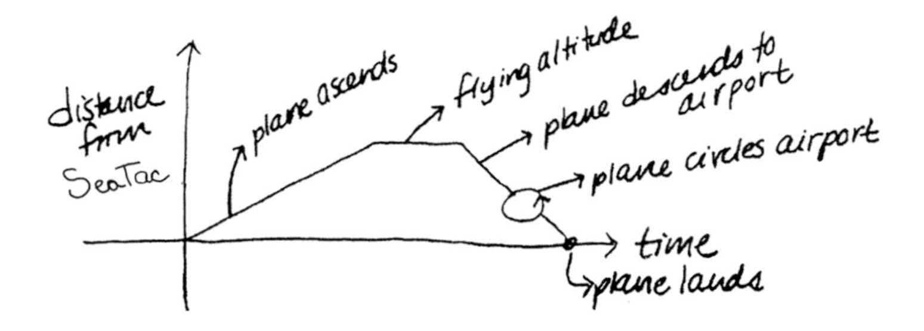

This representation indicates a number of interesting and not uncommon misunderstandings, in which literal features of the story (the plane flying at constant height or circling around the airport) are converted inappropriately into features of the graph (Dugdale 1993; Leinhardt, Zaslavsky, and Stein 1990). Representations of this type can provoke interesting classroom conversations, revealing what the students really understand about graphing. This revelation puts the teacher in a better position to move the class toward a more nearly accurate representation, as sketched in figure 7.41.

Fig. **7.41.**

A more nearly accurate representation of the airplane's distance from its take-off point against the time from takeoff to landing

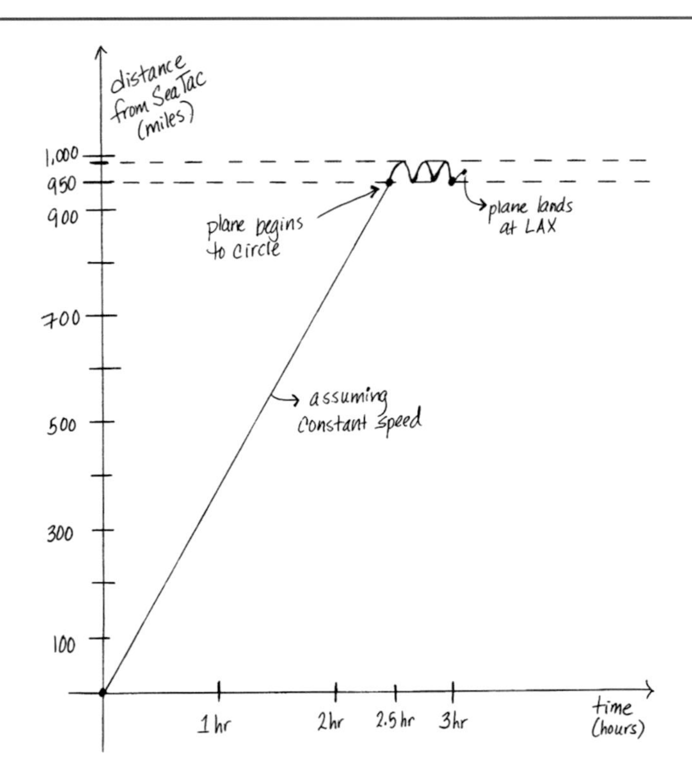

Mathematics is one of humankind's greatest cultural achievements. It is the "language of science," providing a means by which the world around us can be represented and understood. The mathematical representations that high school students learn afford them the opportunity to understand the power and beauty of mathematics and equip them to use representations in their personal lives, in the workplace, and in further study.

Making the vision of mathematics teaching and learning a reality requires a strong system of support at both the local and the national levels.

Working Together to Achieve the Vision

Imagine a classroom in which the Principles and Standards described in this volume have come to life. Students of varied backgrounds and abilities are working with their teachers to learn important mathematical ideas. Expectations are high for all students, including those who need extra support to learn mathematics well. Students are engaged by the mathematics they are learning, study it every year they are in school, and accept responsibility for their own mathematics learning. The classroom environment is equitable, challenging, and technologically equipped for the twenty-first century.

This vision of mathematics education—introduced in chapter 1, given focus in the Principles described in chapter 2, and more fully elaborated in the discussions of the Standards in chapters 3–7—is enticing. But what would it take to realize this vision? Let us look beyond the classroom to a broader context:

Imagine that all mathematics teachers continue to learn new mathematics content and keep current on education research. They collaborate on problems of mathematics teaching and regularly visit one another's classrooms to learn from, and critique, colleagues' teaching. In every school and district, mathematics teacher-leaders are available, serving as expert mentors to their colleagues, recommending resources, orchestrating interaction among teachers, and advising administrators. Education administrators and policymakers at all levels understand the nature of mathematical thinking and learning, help create professional and instructional climates that support students' and teachers' growth, understand the importance of mathematics learning, and provide the time and resources for teachers to teach and students to learn mathematics well. Institutions of higher learning collaborate with schools to study mathematics education and to improve teacher preparation and professional development. Professional mathematicians take an interest in, and contribute constructively to, setting the content goals for mathematics in grades K–12 and for developing teachers' mathematical knowledge. Professional organizations, such as the National Council of Teachers of Mathematics, provide leadership, resources, and professional development opportunities to improve mathematics education. And families, politicians, business and community leaders, and other stakeholders in the system are informed about education issues and serve as valuable resources for schools and children.

Making the vision of mathematics teaching and learning a reality requires a strong system of support at both the local and the national levels. The National Council of Teachers of Mathematics proposes that the Principles and Standards—grounded in decades of research and practice and refined in an extensive, collaborative process of review and revision—serve as a basis for realizing the vision.

Putting the Principles into Action

The Principles in chapter 2 offer perspectives that can guide decision makers in mathematics education. If essential supports of good mathematics classrooms are missing, not all students can learn the mathematics they need. Teachers need to work in environments where they can act, and continue to develop, as professionals. Mathematics teaching and learning should take place in a broader context that embraces and supports high-quality mathematics instruction. The following sections highlight each of the Principles in turn, showing how they can shape answers to important questions in mathematics education.

How can all students have access to high-quality mathematics education?

A major area of policy that affects students' access to mathematics education is "tracking," which is the long-term, often permanent, placement of students in classes, courses, or groups that offer different curricula according to the students' perceived mathematical abilities. Historically, tracking has consistently resulted in a select group of students being enrolled in mathematics courses that challenge and enrich them while others—often poor or minority students—are placed in mathematics classes that concentrate on remediation or do not offer significant mathematical substance (Wheelock 1992). For example, many middle-grades and high school students have been excluded from experiences in which they could learn significant amounts of algebra, instead spending much of their time reviewing the mathematics content studied in the elementary grades. As a result, these students are unable to experience a full program of high school mathematics across the range of content areas. *Principles and Standards* takes a strong stance: All students should have a common foundation of challenging mathematics, whether those students will enter the workplace after high school or pursue further study in mathematics and science.

The Equity Principle

Excellence in mathematics education requires equity—high expectations and strong support for all students.

Taking this stance necessitates addressing the unique mathematical needs of all students. Students with exceptional promise in mathematics and deep interest in advanced mathematical study need appropriate opportunities to pursue their interests. Students with special learning needs in mathematics must be supported both by their classroom teachers and by special education staff. Special-needs educators responsible for mathematics instruction should participate in mathematics professional development, which will allow them to collaborate with classroom teachers in assessing and analyzing students' work in order to plan instruction.

Teachers and school and district leaders face complex decisions about how best to structure different curricular options. One traditional way for students to learn additional mathematics in which they have a particular interest is differential pacing—allowing some students to move rapidly through the mathematical content expected of all so that they can go on to additional areas. However, some alternatives to differential pacing may prove advantageous. For example, curricula can be offered in which students can explore mathematics more deeply rather than more rapidly. This model allows them to develop deep insights into important concepts that prepare them well for later experiences instead of experiencing a more cursory treatment of a broader range of topics. Or schools can offer supplementary mathematics opportunities in areas not studied by all students or in extracurricular activities such as mathematics clubs or competitions.

Schools face difficult decisions about grouping—whether students should be offered mathematics instruction in homogeneous or heterogeneous groups. Students can effectively learn mathematics in heterogeneous groups if structures are developed to provide appropriate, differentiated support for a range of students. Structures that exclude certain groups of students from a challenging, comprehensive mathematics program should be dismantled. All such efforts should be monitored and evaluated to ensure that students are well served.

Are good instructional materials chosen, used, and accepted?

Choices of mathematics instructional materials can be controversial. Teachers should be prepared to work with new curricular materials, and they need considerable time to "live with" curricula in order to discover their strengths and weaknesses. Only then can they develop the kinds of knowledge necessary to make materials work well in particular contexts. The selection of curriculum and materials, therefore, needs to be a long-term collaborative process involving teachers, teacher-leaders, and administrators. Extensive field-testing should be conducted, with information and interactions at the district level so that choices are made wisely and support structures are put in place.

If instructional materials are not consistent with the expectations of families and community members or do not seem reasonable to them, serious difficulties can arise. For that reason, teachers and administrators should help families understand the goals and content of curricular materials, and community members should be consulted and informed about decisions regarding curricula and materials. Choices of instructional materials should be based on a community's agreed-on goals for

The Curriculum Principle

A curriculum is more than a collection of activities: it must be coherent, focused on important mathematics, and well articulated across the grades.

mathematics education. *Principles and Standards*, together with province, state, and local frameworks, offers proposals for such goals.

Developers of instructional materials and frameworks should draw on research in their efforts to implement the ideas of standards. We urge them to use *Principles and Standards* as a guide when making the many decisions involved in creating curricula. Similarly, through the evaluation and study of curricular efforts and through discussions of the ideas in this document, the mathematics education community can continue to develop a base of knowledge to guide the direction of mathematics education in prekindergarten through grade 12.

How can teachers learn what they need to know?

Teachers need to know and use "mathematics for teaching" that combines mathematical knowledge and pedagogical knowledge. They must be information providers, planners, consultants, and explorers of uncharted mathematical territory. They must adjust their practices and extend their knowledge to reflect changing curricula and technologies and to incorporate new knowledge about how students learn mathematics. They also must be able to describe and explain why they are aiming for particular goals.

Preservice preparation is the foundation for mathematics teaching, but it gives teachers only a small part of what they will need to know and understand throughout their careers. No matter how well prepared teachers are when they enter the profession, they need sustained, ongoing professional development in order to offer students a high-quality mathematics education. They must continue to learn new or additional mathematics content, study how students learn mathematics, analyze issues in teaching mathematics, and use new materials and technology. Teachers must develop their own professional knowledge using research, the knowledge base of the profession, and their own experiences as resources. Preservice education, therefore, needs to prepare teachers to learn from their own teaching, from their students, from curriculum materials, from colleagues, and from other experts.

Unfortunately, the preparation today's teachers have received is in many instances inadequate for the needs of tomorrow. The reality is simple: unless teachers are able to take part in ongoing, sustained professional development, they will be handicapped in providing high-quality mathematics education. The current practice of offering occasional workshops and in-service days does not and will not suffice.

Most mathematics teachers work in relative isolation, with little support for innovation and few incentives to improve their practice. Yet much of teachers' best learning occurs when they examine their teaching practices with colleagues. Research indicates that teachers are better able to help their students learn mathematics when they have opportunities to work together to improve their practice, time for personal reflection, and strong support from colleagues and other qualified professionals (see, e.g., Brown and Smith [1997]; Putnam and Borko [2000]; Margaret Smith [forthcoming]). The educational environment must be characterized by trust and respect for teachers and by patience as they work to develop, analyze, and refine their practice. Too often we place the responsibility for change solely on the shoulders of teachers and then blame them when things do not work as expected. We need

The Teaching Principle

Effective mathematics teaching requires understanding what students know and need to learn and then challenging and supporting them to learn it well.

instead to address issues in a systemic way, providing teachers with the resources they need for professional growth.

Such shifts in the system are feasible. The typical structures of teachers' workdays often inhibit community building, but structures can be changed. In some cultures, shared discussions of students and teaching are the norm. In Japan and China, the workdays of teachers include time for meeting together to analyze recent lessons and to plan for upcoming lessons (Ma 1999; Stigler and Hiebert 1999). During this "lesson study," teachers plan the lesson, teach the lesson with colleagues watching, revise the lesson collaboratively, teach the revised lesson, evaluate and reflect again, and share the results in written form. Part of the planning includes predicting what groups of students will do when presented with particular problems and tasks. The ongoing analysis of practice is thus built into the fabric of teaching, not treated as an added task that teachers must organize themselves. Although this level of professional collaboration may be hard for U.S. and Canadian teachers to imagine within the constraints of the prevailing professional culture and system, it illustrates the potential power of learning communities to improve mathematics teaching and learning. Finding ways to establish such communities should be a primary goal for schools and districts that are serious about improving mathematics education.

Do all students have time and the opportunity to learn?

Learning mathematics with understanding requires consistent access to high-quality mathematics instruction. In the elementary grades, students should study mathematics for at least an hour a day under the guidance of teachers who enjoy mathematics and are prepared to teach it well. Achieving this objective takes thoughtful administrative arrangements, such as orchestrating shared teaching responsibilities or using mathematics specialists.

Every middle-grades and high school student should be required to study the equivalent of a full year of mathematics in each grade. Ways of organizing programs will vary according to local goals and situations. Some schools are using such organizational alternatives as block scheduling. The impact of such alternatives on students' learning needs further study. It is not clear, for example, whether long intervals between periods of intensive study benefit or detract from students' learning or whether mathematics-intensive workplace activity can support students' mathematical engagement and growth. All middle-grades and high school students should be expected to spend a substantial amount of time every day working on mathematics outside of class, in activities ranging from typical homework assignments and projects to problem solving in the workplace.

A significant challenge to realizing the vision portrayed in *Principles and Standards* is disengagement. Too many students disengage from school mathematics, which creates a serious problem not only for their teachers but also for a society that increasingly depends on a quantitatively literate citizenry. Students may become uninvolved for various reasons. Many, for example, find it difficult to sustain the motivation and effort required to learn what can be a challenging school subject. They may find the subject as taught to be uninteresting and irrelevant.

Disengagement is too often reinforced in both overt and subtle ways by the attitudes and actions of adults who have influence with students.

The Learning Principle

Students must learn mathematics with understanding, actively building new knowledge from experience and prior knowledge.

Some parents and other authority figures, as well as societal influences like the media, convey the message that not everyone is expected to be successful in mathematics and thus that disengagement from school mathematics is acceptable. Such societal tolerance makes it less likely that all students will be motivated to sustain the effort needed to learn mathematics, which in turn makes the job of their teachers even more challenging. Some teachers also believe that many students cannot learn mathematics, which supports those students in their beliefs that they cannot learn mathematics, which then leads to further disengagement. Thus, a vicious cycle takes hold. It affects school mathematics in profound ways and is especially prevalent in the middle grades and high school.

Although the challenge presented by disengagement is formidable, it is not insurmountable. Teachers need to uphold high expectations that all children should learn with understanding, including children of minorities or from poor communities. Many teachers have found that if they teach mathematics in ways similar to those advanced in *Principles and Standards*—for example, by approaching traditional topics in ways that emphasize conceptual understanding and problem solving—many apparently uninterested students can become quite engaged.

Are assessments aligned with instructional goals?

High-stakes assessments—from nationally normed achievement tests to state, province, or locally developed measures of students' performance—are a particular concern for educators. If they are not aligned with school and community goals for mathematics education and with the curriculum, teachers and students are left in a precarious position. If teachers are committed to pursuing goals and practices consistent with those in *Principles and Standards*, satisfying the sometimes contradictory requirements of the local, state, or province assessment system is challenging. It is not realistic for teachers simply to ignore the pressure of these tests. Students may be penalized if they do not perform well, staff or school evaluations may depend on demonstrating progress, and decisions about resource allocation and salaries may be tied to test scores. Yet "teaching to the test"—a political reality when the consequences of test scores are significant—can undermine the integrity of instruction. To put teachers in the position of deciding between what they believe best enhances their students' learning and what is required to survive in the educational system puts them in an untenable position. High-stakes assessments must be closely linked to the goals teachers are being asked to achieve; where they are not, teachers must be supported in the decisions they make.

The Assessment Principle

Assessment should support the learning of important mathematics and furnish useful information to both teachers and students.

The assessment of students' understanding can be enhanced by the use of multiple forms of assessment, such as portfolios, group projects, and writing questions. However, students and parents alike may find these forms unfamiliar in the mathematics classroom. Teachers need support from administrators in helping students and parents understand the utility and purpose of such approaches in improving mathematics instruction.

Is technology supporting learning?

To make technology an essential part of classrooms, the technological tools must be selected and used in ways that are compatible with the instructional goals. When technological tools are considered essential instructional materials for all students, then decisions about resources must reflect this view, despite the costs of purchases and upgrades. Schools, districts, or provinces that integrate technology in mathematics teaching and learning face challenging issues of equity. The need for high-quality technology is as great in urban and rural settings as in suburban schools—perhaps greater.

Decisions to incorporate new technology also require that teachers be prepared and supported in using it to serve instructional goals. Teachers must themselves experience how technology can enhance the learning of significant mathematics and explore models for incorporating it in their classroom practice. Moreover, technology must be embedded in the mathematics program rather than be treated as just another flashy add-on. Without coherent, comprehensive implementation plans, the incorporation of new technology is likely to fall short in improving mathematics teaching and learning.

The Technology Principle

Technology is essential in teaching and learning mathematics; it influences the mathematics that is taught and enhances students' learning.

Roles and Responsibilities

The sections that follow discuss the kinds of commitments and actions that various communities must make to realize the vision of *Principles and Standards.* The role of teachers, of course, is central. The choices that mathematics teachers make every day determine the quality and effectiveness of their students' mathematics education. But teachers alone do not make all the decisions—they are part of a complex instructional system. Others—students themselves; mathematics teacher-leaders; school, district, and state or province administrators; higher-education faculty; families, other caregivers, and community members; and professional organizations and policymakers—have resources, influence, and responsibilities that can enable teachers and their students to be successful.

Mathematics Teachers

Mathematics teachers must develop and maintain the mathematical and pedagogical knowledge they need to teach their students well. One way to do this is to collaborate with their colleagues and to create their own learning opportunities where none exist. They should also seek out high-quality professional development opportunities that fit their learning needs. By pursuing sources of information, building communities of colleagues, and participating in professional development, teachers can continue to grow as professionals.

Mathematics teachers generally are responsible for what happens in their own classrooms and can try to ensure that their classrooms support learning by all students. For example, whether or not their school has implemented tracking, teachers must challenge and hold high expectations for all their students, not just those they believe are "gifted." Elementary school students need at least an hour of mathematics instruction each day. The decisions teachers make in the classroom about how to offer all students experiences with important mathematics and how to accommodate the wide-ranging interests, talents, and experiences of

Teachers must help students be confident, engaged mathematics learners.

students are essential to giving all students access to mathematics. Although many matters bearing on their classrooms are beyond teachers' sole control, they need to take the initiative in discussing trends and opportunities in mathematics education with administrators.

Teachers must help students be confident, engaged mathematics learners. In the elementary grades, convincing students that they can do mathematics and helping them enjoy it are important goals. Typically, student disengagement has been a serious problem in the middle grades and high school. Teachers at those levels should work to keep students involved in relevant classroom activities, assign projects that make connections between mathematics and students' daily lives, and allow students multiple avenues to display what they have learned. Learning experiences based in the workplace have also proved effective in motivating students who are at risk of becoming disengaged from school.

Mathematics teachers can foster reinforcement of their efforts by families and other community members by maintaining dialogue aimed at the improvement of mathematics education. Communicating about mathematics goals, students' learning, teaching, and programs helps families and other caregivers understand the kind of mathematics learning in which children are engaged. Giving them opportunities to ask questions, express concerns, and experience classroom activities can be very useful in shaping improvements. Many groups of teachers organize "Math Nights" at least once a year. At such events, usually held during the evening for the convenience of parents, students and their parents work together on engaging mathematics activities. Newsletters, homework assignments that involve family collaboration, and other means can help maintain communication between home and school. To do all of this well, teachers need to understand their mathematical goals and their perspectives on mathematics education and be able to articulate them in compelling ways.

Mathematics teachers ultimately control the range of mathematical ideas made available to their students. They have the responsibility to ensure that a full range of mathematical content and processes, such as those described in this document, are taught and that mathematical emphases fit together into a coherent whole. They can do so by using the available textbooks, support materials, technology, and other instructional resources effectively and tailoring these resources to their particular situations so that their goals for mathematics instruction and their students' needs are met. Teachers need to seek out support and professional development as they implement current or new curricula. They should constantly evaluate curricular materials and offer suggestions to teacher-leaders and administrators, and they should find ways to be involved in choosing the instructional materials for their school or district.

Integrating assessment into instruction and using a variety of sources of evidence to evaluate the learning of each student is challenging. It may be especially difficult in the face of mandated high-stakes assessments. Although teachers may at times feel trapped between their goals for their students' mathematics learning and those of high-stakes tests, they are not powerless. If teachers recognize that required tests are not aligned with meaningful instructional goals, they should voice their concerns to their teacher-leaders and administrators and seek ways to participate in decisions about testing.

Mathematics Students

Learning mathematics is stimulating, rewarding, and at times difficult. Mathematics students, particularly in the middle grades and high school, can do their part by engaging seriously with the material and striving to make mathematical connections that will support their learning. If students are committed to communicating their understandings clearly to their teachers, then teachers are better able to plan instruction and respond to students' difficulties. Productive communication requires that students record and revise their thinking and learn to ask good questions as part of learning mathematics.

Beyond the classroom, students need to build time into their days to work on mathematics. They need to learn how to use resources such as the Internet to pursue their mathematical questions and interests. As students begin to identify potential careers, they can take the initiative in researching the mathematics requirements for those careers and investigate whether their school programs offer the necessary preparation.

Mathematics Teacher-Leaders

There is an urgent and growing need for mathematics teacher-leaders—specialists positioned between classroom teachers and administrators who can assist with the improvement of mathematics education. The kinds of roles and influences that such leaders can have, as well as the nature of their position and responsibility, will vary widely. In recent years, schools have increasingly turned to teachers in the system as potential leaders. Sometimes leaders are teachers on special assignment, released from the classroom for an extended period to work in one or several school buildings. In other situations, leaders are released from a portion of their classroom teaching so that they can work directly with other teachers. No matter what the particular arrangement may be, mathematics teacher-leaders should take responsibility for focusing on mathematics through their work with teachers, administrators, and families and other community members.

Teacher-leaders can have a significant influence by assisting teachers in building their mathematical and pedagogical knowledge. Leaders face the challenge of changing the emphasis of the conversation among teachers from "activities that work" to the analysis of practice. Teacher-leaders in some settings work with their colleagues to design professional development plans for individual teachers, for a school, or for a larger system. They can arrange collaborative investigations or discussion groups with teachers at a school site, encourage participation in workshops at the school or district level, promote attendance at professional conferences, organize the study of professional resources such as *Principles and Standards* and articles in professional journals, recommend Internet sites that discuss mathematics teaching, and provide information about in-service programs or graduate courses. Teachers can benefit greatly from the knowledge and support of peers and mentors as they move in the directions recommended in *Principles and Standards*. Teacher-leaders' support on a day-to-day basis—ranging from conversations in the hall to in-classroom coaching to regular grade-level and departmental seminars focused on how students learn mathematics—can be crucial to a teacher's work life.

Teacher-leaders' support on a day-to-day basis can be crucial to a teacher's work life.

Teacher-leaders should have the knowledge and expertise to play a role in the design of curriculum frameworks and the selection of instructional materials, and they should ensure that teachers are involved in these processes, too. Schools and districts can also rely on mathematics teacher-leaders to organize and lead the piloting and implementation of new instructional materials. They might help teachers in a school begin working through the materials themselves, reading and discussing any commentary about student learning found in the materials, analyzing students' work, and designing and critiquing lessons. In this way, the materials can serve as a means for teachers to learn.

Teacher-leaders also have a role in working with administrators and policymakers to help guide their decisions about the improvement of mathematics education. They can help ensure that teachers and administrators develop and share a common perspective about goals for mathematics teaching and learning. Doing so involves keeping everyone informed of new directions and emphases in mathematics education and facilitating substantive discussions and planning sessions aimed at reaching common goals. Finally, like all teachers, mathematics teacher-leaders must themselves engage in ongoing learning and professional development.

School, District, and State or Province Administrators

Administrators at all levels should ensure that mathematics expertise and leadership are developed in their schools or systems.

Administrators at every level have responsibilities for shaping the instructional mission in their jurisdictions, providing for the professional development of teachers, designing and implementing policies, and allocating resources. To deploy their influence well on behalf of mathematics education, administrators who understand the goals of mathematics instruction—including those described in *Principles and Standards*—can work to create institutions in which teachers have access to human and material resources that will help them attain those goals. In particular, administrators can identify individuals or teams of teachers as leaders in establishing mathematics communities, and they can ensure that such learning communities develop, flourish, and grow beyond a few teachers. When administrators themselves become part of the mathematics learning community, they develop deeper understandings of the goals of mathematics instruction. They can understand better what they are seeing in mathematics classrooms (Nelson 1999) and can make more-informed judgments about the curricular, technological, and pedagogical resources teachers need.

If they are truly committed to the improvement of mathematics education, administrators at all levels should ensure that mathematics expertise and leadership are developed in their schools or systems. Administrators can influence the quality of mathematics education by supporting the professional growth of mathematics teachers. They can arrange for meaningful professional development workshops, provide libraries and Web access to instructional and other materials, and foster cross-school conversations about goals and instructional practices. Administrators can help arrange teachers' work schedules so that meaningful collaboration with colleagues is part of the school day. They can shape work environments so that they are conducive to productive professional interactions, they can include such interactions as part of teachers' work, and they can establish a program of mathematics teacher-leaders within the school or system.

Perhaps the most important sphere of influence for administrators is the area of structures and policies. Through their decisions about hiring, teaching assignments, evaluation, and mentoring for new teachers, administrators have powerful opportunities to strengthen the focus on mathematics. They also have a role in guaranteeing equitable access for students and in arranging time and space for effective mathematics instruction. They can work to align curricular materials, technology, and assessments at all levels with agreed-on goals for mathematics education, such as those represented in *Principles and Standards*.

Administrators can support the improvement of mathematics education by establishing effective processes for the analysis and selection of instructional materials. These processes, whether at the classroom, school, or system level, should involve wide consultation with teachers and teacher-leaders and a deep and careful analysis of the materials. Many districts pilot-test one or more programs before making a final decision. No matter how decisions about the selection of instructional materials are made, they should always be guided by school, state or province, and national goals for mathematics instruction. Furthermore, the adoption of new instructional materials is only a beginning. No matter how well curricular materials may be designed, they are unlikely to lead to the continuing improvement of instruction unless plans for implementation and professional development are formulated along with the plan for adoption.

Finally, to make long-term progress in improving students' learning, administrators and policymakers must carefully consider the impact of high-stakes assessments on the instructional climate in schools, and they must understand what can be learned from assessments and what cannot. If a test focuses primarily on the acquisition of superficial skills rather than on the deep mathematical understandings described in this document, its use in making decisions that promote constructive change will be limited and perhaps even counterproductive. Decisions about placing students in different instructional situations and evaluations of teachers' effectiveness should never be based on a single test score, especially when that test has been designed to measure how well students carry out routine procedures. Implementing and sustaining assessment policies and practices that support high-quality mathematics education is a difficult but essential part of administrators' responsibilities.

Administrators and policymakers must carefully consider the impact of high-stakes assessments on the instructional climate in schools.

Higher-Education Faculty

Faculty in two-year and four-year colleges and universities have a significant impact on school mathematics, primarily through their work with students who will become teachers. They have considerable influence on whether teachers enter the profession with the strong knowledge of the mathematics needed to teach pre-K–12 mathematics, of student learning, and of mathematics teaching. They can also model the effective practices they believe teachers should employ.

The first few years in teachers' careers are critical to their persistence in mathematics teaching and to their dispositions toward continued professional growth and learning. The in-service education and professional development of teachers, especially in content knowledge, are not the exclusive mission of any single type of institution, so a significant leadership role is available for higher education. Teachers need in-service and graduate education that help them grow mathematically and as practitioners.

Faculty members in institutions of higher education should be partners in the development of school-based mathematics communities. Teacher educators, mathematicians, and practicing teachers working together can create a rich intellectual environment that will promote veteran teachers' growth and demonstrate to new teachers the value of learning communities. Mathematics education researchers and teacher educators can collaborate with classroom teachers to investigate research questions based in classroom practice or to look at mathematics as it occurs in classrooms. In such contexts, higher-education faculty can serve as resources to mathematics teachers at all levels and also learn from them.

In recent decades, research in mathematics education has coalesced as a powerful field of intellectual study. The efforts of education and mathematics faculty can result in increased knowledge and the improved preparation of teachers, teacher-leaders, administrators, and researchers. Such ongoing efforts, in collaboration with school personnel, are a major aspect of improving mathematics teaching and learning.

Families, Other Caregivers, and Community Members

Families become advocates for education standards when they understand the importance of a high-quality mathematics education for their children.

Teachers and administrators should invite families, other caregivers, and community members to participate in examining and improving mathematics education. All partners in this enterprise need to understand the changing goals and priorities of school mathematics, as expressed in the Principles in chapter 2. Families need to know what options are available for their children and why an extensive and rigorous mathematics education is important. When parents understand and support the schools' mathematics program, they can be invaluable in convincing their daughters and sons of the need to learn mathematics and to take schooling seriously. Families become advocates for education standards when they understand the importance of a high-quality mathematics education for their children.

Families can establish learning environments at home that enhance the work initiated at school. Respect shown to and for teachers is often carried over from parent to child. By providing a quiet place for a child to read and attend to homework and by monitoring students' work, families can signal that they believe mathematics is important. Such attention and appreciation of mathematics is not lost on students.

If families and other members of the public do not understand the intent of, and rationale for, improvements in mathematics education, they can halt even the most carefully planned initiatives. *Principles and Standards* was written with the hope that the conversations it engenders will ultimately generate a widespread commitment to improving mathematics education. As part of this effort, it is the responsibility of the education community to inform the general public and its elected representatives about the goals and priorities in mathematics education, thereby empowering them to participate knowledgeably in its improvement.

Professional Organizations and Policymakers

Professional organizations can provide national and regional leadership and expertise in support of the continuing improvement of mathematics education. The National Council of Teachers of Mathematics (NCTM), using *Principles and Standards* as a focus, will offer its members many means of professional development, including conferences, classroom resources, research publications, and Web-based materials. NCTM is also engaged in a variety of efforts to educate the public and to support parents and other caregivers as they encourage their children in mathematics. (See the NCTM Web site at www.nctm.org for details.) Professional organizations like the NCTM are positioned to promote policies that support high-quality mathematics education. For instance, they can work to establish certification and accreditation requirements that include high expectations for teachers' knowledge of content, teaching, and students. Through their members, publications, and meetings, professional organizations can focus attention on mathematics education issues.

High-quality mathematics education is vital to the health of mathematics as a discipline, and mathematical professional societies have a clear stake in this enterprise. Mathematics education serves as the pipeline for future mathematicians, statisticians, and mathematics teachers as well as scientists, engineers, and all professionals who use mathematics. The public needs to be well informed about, not intimidated by, mathematics if it is to support ongoing research, development, and funding in mathematics and related fields. The nature of undergraduate mathematics programs is closely intertwined with K–12 mathematics education, and efforts to improve mathematics teaching and learning from kindergarten through graduate school can be coordinated across professional organizations. Issues about the mathematical preparation of teachers are currently being addressed by the Conference Board of the Mathematical Sciences (CBMS) in its "Mathematical Education of Teachers" project (CBMS 2000). All professional organizations concerned with the mathematical sciences can help improve mathematics education through coordinated, collaborative efforts.

Local, state, and provincial curricular frameworks and standards need ongoing examination and revision.

Policymakers at national and local levels are in a unique position to view the broad range of influences on mathematics education and to make decisions that promote improvements in the field. They can allocate the funding and resources needed to continue the study and implementation of improvements. They can also examine teacher-certification standards and accreditation requirements to ensure that teachers have the strong and deep content knowledge needed today. In the same way that standards need ongoing examination and revision, so do local, state, and provincial curricular frameworks and standards. With *Principles and Standards* available as a resource to identify key issues in contemporary mathematics education, policymakers can ensure that that process occurs and can promote programmatic activity that is designed to further address those issues.

Using *Principles and Standards* Effectively

Principles and Standards for School Mathematics has both an immediate and a long-term role in realizing the vision of improved mathematics education. First, it sets out a carefully developed and ambitious but attainable set of expectations for school mathematics. Educators, families, policymakers, and others can use the ideas contained in this volume to guide the decisions they make about mathematics education, from classroom practice to establishing local and state education standards and frameworks. The interest generated by NCTM's original *Curriculum and Evaluation Standards for School Mathematics* (1989) demonstrates the extent to which various groups with a stake in mathematics education are committed to its improvement. *Principles and Standards* is intended to build on and extend that commitment. It represents a collective judgment, based on research, practice, and an extended consultative process, about what students need to learn in order to be prepared for the future.

Principles and Standards is also a tool for better understanding the issues and challenges involved in improving mathematics education. It offers information and ideas that those with responsibility for mathematics education—whether at the local, state or provincial, or national level—need in order to engage in constructive dialogues about mathematics teaching, curricula, and assessment. Any vision of school mathematics teaching and learning needs ongoing examination; it needs to be refined continually in light of the greater understanding achieved through practice, research, and evidence-based critiques. The process that NCTM put in place for developing *Principles and Standards* reflects a commitment to ongoing discussion and reflection. This document, therefore, should be seen as part of a work in progress that can help guide decision makers in developing excellent mathematics programs, not as a prescription to be rigidly imposed on others. (See Kilpatrick and Silver [2000] and Ferrini-Mundy [2000] for additional discussion.)

Conclusion

Achieving high standards in mathematics education calls for clear goals. It calls for the active participation of teachers, administrators, policymakers, higher-education faculty, curriculum developers, researchers, families, students, and community members.

Principles and Standards is provided as a catalyst for the continued improvement of mathematics education. It represents our best current understanding of mathematics teaching and learning and the contextual factors that shape them. It was created with the input and collaboration of members of all the communities mentioned above. It articulates high but attainable standards.

Realizing the vision of mathematics education described in this document requires the continued creation of high-quality instructional materials and technology. It requires enhanced preparation for teachers and increased opportunities for professional growth. It requires the creation of assessments aligned with curricular goals. Realizing the vision depends on the participation of all the constituencies mentioned above in reflecting on, supporting, and improving educational practice. We should not underestimate the difficulty of the task, but it can be done. Now is the time to undertake the collaborative efforts that will make the vision come alive. We owe our children nothing less.

Principles and Standards *is provided as a catalyst for the continued improvement of mathematics education.*

References

Andrews, Angela Giglio. "Developing Spatial Sense—a Moving Experience." *Teaching Children Mathematics* 2 (January 1996): 290–93.

———. "Solving Geometric Problems by Using Unit Blocks." *Teaching Children Mathematics* 6 (February 1999): 318–23.

Balanced Assessment for the Mathematics Curriculum. *High School Assessment Package 2*. White Plains, N.Y.: Dale Seymour Publications, 1999b.

———. *Middle Grades Assessment Package 1*. White Plains, N.Y.: Dale Seymour Publications, 1999a.

Ball, Deborah Loewenberg, and Hyman Bass. "Making Believe: The Collective Construction of Public Mathematical Knowledge in the Elementary Classroom." In *Constructivism in Education*, edited by D. Phillips, Yearbook of the National Society for the Study of Education. Chicago, Ill.: University of Chicago Press, forthcoming.

Banchoff, Thomas F. "Dimension." *On the Shoulders of Giants: New Approaches to Numeracy*, edited by Lynn Arthur Steen, pp. 11–59. Washington, D.C.: National Academy Press, 1990.

Baroody, Arthur J. "The Development of Basic Counting, Number, and Arithmetic Knowledge among Children Classified as Mentally Handicapped." In *International Review of Research in Mental Retardation*, edited by Laraine Masters Glidden, pp. 51–103. New York: Academic Press, 1999.

———. "The Development of Preschoolers' Counting Skills and Principles." In *Pathways to Number: Children's Developing Numerical Abilities*, edited by Jacqueline Bideaud, Claire Meljac, and Jean-Paul Fischer, pp. 99–126. Hillsdale, N.J.: Lawrence Erlbaum Associates, 1992.

———. "An Investigative Approach to the Mathematics Instruction of Children Classified as Learning Disabled." In *Cognitive Approaches to Learning Disabilities*, edited by D. Kim Reid, Wayne P. Hresko, and H. Lee Swanson, pp. 547–615. Austin, Tex.: Pro-Ed, 1996.

———. *3-D Geometry: Seeing Solids and Silhouettes*. Investigations in Number, Data, and Space. Palo Alto, Calif.: Dale Seymour Publications, 1995.

Battista, Michael T., Douglas H. Clements, Judy Arnoff, Kathryn Battista, and Caroline Van Auken Borrow. "Students' Spatial Structuring of 2D Arrays of Squares." *Journal for Research in Mathematics Education* 29 (November 1998): 503–32.

Beaton, Albert E., Ina V. S. Mullis, Michael O. Martin, Eugenio J. Gonzalez, Dana L. Kelly, and Teresa A. Smith. *Mathematics Achievement in the Middle School Years: IEA's Third International Mathematics and Science Study (TIMSS)*. Chestnut Hill, Mass.: Boston College, TIMSS International Study Center, 1996.

Behr, Merlyn, Stanley Erlwanger, and Eugene Nichols. *How Children View Equality Sentences.* PMDC Technical Report no. 3. Tallahassee, Fla.: Florida State University, 1976. (ERIC Document Reproduction Service no. ED144802).

Bennett, Albert, Eugene Maier, and Ted Nelson. *Math and the Mind's Eye*. Portland, Oreg.: The Math Learning Center, 1988.

Black, Paul, and Dylan Wiliam. "Inside the Black Box: Raising Standards through Classroom Assessment." *Phi Delta Kappan* (October 1998): 139–48.

Boers-van Oosterum, Monique Agnes Maria. "Understanding of Variables and Their Uses Acquired by Students in Traditional and Computer-Intensive Algebra." Ph.D. diss., University of Maryland College Park, 1990.

Borasi, Raffaella. *Learning through Inquiry*. Portsmouth, N.H.: Heinemann, 1992.

Borko, Hilda, and Ralph T. Putman. "Learning to Teach." In *Handbook of Educational Psychology*, edited by David C. Berliner and Robert C. Calfee, pp. 673–708. New York: Macmillan, 1996.

Bransford, John D., Ann L. Brown, and Rodney R. Cocking, eds. *How People Learn: Brain, Mind, Experience, and School.* Washington, D.C.: National Academy Press, 1999.

Brown, Catherine A., and Margaret S. Smith. "Supporting the Development of Mathematical Pedagogy." *Mathematics Teacher* 90 (February 1997): 138–43.

Brown, Stephen I., and Marion Walter. *The Art of Problem Posing*. Hillsdale, N.J.: Lawrence Erlbaum Associates, 1983.

Brownell, William A. "The Place of Meaning in the Teaching of Arithmetic." *Elementary School Journal* 47 (January 1947): 256–65.

Burger, William F., and J. Michael Shaughnessy. "Characterizing the van Hiele Levels of Development in Geometry." *Journal for Research in Mathematics Education* 17 (January 1986): 31–48.

Burns, Marilyn. *Fifty Problem-Solving Lessons, Grades 1–6.* Sausalito, Calif.: Math Solutions Publications, 1996.

Burns, Marilyn, and Cathy McLaughlin. *A Collection of Math Lessons from Grades 6 through 8.* Sausalito, Calif.: Math Solutions Publications, 1990.

Burrill, Gail F., Jack C. Burrill, Patrick W. Hopfensperger, and James M. Landwehr. *Exploring Regression: Data-Driven Mathematics.* White Plains, N.Y.: Dale Seymour Publications, 1999.

Burton, Grace, Douglas Clements, Terrence Coburn, John Del Grande, John Firkins, Jeane Joyner, Miriam A. Leiva, Mary M. Lindquist, and Lorna Morrow. *Third-Grade Book. Curriculum and Evaluation Standards for School Mathematics* Addenda Series, Grades K–6, edited by Miriam A. Leiva. Reston, Va.: National Council of Teachers of Mathematics, 1992.

California State Department of Education. *A Question of Thinking: A First Look at Students' Performance on Open-Ended Questions in Mathematics*. Sacramento, Calif.: California Assessment Program, California State Department of Education, 1989.

Campbell, Patricia. *Project IMPACT: Increasing Mathematics Power for All Children and Teachers.* Phase 1, Final Report. College Park, Md.: Center for Mathematics Education, University of Maryland, 1995.

Carle, Eric. *Rooster's Off to See the World.* Natick, Mass.: Picture Book Studio, 1971.

Carlson, Marilyn P. "A Cross-Sectional Investigation of the Development of the Function Concept." In *Research in Collegiate Mathematics Education 3*, edited by Alan H. Schoenfeld, Jim Kaput, and Ed Dubinsky, pp. 114–62. Washington, D.C.: Conference Board of the Mathematical Sciences, 1998.

Carnegie Corporation of New York. *Years of Promise: A Comprehensive Learning Strategy for America's Children, Executive Summary*. June 1999. Available online at www.carnegie.org//execsum.html.

Carpenter, Thomas P., Mary Kay Corbitt, Henry S. Kepner, Jr., Mary Montgomery Lindquist, and Robert E. Reys. *Results from the Second Mathematics Assessment of the National Assessment of Educational Progress.* Reston, Va.: National Council of Teachers of Mathematics, 1981.

Carpenter, Thomas P., Elizabeth Fennema, Penelope L. Peterson, Chi-Pang Chiang, and Megan Loef. "Using Knowledge of Children's Mathematics Thinking in Classroom Teaching: An Experimental Study." *American Education Research Journal* 26 (1989): 499–531.

Carpenter, Thomas P., Megan L. Franke, Victoria R. Jacobs, Elizabeth Fennema, and Susan B. Empson. "A Longitudinal Study of Invention and Understanding in Children's Multidigit Addition and Subtraction." *Journal for Research in Mathematics Education* 29 (January 1998): 3–20.

Carpenter, Thomas P., and Linda Levi. "Developing Conceptions of Algebraic Reasoning in the Primary Grades." Paper presented at the annual meeting of the American Educational Research Association, Montreal, April 1999.

Carpenter, Thomas P., and James M. Moser. "The Acquisition of Addition and Subtraction Concepts in Grades One through Three." *Journal for Research in Mathematics Education* 15 (May 1984): 179–202.

Clement, Jennie, Erin DiPerna, Jean Gavin, Sally Hanner, Eric James, Angie Putz, Mark Rohlfing, and Susan Wainwright. *Children's Work with Data.* Madison, Wis.: Wisconsin Center for Education Research, University of Wisconsin—Madison, 1997.

Clements, Douglas H. "Geometric and Spatial Thinking in Young Children." In *Mathematics in the Early Years*, edited by Juanita V. Copley, pp. 66–79. Reston, Va.: National Council of Teachers of Mathematics, 1999b.

———. "Training Effects on the Development and Generalization of Piagetian Logical Operations and Knowledge of Number." *Journal of Educational Psychology* 76, no. 5 (1984): 766–76.

———. "Young Children and Technology." In *Dialogue on Early Childhood Science, Mathematics, and Technology Education*, edited by G. D. Nelson, pp. 92–105. Washington, D.C.: American Association for the Advancement of Science, 1999a.

Clements, Douglas H., Michael T. Battista, Julie Sarama, and Sudha Swaminathan. "Development of Students' Spatial Thinking in a Unit on Geometric Motions and Area." *Elementary School Journal* 98, no. 2 (1997): 171–86.

Clements, Douglas H., Sudha Swaminathan, Mary Anne Zeitler Hannibal, and Julie Sarama. "Young Children's Concepts of Shape." *Journal for Research in Mathematics Education* 30 (March 1999): 192–212.

Cobb, Paul, and Grayson Wheatley. "Children's Initial Understandings of Ten." *Focus on Learning Problems in Mathematics* 10, no. 3 (1988): 1–28.

Cobb, Paul, Terry Wood, and Erna Yackel. "Discourse, Mathematical Thinking, and Classroom Practice." In *Contexts for Learning: Sociocultural Dynamics in Children's Development*. New York: Oxford University Press, 1994.

Cobb, Paul, Terry Wood, Erna Yackel, John Nicholls, Grayson Wheatley, Beatriz Trigatti, and Marcella Perlwitz. "Assessment of a Problem-Centered Second-Grade Mathematics Project." *Journal for Research in Mathematics Education* 22 (January 1991): 3–29.

Cobb, Paul, Erna Yackel, Terry Wood, Grayson Wheatley, and Graceann Merkel. "Creating a Problem-Solving Atmosphere." *Arithmetic Teacher* 36 (September 1988): 46–47.

Collins, Alan, John Seely Brown, and Susan Newman. "Cognitive Apprenticeship: Teaching the Crafts of Reading, Writing, and Mathematics." In *Knowing, Learning, and Instruction: Essays in Honor of Robert Glaser*, edited by Lauren B. Resnick, pp. 453–94. Hillsdale, N.J.: Lawrence Erlbaum Associates, 1989.

Conference Board of the Mathematical Sciences. *CBMS Mathematical Education of Teachers Project Draft Report*. Washington, D.C.: Conference Board of the Mathematical Sciences, March 2000. Available on the Web at www.maa.org/cbms.

Confrey, Jere. "Learning to Listen: A Student's Understanding of Powers of Ten." In *Radical Constructivism in Mathematics Education*, edited by Ernst von Glasersfeld, pp. 111–38. Boston: Kluwer Academic Publishers, 1991.

———. "A Review of the Research on Student Conceptions in Mathematics, Science, and Programming." In *Review of Research in Education*, vol.16, edited by Courtney Cazden, pp. 3–56. Washington, D.C.: American Educational Research Association, 1990.

Consortium for Mathematics and Its Applications. *Mathematics: Modeling Our World.* Cincinnati, Ohio: South-Western Educational Publishing, 1999.

Council of Chief State School Officers. *State Curriculum Frameworks in Mathematics and Science: How Are They Changing across the States?* Washington, D.C.: Council of Chief State School Officers, 1995.

Coxford, Arthur F., James T. Fey, Christian R. Hirsch, Harold L. Schoen, Gail Burrill, Eric W. Hart, and Ann E. Watkins, with Mary Jo Messenger and Beth Ritsema. *Contemporary Mathematics in Context: A Unified Approach*, Course, 2, Part B. Core-Plus Mathematics Project. Chicago: Everyday Learning Corp., 1998.

Curcio, Frances R., and Nadine S. Bezuk, with others. *Understanding Rational Numbers and Proportions. Curriculum and Evaluation Standards for School Mathematics* Addenda Series, Grades 5–8. Reston, Va.: National Council of Teachers of Mathematics, 1994.

Curriculum Development Institute of Singapore. *Primary Mathematics 6B*. Singapore: Federal Publications, 1997.

Demana, Franklin, and Joan Leitzel. "Establishing Fundamental Concepts through Numerical Problem Solving." In *The Ideas of Algebra, K–12*, 1988 Yearbook of the National Council of Teachers of Mathematics, edited by Arthur F. Coxford, pp. 61–68. Reston, Va.: National Council of Teachers of Mathematics, 1988.

Dossey, John A., Ina V. S. Mullis, and Chancey O. Jones. *Can Students Do Mathematical Problem Solving? Results from Constructed-Response Questions in NAEP's 1992 Mathematics Assessment.* 23-FR-01. Washington, D.C.: National Center for Education Statistics, August 1993.

Dugdale, Sharon. "Functions and Graphs: Perspectives on Student Thinking." In *Integrating Research on the Graphical Representation of Functions*, edited by Thomas Romberg, Elizabeth Fennema, and Thomas Carpenter, pp. 101–29. Hillsdale, N.J.: Lawrence Erlbaum Associates, 1993.

Dunham, Penelope H., and Thomas P. Dick. "Research on Graphing Calculators." *Mathematics Teacher* 87 (September 1994): 440–45.

Education Development Center, Inc. *Using Algebraic Thinking: Patterns, Numbers, and Shapes.* Mathscape: Seeing and Thinking Mathematically series. Mountain View, Calif.: Creative Publications, 1998.

Edwards, Carolyn, Lella Gandini, and George Forman. *The Hundred Languages of Children: The Reggio Emilia Approach to Early Childhood Education.* Norwood, N.J.: Ablex Publishing Corp., 1993.

Encyclopaedia Britannica Educational Corporation. *Insights into Data*. Mathematics in Context. Chicago: Encyclopaedia Britannica Education Corp., 1998.

English, Lyn D., and Elizabeth A. Warren. "Introducing the Variable through Pattern Exploration." *Mathematics Teacher* 91 (February 1998): 166–70.

Ferrini-Mundy, Joan. "The Standards Movement in Mathematics Education: Reflections and Hopes." In *Learning Mathematics for a New Century*, edited by Maurice J. Burke, pp. 37–50. Reston, Va.: National Council of Teachers of Mathematics, 2000.

Ferrini-Mundy, Joan, Glenda Lappan, and Elizabeth Phillips. "Experiences with Patterning." *Teaching Children Mathematics* 3 (February 1997): 282–88.

Ferrini-Mundy, Joan, and Thomas Schram, eds. *The Recognizing and Recording Reform in Mathematics Education Project: Insights, Issues, and Implications. Journal for Research in Mathematics Education* Monograph No. 8. Reston, Va.: National Council of Teachers of Mathematics, 1997.

Flores, Alfinio. "Sí Se Puede, 'It Can Be Done': Quality Mathematics in More than One Language." In *Multicultural and Gender Equity in the Mathematics Classroom: The Gift of Diversity*, 1997 Yearbook of the National Council of Teachers of Mathematics, edited by Janet Trentacosta, pp. 81–91. Reston, Va.: National Council of Teachers of Mathematics, 1997.

Fuson, Karen C. *Children's Counting and Concepts of Number*. New York: Springer-Verlag New York, 1988.

———. "Number and Operation." In *A Research Companion to NCTM's Standards*, edited by Jeremy Kilpatrick, W. Gary Martin, and Deborah Schifter. Reston, Va.: National Council of Teachers of Mathematics, forthcoming.

———. "Research on Learning and Teaching Addition and Subtraction of Whole Numbers." In *Analysis of Arithmetic for Mathematics Teaching*, edited by Gaea Leinhardt, Ralph Putnam, and Rosemary A. Hattrup, pp. 53–188. Hillsdale, N.J.: Lawrence Erlbaum Associates, 1992.

Fuson, Karen C., Diana Wearne, James C. Hiebert, Hanlie G. Murray, Pieter G. Human, Alwyn I. Olivier, Thomas P. Carpenter, and Elizabeth Fennema. "Children's Conceptual Structures for Multidigit Numbers and Methods of Multidigit Addition and Subtraction." *Journal for Research in Mathematics Education* 28 (March 1997): 130–62.

Fuys, David, Dorothy Geddes, and Rosamond Tischler. *The van Hiele Model of Thinking in Geometry among Adolescents. Journal for Research in Mathematics Education* Monograph No. 3. Reston, Va.: National Council of Teachers of Mathematics, 1988.

Garofalo, Joe, and Frank K. Lester, Jr. "Metacognition, Cognitive Monitoring, and Mathematical Performance." *Journal for Research in Mathematics Education* 16 (May 1985): 163–76.

Gelfand, Israel M., and Alexander Shen. *Algebra*. Boston: Birkhäuser Boston, 1993.

Gelman, Rochel. "Constructivism and Supporting Environments." In *Implicit and Explicit Knowledge: An Educational Approach*, edited by Dina Tirosh, pp. 55–82, vol. 6 of *Human Development*, edited by S. Strauss. Norwood, N.J.: Ablex Publishing Corp., 1994.

Gelman, Rochel, and C. R. Gallistel. *The Child's Understanding of Number.* Cambridge: Harvard University Press, 1978.

Ginsburg, Herbert P., Alice Klein, and Prentice Starkey. "The Development of Children's Mathematical Thinking: Connecting Research with Practice." In *Child Psychology in Practice*, edited by Irving E. Sigel and K. Ann Renninger, pp. 401–76, vol. 4 of *Handbook of Child Psychology*, edited by William Damon. New York: John Wiley & Sons, 1998.

Goldberg, Howard. "Evaporation I." *TIMS Laboratory Investigations.* Dubuque, Iowa: Kendall/Hunt Publishing Co., 1997.

Goldenberg, Paul, Philip Lewis, and James O'Keefe. "Dynamic Representation and the Development of a Process Understanding of Function." In *The Concept of Function: Aspects of Epistemology and Pedagogy*, edited by Ed Dubinsky and Guershon Harel, pp. 235–60. MAA notes no. 25. Washington, D.C.: Mathematical Association of America, 1992.

Gordon, Sheldon P., Deborah Hughes-Hallett, Arnold Ostebee, and Zalman Usiskin. "Report of the Content Workshops." In *Preparing for a New Calculus: Conference Proceedings*, MAA Notes no. 36, edited by Anita E. Solow, pp. 55–59. Washington, D.C.: Mathematical Association of America, 1994.

Graeber, Anna O., and Patricia F. Campbell. "Misconceptions about Multiplication and Division." *Arithmetic Teacher* 40 (March 1993): 408–11.

Graeber, Anna O., and Elaine Tanenhaus. "Multiplication and Division: From Whole Numbers to Rational Numbers." In *Research Ideas for the Classroom, Middle Grades Mathematics*, National Council of Teachers of Mathematics Research Interpretation Project, edited by Douglas T. Owens, pp. 99–117. New York: Macmillan Publishing Co., 1993.

Gravemeijer, Koeno Pay Eskelhoff. *Developing Realistic Mathematics Instruction.* Utrecht, Netherlands: Freudenthal Institute, 1994.

Greeno, James G., and Roger B. Hall. "Practicing Representation: Learning with and about Representational Forms." *Phi Delta Kappan* (January 1997): 361–67.

Greer, Brian. "Multiplication and Division as Models of Situations." In *Handbook of Research on Mathematics Teaching and Learning*, edited by Douglas A. Grouws, pp. 276–95. New York: Macmillan Publishing Co., 1992.

Griffin, Sharon, and Robbie Case. "Re-thinking the Primary School Math Curriculum: An Approach Based on Cognitive Science." *Issues in Education* 3, no. 1 (1997): 1–49.

Griffin, Sharon A., Robbie Case, and Robert S. Siegler. "Rightstart: Providing the Central Conceptual Prerequisites for First Formal Learning of Arithmetic to Students at Risk for School Failure." In *Classroom Lessons: Integrating Cognitive Theory and Classroom Practice*, edited by Kate McGilly, pp. 25–49. Cambridge: MIT Press, 1994.

Groves, Susie. "Calculators: A Learning Environment to Promote Number Sense." Paper presented at the annual meeting of the American Educational Research Association, New Orleans, April 1994.

Hall, Rogers, Dennis Kibler, Etienne Wenger, and Chris Truxaw. "Exploring the Episodic Structure of Algebra Story Problem Solving." *Cognition and Instruction* 6, no. 3 (1989): 223–83.

Halmos, P. R. "The Heart of Mathematics." *American Mathematical Monthly* 87, no. 7 (1980): 519–24.

Hamilton, Johnny E., and Margaret S. Hamilton. *Math to Build On: A Book for Those Who Build.* Clinton, N.C.: Construction Trades Press, 1993.

Hancock, Chris, James J. Kaput, and Lynn T. Goldsmith. "Authentic Inquiry with Data: Critical Barriers to Classroom Implementation." *Educational Psychologist* 27, no. 3 (1992): 337–64.

Hanna, Gila, and Erna Yackel. "Reasoning and Proof." In *A Research Companion to NCTM's Standards*, edited by Jeremy Kilpatrick, W. Gary Martin, and Deborah Schifter. Reston, Va.: National Council of Teachers of Mathematics, forthcoming.

Hatano, Giyoo, and Kayoko Inagaki. "Sharing Cognition through Collective Comprehension Activity." In *Perspectives on Socially Shared Cognition*, edited by Lauren B. Resnick, John M. Levine, and Stephanie D. Teasley, pp. 331–48. Washington, D.C.: American Psychological Association, 1991.

Hiebert, James. "Relationships between Research and the NCTM Standards." *Journal for Research in Mathematics Education* 30 (January 1999): 3–19.

Hiebert, James, and Thomas P. Carpenter. "Learning and Teaching with Understanding." In *Handbook of Research on Mathematics Teaching and Learning*, edited by Douglas A. Grouws, pp. 65–97. New York: Macmillan Publishing Co., 1992.

Hiebert, James, Thomas P. Carpenter, Elizabeth Fennema, Karen C. Fuson, Diana Wearne, Hanlie Murray, Alwyn Olivier, and Piet Human. *Making Sense: Teaching and Learning Mathematics with Understanding*. Portsmouth, N.H.: Heinemann, 1997.

Hiebert, James, and Mary Lindquist. "Developing Mathematical Knowledge in the Young Child." In *Mathematics for the Young Child*, edited by Joseph N. Payne, pp. 17–36. Reston, Va.: National Council of Teachers of Mathematics, 1990.

Hoachlander, Gary. "Organizing Mathematics Education around Work." In *Why Numbers Count: Quantitative Literacy for Tomorrow's America*, edited by Lynn Arthur Steen, pp. 122–36. New York: College Entrance Examination Board, 1997.

Hughes-Hallett, Deborah, Andrew M. Gleason, Daniel E. Flath, Sheldon P. Gordon, David O. Lomen, David Lovelock, William G. McCallum, Brad G. Osgood, Andrew Pasquale, Jeff Tecosky-Feldman, Joe B. Thrash, Karen R. Thrash, and Thomas W. Tucker, with the assistance of Otto K. Bretscher. *Calculus.* New York: John Wiley & Sons, 1994.

Hutchins, Pat. *1 Hunter*. New York: Greenwillow Books, 1982.

Isaacs, Andrew C., and William M. Carroll. "Strategies for Basic-Facts Instruction." *Teaching Children Mathematics* 5 (May 1999): 508–15.

Kamii, Constance, and Ann Dominick. "The Harmful Effects of Algorithms in Grades 1–4," In *The Teaching and Learning of Algorithms in School Mathematics*, 1998 Yearbook of the National Council of Teachers of Mathematics, edited by Lorna J. Morrow, pp. 130–40. Reston, Va.: National Council of Teachers of Mathematics, 1998.

Kamii, Constance, Barbara A. Lewis, and Sally Jones Livingston. "Primary Arithmetic: Children Inventing Their Own Procedures." *Arithmetic Teacher* 41 (December 1993): 200–203.

Kamii, Constance K., with Leslie Baker Housman. *Young Children Reinvent Arithmetic: Implications of Piaget's Theory.* 2nd ed. New York: Teachers College Press, 2000.

Kazemi, Elham. "Discourse That Promotes Conceptual Understanding." *Teaching Children Mathematics* 4 (March 1998): 410–14.

Kenney, Patricia Ann, and Vicky L. Kouba. "What Do Students Know about Measurement?" In *Results from the Sixth Mathematics Assessment of the National Assessment of Educational Progress*, edited by Patricia Ann Kenney and Edward A. Silver, pp. 141–63. Reston, Va.: National Council of Teachers of Mathematics, 1997.

Kenney, Patricia Ann, and Edward A. Silver, eds. *Results from the Sixth Mathematics Assessment of the National Assessment of Educational Progress.* Reston, Va.: National Council of Teachers of Mathematics, 1997.

Keynes, Harvey B. "Preparing Students for Postsecondary Education." In *High School Mathematics at Work: Essays and Examples for the Education of All Students*, edited by National Research Council, Mathematical Sciences Education Board, pp. 107–10. Washington, D.C.: National Academy Press, 1998.

Kieran, Carolyn. "Concepts Associated with the Equality Symbol." *Educational Studies in Mathematics* 12 (August 1981): 317–26.

———. "Relationships between Novices' Views of Algebraic Letters and Their Use of Symmetric and Asymmetric Equation-Solving Procedures." In *Proceedings of the Fifth Annual Meeting of PME-NA*, vol. 1, edited by J. C. Bergeron and N. Herscovics, pp. 161–68. Montreal: Université de Montréal, 1983.

Kilpatrick, Jeremy, and Edward A. Silver. "Unfinished Business: Challenges for Mathematics Educators in the Next Decades." In *Learning Mathematics for a New Century*, edited by Maurice J. Burke, pp. 223–35. Reston, Va.: National Council of Teachers of Mathematics, 2000.

Klein, Alice, Prentice Starkey, and Ann Wakeley. "Enhancing Pre-kindergarten Children's Readiness for School Mathematics." Paper presented at the annual meeting of the American Educational Research Association, Montreal, April 1999.

Knapp, Michael S., Nancy E. Adelman, Camille Marder, Heather McCollum, Margaret C. Needels, Christine Padilla, Patrick M. Shields, Brenda J. Turnbull, and Andrew A. Zucker. *Teaching for Meaning in High-Poverty Schools.* New York: Teachers College Press, 1995.

Konold, Clifford. "Informal Conceptions of Probability." *Cognition and Instruction* 6, no. 1 (1989): 59–98.

———. "Probability, Statistics, and Data Analysis." In *A Research Companion to NCTM's Standards*, edited by Jeremy Kilpatrick, W. Gary Martin, and Deborah Schifter. Reston, Va.: National Council of Teachers of Mathematics, forthcoming.

Kouba, Vicky L., Thomas P. Carpenter, and Jane O. Swafford. "Number and Operations." In *Results from the Fourth Mathematics Assessment of the National Assessment of Educational Progress*, edited by Mary Montgomery Lindquist, pp. 64–93. Reston, Va.: National Council of Teachers of Mathematics, 1989.

Kouba, Vicki L., Judith S. Zawojewski, and Marilyn E. Strutchens. "What Do Students Know about Numbers and Operations?" In *Results from the Sixth Mathematics Assessment of the National Assessment of Educational Progress*, edited by Patricia Ann Kenney and Edward A. Silver, pp. 87–140. Reston, Va.: National Council of Teachers of Mathematics, 1997.

Kroll, Diana Lambdin, and Tammy Miller. "Insights from Research on Mathematical Problem Solving in the Middle Grades." In *Research Ideas for the Classroom, Middle Grades Mathematics*, National Council of Teachers of Mathematics Research Interpretation Project, edited by Douglas T. Owens, 58–77. New York: Macmillan Publishing Co., 1993.

Krutetskii, V. A. *The Psychology of Mathematical Abilities in Schoolchildren.* Chicago: University of Chicago Press, 1976.

Küchemann, D. "Children's Understanding of Numerical Variables." *Mathematics in School* 7, no. 4 (1978): 23–26.

Lampert, Magdalene. "Arithmetic as Problem Solving." *Arithmetic Teacher* 36 (March 1989): 34–36.

———. "Teaching Multiplication." *Journal of Mathematical Behavior* 5, no. 3 (December 1986): 241–80.

———. "When the Problem Is Not the Question and the Solution Is Not the Answer: Mathematical Knowing and Teaching." *American Educational Research Journal* 27, no. 1 (Spring 1990): 29–63.

Lampert, Magdalene, and Paul Cobb. "Communications and Language." In *A Research Companion to NCTM's Standards*, edited by Jeremy Kilpatrick, W. Gary Martin, and Deborah Schifter. Reston, Va.: National Council of Teachers of Mathematics, forthcoming.

Lappan, Glenda, James T. Fey, William M. Fitzgerald, Susan N. Friel, and Elizabeth Difanis Phillips. *Comparing and Scaling: Ratio, Proportion, and Percents.* Connected Mathematics series. Palo Alto, Calif.: Dale Seymour Publications, 1998.

Lehrer, Richard. "Developing Understanding of Measurement." In *A Research Companion to NCTM's Standards*, edited by Jeremy Kilpatrick, W. Gary Martin, and Deborah Schifter. Reston, Va.: National Council of Teachers of Mathematics, forthcoming.

Lehrer, Richard, Michael Jenkins, and Helen Osana. "Longitudinal Study of Children's Reasoning about Space and Geometry." In *Designing Learning Environments for Developing Understanding of Geometry and Space*, edited by Richard Lehrer and Daniel Chazan, pp. 137–67. Mahwah, N.J.: Lawrence Erlbaum Associates, 1998.

Leinhardt, Gaea, Orit Zaslavsky, and Mary Kay Stein. "Functions, Graphs, and Graphing: Tasks, Learning, and Teaching." *Review of Educational Research* 60, no. 1 (1990): 1–64.

Liben, Lynn S., and Roger M. Downs. "Understanding Maps as Symbols: The Development of Map Concepts in Children." In *Advances in Child Development and Behavior*, vol. 22, edited by Hayne W. Reese, pp. 145–201. San Diego, Calif.: Academic Press, 1989.

Lindquist, Mary M., and Vicky L. Kouba. "Measurement." In *Results from the Fourth Mathematics Assessment of the National Assessment of Educational Progress*, edited by Mary Montgomery Lindquist, pp. 35–43. Reston, Va.: National Council of Teachers of Mathematics, 1989.

Ma, Liping. *Knowing and Teaching Elementary Mathematics: Teachers' Understanding of Fundamental Mathematics in China and the United States.* Mahwah, N.J.: Lawrence Erlbaum Associates, 1999.

Mack, Nancy K. "Learning Fractions with Understanding: Building on Informal Knowledge." *Journal for Research in Mathematics Education* 21 (January 1990): 16–32.

Madell, Rob. "Children's Natural Processes." *Arithmetic Teacher* 32 (March 1985): 20–22.

Maher, Carolyn A., and Amy M. Martino. "The Development of the Idea of Mathematical Proof: A Five-Year Case Study." *Journal for Research in Mathematics Education* 27 (March 1996): 194–214.

Malloy, Carol E. "Mathematics Projects Promote Students' Algebraic Thinking." *Mathematics Teaching in the Middle School* 2 (February 1997): 282–88.

Mathematical Association of America, Committee of the Mathematical Education of Teachers. *A Call for Change: Recommendations for the Mathematical Preparation of Teachers of Mathematics*, edited by James R. C. Leitzel. Washington, D.C.: Mathematical Association of America, 1991.

McClain, Kay. "Reflecting on Students' Understanding of Data." *Mathematics Teaching in the Middle School* 4 (March 1999): 374–80.

McClain, Kay, Paul Cobb, and Janet Bowers. "A Contextual Investigation of Three-Digit Addition and Subtraction." In *The Teaching and Learning of Algorithms in School Mathematics*, 1998 Yearbook of the National Council of Teachers of Mathematics, edited by Lorna J. Morrow, pp. 141–50. Reston, Va.: National Council of Teachers of Mathematics, 1998.

Mokros, Jan, and Susan Jo Russell. "Children's Concepts of Average and Representativeness." *Journal for Research in Mathematics Education* 26 (January 1995): 20–39.

Moore, David S. "Uncertainty." *On the Shoulders of Giants: New Approaches to Numeracy*, edited by Lynn Arthur Steen, pp. 95–137. Washington, D.C.: National Academy Press, 1990.

Moore, R. C. "Making the Transition to Formal Proof." *Educational Studies in Mathematics* 27, no. 3 (1994): 249–66.

Moschkovich, Judit N. "Resources for Refining Mathematical Conceptions: Case Studies in Learning about Linear Functions." *Journal of the Learning Sciences* 7, no. 2 (1998): 209–37.

Moschkovich, Judit, Alan H. Schoenfeld, and Abraham Arcavi. "Aspects of Understanding: On Multiple Perspectives and Representations of Linear Relations and Connections among Them." In *Integrating Research on the Graphical Representation of Functions*, edited by Thomas A. Romberg, Elizabeth Fennema, and Thomas P. Carpenter, pp. 69–100. Hillsdale, N.J.: Lawrence Erlbaum Associates, 1993.

Mullis, Ina V. S., Michael O. Martin, Albert E. Beaton, Eugenio J. Gonzales, Dana L. Kelly, and Teresa A. Smith. *Mathematics Achievement in the Primary School Years: IEA's Third International Mathematics and Science Study (TIMSS).* Chestnut Hill, Mass.: Boston College, TIMSS International Study Center, 1997.

National Commission on Teaching and America's Future. *What Matters Most: Teaching for America's Future.* New York: National Commission on Teaching and America's Future, 1996.

National Council of Teachers of Mathematics. *Assessment Standards for School Mathematics.* Reston, Va.: National Council of Teachers of Mathematics, 1995.

———. *Curriculum and Evaluation Standards for School Mathematics.* Reston, Va.: National Council of Teachers of Mathematics, 1989.

———. *Professional Standards for Teaching Mathematics.* Reston, Va.: National Council of Teachers of Mathematics, 1991.

National Council of Teachers of Mathematics, Algebra Working Group. "A Framework for Constructing a Vision of Algebra: A Discussion Document." Appendix E in *The Nature and Role of Algebra in the K–14 Curriculum: Proceedings of a National Symposium, May 27 and 28, 1997*, edited by the National Council of Teacher of Mathematics and the National Research Council, pp. 145–90. Washington, D.C.: National Academy Press, 1998. Available on the Web at books.nap.edu/books/0309061474/html/R1.html.

National Research Council. *High School Mathematics at Work: Essays and Examples for the Education of All Students.* Washington, D.C.: National Academy Press, 1998.

———. *National Science Education Standards.* Washington, D.C.: National Academy Press, 1996.

National Research Council, Mathematical Sciences Education Board. *Everybody Counts: A Report to the Nation on the Future of Mathematics Education.* Washington, D.C.: National Academy Press, 1989.

Nelson, Barbara S. *Building New Knowledge by Thinking: How Administrators Can Learn What They Need to Know about Mathematics Education Reform.* Newton, Mass.: Center for the Development of Teaching, Education Development Center, 1999.

Pajitnov, Alexeyo. Tetris. San Francisco, Calif.: Blue Planet Software, 1996.

Pólya, George. *How to Solve It: A New Aspect of Mathematical Method.* 2nd ed. Princeton, N.J.: Princeton University Press, 1957.

Price, Glenda. "Quantitative Literacy across the Curriculum." In *Why Numbers Count: Quantitative Literacy for Tomorrow's America*, edited by Lynn Arthur Steen, pp. 155–60. New York: College Entrance Examination Board, 1997.

Putnam, Ralph T., and Hilda Borko. "What Do New Views of Knowledge and Thinking Have to Say about Research on Teacher Learning?" *Educational Researcher* 29 (January-February 2000): 4–15.

Raimi, Ralph A., and Lawrence S. Braden. *State Mathematics Standards: An Appraisal of Math Standards in Forty-six States, the District of Columbia, and Japan.* Washington, D.C.: Thomas B. Fordham Foundation, 1998.

Razel, Micha, and Bat-Sheva Eylon. "Developing Mathematics Readiness in Young Children with the Agam Program." Paper presented at the Fifteenth Conference of the International Group for the Psychology of Mathematics Education, Genoa, Italy, 1991.

Resnick, Lauren B. *Education and Learning to Think.* Washington, D.C.: National Academy Press, 1987.

Resnick, Lauren B., Pearla Nesher, Francois Leonard, Maria Magone, Susan Omanson, and Irit Peled. "Conceptual Bases of Arithmetic Errors: The Case of Decimal Fractions." *Journal for Research in Mathematics Education* 20 (January 1989): 8–27.

Reys, Robert E., and Der-Ching Yang. "Relationship between Computational Performance and Number Sense among Sixth- and Eighth-Grade Students in Taiwan." *Journal for Research in Mathematics Education* 29 (March 1998): 225–37.

Reys, Robert E., and Nobuhiko Nohda, eds. *Computational Alternatives for the Twenty-first Century: Cross-Cultural Perspectives from Japan and the United States.* Reston, Va.: National Council of Teachers of Mathematics, 1994.

Riedesel, C. Alan. *Teaching Elementary School Mathematics.* 3rd ed. Englewood Cliffs, N.J.: Prentice-Hall, 1980.

Roberts, Fred S. "The Role of Applications in Teaching Discrete Mathematics." In *Discrete Mathematics in the Schools*, edited by Joseph G. Rosenstein, Deborah S. Franzblau, and Fred S. Roberts, pp. 105–17. DIMACS Series in Discrete Mathematics and Theoretical Computer Science, vol. 36. Providence, R.I.: American Mathematical Society and National Council of Teachers of Mathematics, 1997.

Roche, Robert. "Cranberry Estimation." In *Estimation*, produced by WGBH Boston. Teaching Math, a Video Library, K–4. Boston: WGBH Boston, 1996.

Rojano, Teresa. "Developing Algebraic Aspects of Problem Solving within a Spreadsheet Environment." In *Approaches to Algebra: Perspectives for Research and Teaching*, edited by Nadine Bednarz, Carolyn Kieran, and Lesley Lee. Boston: Kluwer Academic Publishers, 1996.

Russell, Susan Jo, Douglas H. Clements, and Julie Sarama. "Building a Town." In *Quilt Squares and Block Towns: 2-D and 3-D Geometry*, edited by Catherine Anderson and Beverly Cory, pp. 113–18. Menlo Park, Calif.: Dale Seymour Publications, 1998.

Russell, Susan Jo, Deborah Schifter, and Virginia Bastable. *Working with Data Casebook*. Pilot-test draft. Developing Mathematical Ideas. Newton, Mass.: Education Development Center, 1999.

Sarmiento, Jorge L. "Thousand-Year Record Documents Rise in Atmospheric CO_2." *Chemical and Engineering News*, 31 May 1993, pp. 30–43.

Schifter, Deborah. "Reasoning about Operations: Early Algebraic Thinking in Grades K–6." In *Developing Mathematical Reasoning in Grades K–12*, 1999 Yearbook of the National Council of Teachers of Mathematics, edited by Lee V. Stiff, pp. 62–81. Reston, Va.: National Council of Teachers of Mathematics, 1999.

Schifter, Deborah, Virginia Bastable, and Susan Jo Russell. *Building a System of Tens Casebook*. Developing Mathematical Ideas: Number and Operations, Part 1. Parsippany, N.J.: Dale Seymour Publications, 1999.

Schoenfeld, Alan H. *Mathematical Problem Solving*. Orlando, Fla.: Academic Press, 1985.

———. "What's All the Fuss about Metacognition?" In *Cognitive Science and Mathematics Education*, edited by Alan H. Schoenfeld, pp. 189–215. Hillsdale, N.J.: Lawrence Erlbaum Associates, 1987.

———. "When Good Teaching Leads to Bad Results: The Disasters of Well Taught Mathematics Classes." *Educational Psychologist* 23 (Spring 1988): 145–66.

Schoenfeld, Alan H., and Abraham Arcavi. "On the Meaning of Variable." *Mathematics Teacher* 81 (September 1988): 420–27.

Schroeder, Thomas L., and Frank K. Lester, Jr. "Developing Understanding in Mathematics via Problem Solving." In *New Directions for Elementary School Mathematics*, 1989 Yearbook of the National Council of Teachers of Mathematics, edited by Paul R. Trafton, pp. 31–42. Reston, Va.: National Council of Teachers of Mathematics, 1989.

Schwartz, Daniel L., Susan R. Goldman, Nancy J. Vye, and Brigid J. Barron. "Aligning Everyday and Mathematical Reasoning: The Case of Sampling Assumptions." In *Reflections on Statistics: Learning, Teaching, and Assessment in Grades K–12* , edited by Suzanne P. Lajoie, pp. 233–73. Mahwah, N.J.: Lawrence Erlbaum Associates, 1998.

Senk, Sharon L. "Van Hiele Levels and Achievement in Writing Geometry Proofs." *Journal for Research in Mathematics Education* 20 (May 1989): 309–21.

Senk, Sharon L., Denisse R. Thompson, Steven S. Viktora, Rheta Rubenstein, Judy Halvorson, James Flanders, Natalie Jakucyn, Gerald Pillsbury, and Zalman Usiskin. *UCSMP Advanced Algebra*. 2nd ed. Glenview, Ill.: Scott, Foresman & Co., 1996.

Sfard, Ana. "On the Dual Nature of Mathematical Conceptions: Reflections on Processes and Objects as Different Sides of the Same Coin." *Educational Studies in Mathematics* 22 (1991): 1–36.

Sheets, Charlene. "Effects of Computer Learning and Problem-Solving Tools on the Development of Secondary School Students' Understanding of Mathematical Functions." Ph.D. diss., University of Maryland College Park, 1993.

Siegler, Robert S. *Emerging Minds: The Process of Change in Children's Thinking*. New York: Oxford University Press, 1996.

Silver, Edward A., Jeremy Kilpatrick, and Beth G. Schlesinger. *Thinking through Mathematics: Fostering Inquiry and Communication in Mathematics Classrooms*. New York: College Entrance Examination Board, 1990.

Silver, Edward A., and Margaret S. Smith. "Implementing Reform in the Mathematics Classroom: Creating Mathematical Discourse Communities." In *Reform in Math and Science Education: Issues for Teachers*. Columbus, Ohio: Eisenhower National Clearinghouse for Mathematics and Science Education, 1997. CD-ROM.

Silver, Edward A., Margaret Schwan Smith, and Barbara Scott Nelson. "The QUASAR Project: Equity Concerns Meet Mathematics Education Reform in the Middle School." In *New Directions for Equity in Mathematics Education*, edited by Walter G. Secada, Elizabeth Fennema, and Lisa Byrd Adajian, pp. 9–56. New York: Cambridge University Press, 1995.

Silver, Edward A., and Mary Kay Stein. "The QUASAR Project: The 'Revolution of the Possible' in Mathematics Instructional Reform in Urban Middle Schools." *Urban Education* 30 (January1996): 476–521.

Silver, Edward A., Marilyn E. Strutchens, and Judith S. Zawojewski. "NAEP Findings regarding Race/Ethnicity and Gender: Affective Issues, Mathematics Performance, and Instructional Context." In *Results from the Sixth Mathematics Assessment of the National Assessment of Educational Progress*, edited by Patricia Ann Kenney and Edward A. Silver, pp. 33–59. Reston, Va.: National Council of Teachers of Mathematics, 1997.

Silverstein, Shel. "One Inch Tall." In *Where the Sidewalk Ends: The Poems and Drawings of Shel Silverstein*, p. 55. New York: Harper and Row, 1974.

Skemp, Richard R. "Relational Understanding and Instrumental Understanding." *Mathematics Teaching* 77 (December 1976): 20–26. Reprinted in *Arithmetic Teacher* (November 1978): 9–15.

Smith, John P. III, Andrea A. diSessa, and Jeremy Roschelle. "Misconceptions Reconceived: A Constructivist Analysis of Knowledge in Transition." *Journal of the Learning Sciences* 3, no. 2 (1993): 115–63.

Smith, Erick. "Patterns, Functions, and Algebra." In *A Research Companion to NCTM's Standards*, edited by Jeremy Kilpatrick, W. Gary Martin, and Deborah Schifter. Reston, Va.: National Council of Teachers of Mathematics, forthcoming.

Smith, Margaret S. "Balancing on a Sharp, Thin Edge: A Study of Teacher Learning in the Context of Mathematics Instructional Reform." *Elementary School Journal* (forthcoming).

Smith, Ray. *An Introduction to Perspective*. New York: D K Publishing, 1995.

Society for Industrial and Applied Mathematics. *SIAM Report on Mathematics in Industry*. Philadelphia: Society for Industrial and Applied Mathematics, 1996.

Sowder, Judith T. "Making Sense of Numbers in School Mathematics." In *Analysis of Arithmetic for Mathematics Teaching*, edited by Gaea Leinhardt, Ralph Putman, and Rosemary A. Hattrup, pp. 1–51. Hillsdale, N.J.: Lawrence Erlbaum Associates, 1992.

Starkey, Prentice, and Robert G. Cooper, Jr. "Perception of Numbers by Human Infants." *Science* 210 (November 1980): 1033–35.

Steen, Lynn Arthur. "Preface: The New Literacy." In *Why Numbers Count: Quantitative Literacy for Tomorrow's America*, edited by Lynn Arthur Steen, pp. xv–xxviii. New York: College Entrance Examination Board, 1997.

———. "The Science of Patterns." *Science* 240 (April 1988): pp. 611–16.

Steffe, Leslie P. "Children's Multiplying Schemes." In *The Development of Multiplicative Reasoning in the Learning of Mathematics*, edited by Guershon Harel and Jere Confrey, pp. 3–39. Albany, N.Y.: State University of New York Press, 1994.

Steffe, Leslie P., and Paul Cobb. *Construction of Arithmetical Meanings and Strategies*. New York: Springer-Verlag, 1988.

Stigler, James W., and James Hiebert. *The Teaching Gap: Best Ideas from the World's Teachers for Improving Education in the Classroom*. New York: The Free Press, 1999.

Thompson, Alba G. "Teachers' Beliefs and Conceptions: A Synthesis of the Research." In *Handbook of Research on Mathematics Teaching and Learning*, edited by Douglas A. Grouws, pp. 127–46. New York: Macmillan Publishing Co., 1992.

Thompson, Patrick W. "To Understand Post-counting Numbers and Operations." In *A Research Companion to NCTM's Standards*, edited by Jeremy Kilpatrick, W. Gary Martin, and Deborah Schifter. Reston, Va.: National Council of Teachers of Mathematics, forthcoming.

Thornton, Carol A. "Strategies for the Basic Facts." In *Mathematics for the Young Child*, edited by Joseph N. Payne, pp. 133–51. Reston, Va.: National Council of Teachers of Mathematics, 1990.

Tierney, Cornelia C., and Mary Berle-Carman. *Fractions: Fair Shares*. Investigations in Number, Data, and Space. Palo Alto, Calif.: Dale Seymour Publications, 1995.

Trafton, Paul R., and Christina L. Hartman. "Developing Number Sense and Computational Strategies in Problem-Centered Classrooms." *Teaching Children Mathematics* 4 (December 1997): 230–33.

Tucker, Alan. "New GRE Mathematical Reasoning Test." *Focus* 15 (February 1995): 14–16.

Uccellini, John C. "Teaching the Mean Meaningfully." *Mathematics Teaching in the Middle School* 2 (November–December 1996): 112–15.

University of North Carolina Mathematics and Science Education Network. *Teach-Stat Activities: Statistical Investigations for Grades 3 through 6*. Palo Alto, Calif.: Dale Seymour Publications, 1997.

U.S. Department of Education. *Exemplary and Promising Mathematics Programs*. Washington, D.C.: U.S. Department of Education, 1999.

U.S. Department of Labor, Secretary's Commission on Achieving Necessary Skills. *What Work Requires of Schools: A SCANS Report for America 2000*. Washington, D.C.: U.S. Department of Labor, 1991.

van Hiele, Pierre M. *Structure and Insight: A Theory of Mathematics Education*. Orlando, Fla.: Academic Press, 1986.

Vinner, Shlomo, and Tommy Dreyfus. "Images and Definitions for the Concept of Function." *Journal for Research in Mathematics Education* 20 (July 1989): 356–66.

Wagner, Sigrid, and Sheila Parker. "Advancing Algebra." In *Research Ideas for the Classroom, High School Mathematics*, edited by Patricia S. Wilson, pp. 119–39. New York: Macmillan Publishing Co., 1993.

Waits, Bert, and Franklin Demana. "The Role of Hand-Held Computer Symbolic Algebra in Mathematics Education in the Twenty-first Century: A Call for Action!" Paper presented at the Technology and NCTM Standards 2000 Conference, Arlington, Va., June 1998.

Wheelock, Anne. *Crossing the Tracks: How "Untracking" Can Save America's Schools*. New York: New Press, 1992.

Wilson, Linda, and Patricia Ann Kenney. "Assessment." In *A Research Companion to NCTM's Standards*, edited by Jeremy Kilpatrick, W. Gary Martin, and Deborah Schifter. Reston, Va.: National Council of Teachers of Mathematics, forthcoming.

Wilson, S. M., L. S. Shulman, and A. Richert. "One Hundred Fifty Ways of Knowing." In *Exploring Teachers' Thinking*, edited by James Calderhead, pp. 104–24. London: Cassell, 1987.

Wright, June L., and Daniel D. Shade. *Young Children: Active Learners in a Technological Age*. Washington, D.C.: National Association for the Education of Young Children, 1994.

Yackel, Erna, and Paul Cobb. "The Development of Young Children's Understanding of Mathematical Argumentation." Paper presented at the annual meeting of the American Educational Research Association, New Orleans, 1994.

———. "Sociomathematical Norms, Argumentation, and Autonomy in Mathematics." *Journal for Research in Mathematics Education* 27 (July 1996): 458–77.

Yackel, Erna, and Grayson H. Wheatley. "Promoting Visual Imagery in Young Pupils." *Arithmetic Teacher* 37 (February 1990): 52–58.

Yates, Billy Charles. "The Computer as an Instructional Aid and Problem Solving Tool: An Experimental Analysis of Two Instructional Methods for Teaching Spatial Skills to Junior High School Students." (Doctoral diss., University of Oregon.) *Dissertation Abstracts International* 49 (1988): 3612-A. (University Microfilms no. DA8903857).

Yerushalmy, Michal, and Judah L. Schwartz. "Seizing the Opportunity to Make Algebra Mathematically and Pedagogically Interesting." In *Integrating Research on the Graphical Representation of Functions*, edited by Thomas A. Romberg, Elizabeth Fennema, and Thomas P. Carpenter, pp. 41–68. Hillsdale, N.J.: Lawrence Erlbaum Associates, 1993.

Zbiek, Rose Mary. "The Pentagon Problem: Geometric Reasoning with Technology." *Mathematics Teacher* 89 (February 1996): 86–90.

Zbiek, Rose Mary, and B. Glass. "Conjecturing and Formal Reasoning about Functions in a Dynamic Environment." *Proceedings of the Twelfth International Conference on Technology in Collegiate Mathematics*, edited by G. Goodell. Reading, Mass.: Addison Wesley Longman, forthcoming.

Zbiek, Rose Mary, and M. Kathleen Heid. "The Skateboard Experiment: Mathematical Modeling for Beginning Algebra." *Computing Teacher* 18, no. 2 (1990): 32–36.

Zeisel, Hans. "Statistics as Legal Evidence." In *International Encyclopedia of the Social Sciences*, vol. 15, edited by David L. Sills, pp. 246–50. New York: Crowell Collier and Macmillan, 1968.

Appendix

Table of Standards and Expectations

Number and Operations

STANDARD *Instructional programs from prekindergarten through grade 12 should enable all students to—*	**Pre-K–2** **Expectations** In prekindergarten through grade 2 all students should–	**Grades 3–5** **Expectations** In grades 3–5 all students should–
Understand numbers, ways of representing numbers, relationships among numbers, and number systems	• count with understanding and recognize "how many" in sets of objects; • use multiple models to develop initial understandings of place value and the base-ten number system; • develop understanding of the relative position and magnitude of whole numbers and of ordinal and cardinal numbers and their connections; • develop a sense of whole numbers and represent and use them in flexible ways, including relating, composing, and decomposing numbers; • connect number words and numerals to the quantities they represent, using various physical models and representations; • understand and represent commonly used fractions, such as 1/4, 1/3, and 1/2.	• understand the place-value structure of the base-ten number system and be able to represent and compare whole numbers and decimals; • recognize equivalent representations for the same number and generate them by decomposing and composing numbers; • develop understanding of fractions as parts of unit wholes, as parts of a collection, as locations on number lines, and as divisions of whole numbers; • use models, benchmarks, and equivalent forms to judge the size of fractions; • recognize and generate equivalent forms of commonly used fractions, decimals, and percents; • explore numbers less than 0 by extending the number line and through familiar applications; • describe classes of numbers according to characteristics such as the nature of their factors.
Understand meanings of operations and how they relate to one another	• understand various meanings of addition and subtraction of whole numbers and the relationship between the two operations; • understand the effects of adding and subtracting whole numbers; • understand situations that entail multiplication and division, such as equal groupings of objects and sharing equally.	• understand various meanings of multiplication and division; • understand the effects of multiplying and dividing whole numbers; • identify and use relationships between operations, such as division as the inverse of multiplication, to solve problems; • understand and use properties of operations, such as the distributivity of multiplication over addition.
Compute fluently and make reasonable estimates	• develop and use strategies for whole-number computations, with a focus on addition and subtraction; • develop fluency with basic number combinations for addition and subtraction; • use a variety of methods and tools to compute, including objects, mental computation, estimation, paper and pencil, and calculators.	• develop fluency with basic number combinations for multiplication and division and use these combinations to mentally compute related problems, such as 30×50; • develop fluency in adding, subtracting, multiplying, and dividing whole numbers; • develop and use strategies to estimate the results of whole-number computations and to judge the reasonableness of such results; • develop and use strategies to estimate computations involving fractions and decimals in situations relevant to students' experience; • use visual models, benchmarks, and equivalent forms to add and subtract commonly used fractions and decimals; • select appropriate methods and tools for computing with whole numbers from among mental computation, estimation, calculators, and paper and pencil according to the context and nature of the computation and use the selected method or tool.

Number and Operations

Standard *Instructional programs from prekindergarten through grade 12 should enable all students to—*	Grades 6–8 **Expectations** In grades 6–8 all students should–	Grades 9–12 **Expectations** In grades 9–12 all students should–
Understand numbers, ways of representing numbers, relationships among numbers, and number systems	• work flexibly with fractions, decimals, and percents to solve problems; • compare and order fractions, decimals, and percents efficiently and find their approximate locations on a number line; • develop meaning for percents greater than 100 and less than 1; • understand and use ratios and proportions to represent quantitative relationships; • develop an understanding of large numbers and recognize and appropriately use exponential, scientific, and calculator notation; • use factors, multiples, prime factorization, and relatively prime numbers to solve problems; • develop meaning for integers and represent and compare quantities with them.	• develop a deeper understanding of very large and very small numbers and of various representations of them; • compare and contrast the properties of numbers and number systems, including the rational and real numbers, and understand complex numbers as solutions to quadratic equations that do not have real solutions; • understand vectors and matrices as systems that have some of the properties of the real-number system; • use number-theory arguments to justify relationships involving whole numbers.
Understand meanings of operations and how they relate to one another	• understand the meaning and effects of arithmetic operations with fractions, decimals, and integers; • use the associative and commutative properties of addition and multiplication and the distributive property of multiplication over addition to simplify computations with integers, fractions, and decimals; • understand and use the inverse relationships of addition and subtraction, multiplication and division, and squaring and finding square roots to simplify computations and solve problems.	• judge the effects of such operations as multiplication, division, and computing powers and roots on the magnitudes of quantities; • develop an understanding of properties of, and representations for, the addition and multiplication of vectors and matrices; • develop an understanding of permutations and combinations as counting techniques.
Compute fluently and make reasonable estimates	• select appropriate methods and tools for computing with fractions and decimals from among mental computation, estimation, calculators or computers, and paper and pencil, depending on the situation, and apply the selected methods; • develop and analyze algorithms for computing with fractions, decimals, and integers and develop fluency in their use; • develop and use strategies to estimate the results of rational-number computations and judge the reasonableness of the results; • develop, analyze, and explain methods for solving problems involving proportions, such as scaling and finding equivalent ratios.	• develop fluency in operations with real numbers, vectors, and matrices, using mental computation or paper-and-pencil calculations for simple cases and technology for more-complicated cases. • judge the reasonableness of numerical computations and their results.

Algebra
Standard

Instructional programs from prekindergarten through grade 12 should enable all students to—	**Pre-K–2** **Expectations** In prekindergarten through grade 2 all students should–	**Grades 3–5** **Expectations** In grades 3–5 all students should–
Understand patterns, relations, and functions	• sort, classify, and order objects by size, number, and other properties; • recognize, describe, and extend patterns such as sequences of sounds and shapes or simple numeric patterns and translate from one representation to another; • analyze how both repeating and growing patterns are generated.	• describe, extend, and make generalizations about geometric and numeric patterns; • represent and analyze patterns and functions, using words, tables, and graphs.
Represent and analyze mathematical situations and structures using algebraic symbols	• illustrate general principles and properties of operations, such as commutativity, using specific numbers; • use concrete, pictorial, and verbal representations to develop an understanding of invented and conventional symbolic notations.	• identify such properties as commutativity, associativity, and distributivity and use them to compute with whole numbers; • represent the idea of a variable as an unknown quantity using a letter or a symbol; • express mathematical relationships using equations.
Use mathematical models to represent and understand quantitative relationships	• model situations that involve the addition and subtraction of whole numbers, using objects, pictures, and symbols.	• model problem situations with objects and use representations such as graphs, tables, and equations to draw conclusions.
Analyze change in various contexts	• describe qualitative change, such as a student's growing taller; • describe quantitative change, such as a student's growing two inches in one year.	• investigate how a change in one variable relates to a change in a second variable; • identify and describe situations with constant or varying rates of change and compare them.

Algebra
Standard

Instructional programs from prekindergarten through grade 12 should enable all students to—	**Grades 6–8** **Expectations** In grades 6–8 all students should–	**Grades 9–12** **Expectations** In grades 9–12 all students should–
Understand patterns, relations, and functions	• represent, analyze, and generalize a variety of patterns with tables, graphs, words, and, when possible, symbolic rules; • relate and compare different forms of representation for a relationship; • identify functions as linear or nonlinear and contrast their properties from tables, graphs, or equations.	• generalize patterns using explicitly defined and recursively defined functions; • understand relations and functions and select, convert flexibly among, and use various representations for them; • analyze functions of one variable by investigating rates of change, intercepts, zeros, asymptotes, and local and global behavior; • understand and perform transformations such as arithmetically combining, composing, and inverting commonly used functions, using technology to perform such operations on more-complicated symbolic expressions; • understand and compare the properties of classes of functions, including exponential, polynomial, rational, logarithmic, and periodic functions; • interpret representations of functions of two variables.
Represent and analyze mathematical situations and structures using algebraic symbols	• develop an initial conceptual understanding of different uses of variables; • explore relationships between symbolic expressions and graphs of lines, paying particular attention to the meaning of intercept and slope; • use symbolic algebra to represent situations and to solve problems, especially those that involve linear relationships; • recognize and generate equivalent forms for simple algebraic expressions and solve linear equations.	• understand the meaning of equivalent forms of expressions, equations, inequalities, and relations; • write equivalent forms of equations, inequalities, and systems of equations and solve them with fluency–mentally or with paper and pencil in simple cases and using technology in all cases; • use symbolic algebra to represent and explain mathematical relationships; • use a variety of symbolic representations, including recursive and parametric equations, for functions and relations; • judge the meaning, utility, and reasonableness of the results of symbol manipulations, including those carried out by technology.
Use mathematical models to represent and understand quantitative relationships	• model and solve contextualized problems using various representations, such as graphs, tables, and equations.	• identify essential quantitative relationships in a situation and determine the class or classes of functions that might model the relationships; • use symbolic expressions, including iterative and recursive forms, to represent relationships arising from various contexts; • draw reasonable conclusions about a situation being modeled.
Analyze change in various contexts	• use graphs to analyze the nature of changes in quantities in linear relationships.	• approximate and interpret rates of change from graphical and numerical data.

Geometry

Standard

Instructional programs from prekindergarten through grade 12 should enable all students to—	**Pre-K–2** **Expectations** In prekindergarten through grade 2 all students should–	**Grades 3–5** **Expectations** In grades 3–5 all students should–
Analyze characteristics and properties of two- and three-dimensional geometric shapes and develop mathematical arguments about geometric relationships	• recognize, name, build, draw, compare, and sort two- and three-dimensional shapes; • describe attributes and parts of two- and three-dimensional shapes; • investigate and predict the results of putting together and taking apart two- and three-dimensional shapes.	• identify, compare, and analyze attributes of two- and three-dimensional shapes and develop vocabulary to describe the attributes; • classify two- and three-dimensional shapes according to their properties and develop definitions of classes of shapes such as triangles and pyramids; • investigate, describe, and reason about the results of subdividing, combining, and transforming shapes; • explore congruence and similarity; • make and test conjectures about geometric properties and relationships and develop logical arguments to justify conclusions.
Specify locations and describe spatial relationships using coordinate geometry and other representational systems	• describe, name, and interpret relative positions in space and apply ideas about relative position; • describe, name, and interpret direction and distance in navigating space and apply ideas about direction and distance; • find and name locations with simple relationships such as "near to" and in coordinate systems such as maps.	• describe location and movement using common language and geometric vocabulary; • make and use coordinate systems to specify locations and to describe paths; • find the distance between points along horizontal and vertical lines of a coordinate system.
Apply transformations and use symmetry to analyze mathematical situations	• recognize and apply slides, flips, and turns; • recognize and create shapes that have symmetry.	• predict and describe the results of sliding, flipping, and turning two-dimensional shapes; • describe a motion or a series of motions that will show that two shapes are congruent; • identify and describe line and rotational symmetry in two- and three-dimensional shapes and designs.
Use visualization, spatial reasoning, and geometric modeling to solve problems	• create mental images of geometric shapes using spatial memory and spatial visualization; • recognize and represent shapes from different perspectives; • relate ideas in geometry to ideas in number and measurement; • recognize geometric shapes and structures in the environment and specify their location	• build and draw geometric objects; • create and describe mental images of objects, patterns, and paths; • identify and build a three-dimensional object from two-dimensional representations of that object; • identify and build a two-dimensional representation of a three-dimensional object; • use geometric models to solve problems in other areas of mathematics, such as number and measurement; • recognize geometric ideas and relationships and apply them to other disciplines and to problems that arise in the classroom or in everyday life.

Geometry
Standard

Instructional programs from prekindergarten through grade 12 should enable all students to—	**Grades 6–8** **Expectations** In grades 6–8 all students should–	**Grades 9–12** **Expectations** In grades 9–12 all students should–
Analyze characteristics and properties of two- and three-dimensional geometric shapes and develop mathematical arguments about geometric relationships	• precisely describe, classify, and understand relationships among types of two- and three-dimensional objects using their defining properties; • understand relationships among the angles, side lengths, perimeters, areas, and volumes of similar objects; • create and critique inductive and deductive arguments concerning geometric ideas and relationships, such as congruence, similarity, and the Pythagorean relationship.	• analyze properties and determine attributes of two- and three-dimensional objects; • explore relationships (including congruence and similarity) among classes of two- and three-dimensional geometric objects, make and test conjectures about them, and solve problems involving them; • establish the validity of geometric conjectures using deduction, prove theorems, and critique arguments made by others; • use trigonometric relationships to determine lengths and angle measures.
Specify locations and describe spatial relationships using coordinate geometry and other representational systems	• use coordinate geometry to represent and examine the properties of geometric shapes; • use coordinate geometry to examine special geometric shapes, such as regular polygons or those with pairs of parallel or perpendicular sides.	• use Cartesian coordinates and other coordinate systems, such as navigational, polar, or spherical systems, to analyze geometric situations; • investigate conjectures and solve problems involving two- and three-dimensional objects represented with Cartesian coordinates.
Apply transformations and use symmetry to analyze mathematical situations	• describe sizes, positions, and orientations of shapes under informal transformations such as flips, turns, slides, and scaling; • examine the congruence, similarity, and line or rotational symmetry of objects using transformations.	• understand and represent translations, reflections, rotations, and dilations of objects in the plane by using sketches, coordinates, vectors, function notation, and matrices; • use various representations to help understand the effects of simple transformations and their compositions.
Use visualization, spatial reasoning, and geometric modeling to solve problems	• draw geometric objects with specified properties, such as side lengths or angle measures; • use two-dimensional representations of three-dimensional objects to visualize and solve problems such as those involving surface area and volume; • use visual tools such as networks to represent and solve problems; • use geometric models to represent and explain numerical and algebraic relationships; • recognize and apply geometric ideas and relationships in areas outside the mathematics classroom, such as art, science, and everyday life.	• draw and construct representations of two- and three-dimensional geometric objects using a variety of tools; • visualize three-dimensional objects from different perspectives and analyze their cross sections; • use vertex-edge graphs to model and solve problems; • use geometric models to gain insights into, and answer questions in, other areas of mathematics; • use geometric ideas to solve problems in, and gain insights into, other disciplines and other areas of interest such as art and architecture.

Measurement

Standard	Pre-K–2	Grades 3–5
Instructional programs from prekindergarten through grade 12 should enable all students to—	**Expectations** In prekindergarten through grade 2 all students should–	**Expectations** In grades 3–5 all students should–
Understand measurable attributes of objects and the units, systems, and processes of measurement	• recognize the attributes of length, volume, weight, area, and time; • compare and order objects according to these attributes; • understand how to measure using nonstandard and standard units; • select an appropriate unit and tool for the attribute being measured.	• understand such attributes as length, area, weight, volume, and size of angle and select the appropriate type of unit for measuring each attribute; • understand the need for measuring with standard units and become familiar with standard units in the customary and metric systems; • carry out simple unit conversions, such as from centimeters to meters, within a system of measurement; • understand that measurements are approximations and understand how differences in units affect precision; • explore what happens to measurements of a two-dimensional shape such as its perimeter and area when the shape is changed in some way.
Apply appropriate techniques, tools, and formulas to determine measurements	• measure with multiple copies of units of the same size, such as paper clips laid end to end; • use repetition of a single unit to measure something larger than the unit, for instance, measuring the length of a room with a single meterstick; • use tools to measure; • develop common referents for measures to make comparisons and estimates.	• develop strategies for estimating the perimeters, areas, and volumes of irregular shapes; • select and apply appropriate standard units and tools to measure length, area, volume, weight, time, temperature, and the size of angles; • select and use benchmarks to estimate measurements; • develop, understand, and use formulas to find the area of rectangles and related triangles and parallelograms; • develop strategies to determine the surface areas and volumes of rectangular solids.

Measurement

STANDARD *Instructional programs from prekindergarten through grade 12 should enable all students to—*	Grades 6–8 **Expectations** In grades 6–8 all students should–	Grades 9–12 **Expectations** In grades 9–12 all students should–
Understand measurable attributes of objects and the units, systems, and processes of measurement	• understand both metric and customary systems of measurement; • understand relationships among units and convert from one unit to another within the same system; • understand, select, and use units of appropriate size and type to measure angles, perimeter, area, surface area, and volume.	• make decisions about units and scales that are appropriate for problem situations involving measurement.
Apply appropriate techniques, tools, and formulas to determine measurements	• use common benchmarks to select appropriate methods for estimating measurements; • select and apply techniques and tools to accurately find length, area, volume, and angle measures to appropriate levels of precision; • develop and use formulas to determine the circumference of circles and the area of triangles, parallelograms, trapezoids, and circles and develop strategies to find the area of more-complex shapes; • develop strategies to determine the surface area and volume of selected prisms, pyramids, and cylinders; • solve problems involving scale factors, using ratio and proportion; • solve simple problems involving rates and derived measurements for such attributes as velocity and density.	• analyze precision, accuracy, and approximate error in measurement situations; • understand and use formulas for the area, surface area, and volume of geometric figures, including cones, spheres, and cylinders; • apply informal concepts of successive approximation, upper and lower bounds, and limit in measurement situations; • use unit analysis to check measurement computations.

Data Analysis and Probability

Standard	Pre-K–2	Grades 3–5
Instructional programs from prekindergarten through grade 12 should enable all students to—	**Expectations** In prekindergarten through grade 2 all students should–	**Expectations** In grades 3–5 all students should–
Formulate questions that can be addressed with data and collect, organize, and display relevant data to answer them	• pose questions and gather data about themselves and their surroundings; • sort and classify objects according to their attributes and organize data about the objects; • represent data using concrete objects, pictures, and graphs.	• design investigations to address a question and consider how data-collection methods affect the nature of the data set; • collect data using observations, surveys, and experiments; • represent data using tables and graphs such as line plots, bar graphs, and line graphs; • recognize the differences in representing categorical and numerical data.
Select and use appropriate statistical methods to analyze data	• describe parts of the data and the set of data as a whole to determine what the data show.	• describe the shape and important features of a set of data and compare related data sets, with an emphasis on how the data are distributed; • use measures of center, focusing on the median, and understand what each does and does not indicate about the data set; • compare different representations of the same data and evaluate how well each representation shows important aspects of the data.
Develop and evaluate inferences and predictions that are based on data	• discuss events related to students' experiences as likely or unlikely.	• propose and justify conclusions and predictions that are based on data and design studies to further investigate the conclusions or predictions.
Understand and apply basic concepts of probability		• describe events as likely or unlikely and discuss the degree of likelihood using such words as *certain*, *equally likely*, and *impossible*; • predict the probability of outcomes of simple experiments and test the predictions; • understand that the measure of the likelihood of an event can be represented by a number from 0 to 1.

Data Analysis and Probability

STANDARD *Instructional programs from prekindergarten through grade 12 should enable all students to—*	Grades 6–8 Expectations In grades 6–8 all students should–	Grades 9–12 Expectations In grades 9–12 all students should–
Formulate questions that can be addressed with data and collect, organize, and display relevant data to answer them	• formulate questions, design studies, and collect data about a characteristic shared by two populations or different characteristics within one population; • select, create, and use appropriate graphical representations of data, including histograms, box plots, and scatterplots.	• understand the differences among various kinds of studies and which types of inferences can legitimately be drawn from each; • know the characteristics of well-designed studies, including the role of randomization in surveys and experiments; • understand the meaning of measurement data and categorical data, of univariate and bivariate data, and of the term variable; • understand histograms, parallel box plots, and scatterplots and use them to display data; • compute basic statistics and understand the distinction between a statistic and a parameter.
Select and use appropriate statistical methods to analyze data	• find, use, and interpret measures of center and spread, including mean and interquartile range; • discuss and understand the correspondence between data sets and their graphical representations, especially histograms, stem-and-leaf plots, box plots, and scatterplots.	• for univariate measurement data, be able to display the distribution, describe its shape, and select and calculate summary statistics; • for bivariate measurement data, be able to display a scatterplot, describe its shape, and determine regression coefficients, regression equations, and correlation coefficients using technological tools; • display and discuss bivariate data where at least one variable is categorical; • recognize how linear transformations of univariate data affect shape, center, and spread; • identify trends in bivariate data and find functions that model the data or transform the data so that they can be modeled.
Develop and evaluate inferences and predictions that are based on data	• use observations about differences between two or more samples to make conjectures about the populations from which the samples were taken; • make conjectures about possible relationships between two characteristics of a sample on the basis of scatterplots of the data and approximate lines of fit; • use conjectures to formulate new questions and plan new studies to answer them.	• use simulations to explore the variability of sample statistics from a known population and to construct sampling distributions; • understand how sample statistics reflect the values of population parameters and use sampling distributions as the basis for informal inference; • evaluate published reports that are based on data by examining the design of the study, the appropriateness of the data analysis, and the validity of conclusions; • understand how basic statistical techniques are used to monitor process characteristics in the workplace.
Understand and apply basic concepts of probability	• understand and use appropriate terminology to describe complementary and mutually exclusive events; • use proportionality and a basic understanding of probability to make and test conjectures about the results of experiments and simulations; • compute probabilities for simple compound events, using such methods as organized lists, tree diagrams, and area models.	• understand the concepts of sample space and probability distribution and construct sample spaces and distributions in simple cases; • use simulations to construct empirical probability distributions; • compute and interpret the expected value of random variables in simple cases; • understand the concepts of conditional probability and independent events; • understand how to compute the probability of a compound event.

Problem Solving

Standard

Instructional programs from prekindergarten through grade 12 should enable all students to—

- Build new mathematical knowledge through problem solving
- Solve problems that arise in mathematics and in other contexts
- Apply and adapt a variety of appropriate strategies to solve problems
- Monitor and reflect on the process of mathematical problem solving

Reasoning and Proof

Standard

Instructional programs from prekindergarten through grade 12 should enable all students to—

- Recognize reasoning and proof as fundamental aspects of mathematics
- Make and investigate mathematical conjectures
- Develop and evaluate mathematical arguments and proofs
- Select and use various types of reasoning and methods of proof

Communication

Standard

Instructional programs from prekindergarten through grade 12 should enable all students to—

- Organize and consolidate their mathematical thinking through communication
- Communicate their mathematical thinking coherently and clearly to peers, teachers, and others
- Analyze and evaluate the mathematical thinking and strategies of others
- Use the language of mathematics to express mathematical ideas precisely

Connections

Standard

Instructional programs from prekindergarten through grade 12 should enable all students to—

- Recognize and use connections among mathematical ideas
- Understand how mathematical ideas interconnect and build on one another to produce a coherent whole
- Recognize and apply mathematics in contexts outside of mathematics

Representation

Standard

Instructional programs from prekindergarten through grade 12 should enable all students to—

- Create and use representations to organize, record, and communicate mathematical ideas
- Select, apply, and translate among mathematical representations to solve problems
- Use representations to model and interpret physical, social, and mathematical phenomena

Printed by
Graphics Arts Center
Portland, Oregon